T0393045

# Principles of
# Communication Engineering

# Principles of Communication Engineering

**M.L. Anand**

Consultant Engineer

CRC Press
Taylor & Francis Group
Boca Raton  London  New York

CRC Press is an imprint of the
Taylor & Francis Group, an **informa** business

Manakin
PRESS

First published 2022
by CRC Press
2 Park Square, Milton Park, Abingdon, Oxon, OX14 4RN

and by CRC Press
6000 Broken Sound Parkway NW, Suite 300, Boca Raton, FL 33487-2742

© 2022 Manakin Press Pvt. Ltd.

*CRC Press is an imprint of Informa UK Limited*

*British Library Cataloguing-in-Publication Data*
A catalogue record for this book is available from the British Library

*Library of Congress Cataloging-in-Publication Data*
A catalog record has been requested

ISBN: 978-1-032-11944-1 (hbk)
ISBN: 978-1-003-22227-9 (ebk)

DOI: 10.1201/9781003222279

Printed in the United Kingdom
by Henry Ling Limited

# Preface

Is there any justification of adding one more book to the already large stock of books on the subject. Perhaps there is. This is the book, in which the subject matter is dealt from elementary to the advance level in a unique manner, which will certainly fascinate the readers.

Three outstanding features can be claimed for the book viz. (*i*) style; the student, while going through the pages would feel as if he is attending a class room. (*ii*) language: that an average student can follow and (*iii*) approach: it takes the student from "known to unknown" and "simple to complex."

The book is reader friendly, thought provoking and stimulating. It helps in clearing cobwebs of the mind. The style is lucid and un-adulterated. Unnecessary mathematics has been avoided. Understandably, it has the language of an average student. What strands out is the stark simplicity, with which the ideas have been portrayed.

Errors might have crept in, inspite of utmost care to avoid them. The author will be grateful if the same are pointed out along with suggestions for improvement of the book.

The author thanks the Publishers for publishing the book and pricing it moderately inspite of heavy cost of paper and printing.

**M.L. ANAND**

A man of true science
uses but few hard words-
and those only, when
None other will answer his purpose-

Whereas a smaller
in science thinks that-
By mouthing hard words,
he understands hard things.

# Brief Contents

# Detail Contents

# 1

# Basic Concepts
# and Signal Analysis

A significant point about communication is that it involves a *sender* (transmitter) and a *receiver.* Only a receiver can complete the process of communication. Therefore dual process of "transmitting and receiving" or "coding and decoding" an information can be called as communication; thus, this is a two way process.

## 1.1 COMMUNICATION

As a general concept, we can say that transfer of information from one place to another is *communication.*

The important elements of a communication system are:

1. Message or information
2. Sender, transmitter or coder
3. Receiver or decoder
4. Code
5. Channel (transmission path)

## 1.2 METHODS OF COMMUNICATION

The communication may be:

(*a*) **Oral Communication:** In this type of communication, the message is sent or transmitted from the sender to the receiver through spoken words *e.g.,* direct talk or through telephone. The communication through hints or face expression also come under this category.

(*b*) **Written Communication:** When the message is sent to the receiver in writing, it is called written communication, *e.g.*, communication through letter, FAX etc.

## 1.3 PROCESS OF COMMUNICATION (SEE FIG. 1.1)

The process of communication involves the following steps:

1. **Encoding the message:** The encoder for transmission encodes the message into suitable words, symbols etc.

2. **Transmission:** After developing the message into suitable code, it may be transmitted through a proper channel.

3. **Reception:** The information is received on the other side by the receiver.

4. **Decoding:** The coded message is decoded into the original form, so that it is easily understood by the person on the receiver side.

**Fig. 1.1**

5. **Use:** The final stage of the communication process is to use the information for the purpose, it has been transmitted.

## 1.4 BRIEF HISTORY OF COMMUNICATION

In earlier days pigeons were used to send message (Information, signal) from one place to another. For this, pigeons were trained so that they could travel hundreds of miles to reach the destination. The message was tied round their neck or fastened in the beak and was flown towards the destination. See Fig. 1.2.

**Fig. 1.2**

Men were also engaged for this job. They were known as *Harkara*, they were collecting *dak* from one place and carried to the destination. Later on, they were provided with horses to speed up the work. See Fig. 1.3.

In 1830, letter boxes were installed in Britain and as India was under British rule, many such **red painted letter** boxes were installed in the localities in India, where the English people resided.

**Fig. 1.3**

We've had postal services of some kind or the other since times beyond memory. Every ruler employed *dak runners* to carry information to and fro from the outposts of his kingdom to the palace. It was during British rule that postal services were linked to the police. A regular police force was set up in l832; and the first Indian postage stamp was issued in 1840. To start with, post offices were located in the same buildings as police stations.

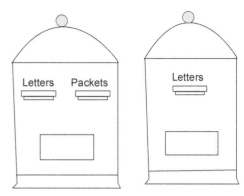

**Fig. 1.4** Typical red painted PO Boxes used all over the country

Then postal services outstripped the police and had to have large buildings like General Post Offices to handle mail, telegrams, money orders, fixed deposits, etc. Now postal services are on the decline. People use telephones, courier services, e-mail and fax. In near future, post offices may become a relic of the past.

## 1.5 ELECTRONIC COMMUNICATION

The function of an electronic communication system is to convey or send a message from one place to another using electronic equipment. The message may be an information or a signal. The information or a signal is obtained from a source, and through an electronic network it is sent to the receiver.

Various communication systems are employed to transmit A.V. (audio-video) signals of telephone, radio, T.V., radar, etc.

## 1.6 STRUCTURE OF AN ELECTRONIC COMMUNICATION SYSTEM

A communication system consists of following parts [Fig. 1.5]:

1. Information Source
2. Transmitter
3. Receiver

1. **The information source:** The information source is the source which generates or produces information or signal.

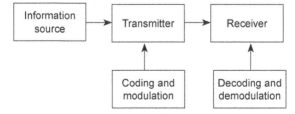

**Fig. 1.5**

2. **Transmitter:** Transmitter is the device which transmits the generated information or signal. It has the following components:

   (*a*)   **Coding:** The transformation of the signal into a suitable form in which it can be transmitted.

   (*b*)   **Modulation:** To superimpose the signal on an H.F. carrier, so that the signal can travel long distances.

3. **Receiver:** It receives the information, *e.g.*, radio receiver, T. V. receiver, telephone receiver, etc. It consists of:

   (*a*)   **Decoding:** It is the reverse of coding, *e.g.*, to regain the original form of the signal.

   (*b*)   **Demodulation:** It is the reverse of modulation, *i.e.*, to separate the original signal from the carrier.

## 1.7  BANDWIDTH REQUIREMENT

By limiting the bandwidth for a signal, more channels can be accommodated. The bandwidth requirement for a signal mainly depends upon the modulating signals. The audio signal occupies a bandwidth upto 15 kHz but when a carrier is modulated by the audio signal, the modulated signal will certainly need more bandwidth. Here we are giving B.W. of few signals (Fig. 1.6).

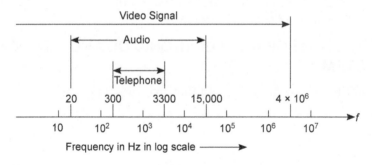

**Fig. 1.6**

1. The audio signal needs B.W. of about 20 Hz to 15 kHz for transmission.

2. The video signal needs a B.W. of about 4 MHz, while a fax signal needs a B.W. of 1 kHz only. In T.V. the picture is scanned in 1/30 seconds while a FAX needs 10 minutes to scan a page.

3. For a telephone, a B.W. of 300 to 3300 Hz is required.

## 1.8 TYPES OF ELECTRONIC COMMUNICATION SYSTEMS

Basically, communication systems are of two types:

1. *Wire communication, i.e.*, where communication is done through wires, *e.g.*, cable T.V., wire telephony, etc.

2. *Wireless or carrier communication, i.e.*, where communication is done without wires. In this system, a carrier wave is used. In other words, modulation is carried out, *e.g.*, radio, T.V., radar, radio telephony, etc.

[A carrier is a high frequency wave which 'carries' the signal, i.e., the signal is 'superimposed' on the carrier.]

Details of frequency band for different types of communication are given in Table 1.1.

**Table 1.1** The Frequency Band for Various Communication Systems

| SI No. | Frequency band | Type of communication |
|--------|----------------|----------------------|
| 1. | Very low frequency (5–30 kHz) | Long distance communication |
| 2. | Low frequency (30–300 kHz) | Radio navigation |
| 3. | Medium frequency (0.3–3 MHz) | Broadcasting marine |
| 4. | High frequency (3–30 MHz) | FM broadcasting television |
| 5. | Ultra high frequency (0.3–3 GHz) | Radar |
| 6. | Super high frequency (3–30 GHz) | Satellite communication |

## 1.9 TRANSMISSION MEDIUMS

The various transmission mediums used in various ranges of the electro magnetic spectrum are shown in Fig. 1.7.

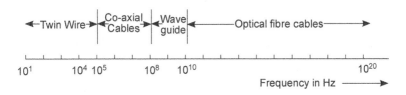

**Fig. 1.7**

These mediums are twin wire lines, co-axial cables, wave guides and optical fibre cables. The various communication systems operating in different ranges of spectrum are: telephone, A.M. and F.M. broadcasting, T.V., satellite etc.

The telephone channels need a bandwidth (B.W.) of 300–3300 Hz, the medium wave broadcasting needs a B.W. of about 10 kHz, the VHF stations need a B.W. of about 150 kHz and T.V. channels need a B.W. of about 8 MHz and soon.

## 1.10  IMPORTANT FACTS ABOUT SOUND AND LIGHT

(*a*) **Radio receivers and other audio devices/circuits deal with sound. Hence we should know important facts about sound.**

   1.   Sound is a form of energy.

   2.   Newton's law for the velocity of sound in a medium is $v = \sqrt{E/p}$, where $E$ is the elasticity and $p$ density of the medium. Velocity of sound in air is 330 m/s, in water 1500 m/s. Velocity of electricity is $10^5$ times more than that of sound.

   3.   A woman's voice is sweet, has more frequency, less wavelength and low pitch. A child can produce sound of even more high frequency than a woman can. The frequency of an adult's voice is low.

   4.   An adult's voice is hoase. He generates a sound of 20 Hz to 20 kHz. Note that a 20 Hz sound has a wavelength of 17 m and a 20 kHz sound has 0.017 m (or 1.7 cm) wavelength.

   5.   As mosquito cannot tolerate more than 25 kHz, the electronic mosquito repellers generate a frequency of more than 40 kHz.

   6.   A bat produces ultrasonic waves (above 100 kHz). By echo the bat finds the presence of an obstacle. By phase difference of echoes coming from different directions, it finds its way. Note that a bat cannot see.

   7.   Best echo is heard from 11 m distance. In cinema halls, absorbers on walls and ceiling do not reflect sound; echo is not produced and we hear the sound clearly.

   8.   A man has two ears. We can therefore find also the direction of a particular sound from phase difference between the sounds coming from different directions of two ears.

   9.   A man can hear a sound of frequency between 20 Hz and 20 kHz (called audio frequency range).

   10.  On moon, we cannot hear as there is no air. The astronauts talk in hints. Note that sound cannot travel without a medium.

   11.  Sound effect remains for 1/10 second and light effect remains for 1/16 second.

12. The pitch of buzzing of mosquito is much that than the roaring of a lion; But loudness of mosquito sound is much less than that of a lion. Loudness does not depend on pitch.

13. Earthquakes produce ultrasonics. Dogs can hear ultrasonic sounds and hence they start barking.

14. The relation between velocity, frequency and wavelength of sound is $V = f\lambda$.

    If a radio station has 1350 kHz frequency ($f$) and 250 m wavelength ($\lambda$), you can calculate the velocity of the radio waves ($V = 3 \times 10^8$ m/s).

15. Music is a sound with simple harmonic motion and regular waves, whereas noise is a sound of combination of irregular waves.

16. All wireless communication occurs through ozone layer in the atmosphere. The ozone layer is only 3 mm thick.

17. Transducers are devices which convert one form of energy into another. A loudspeaker is a transducer which converts sound into electrical waves. A camera converts light (picture) into electrical waves. A picture tube converts electrical waves into picture. Human eye, brain, etc., are also typical transducers.

18. The response of our ears to sound is 'logarithmic' and not linear. If sound becomes 100 times high, our ears feel it as double (log $100 = 2$), not 100 times otherwise ear drum will tear off. This is the reason, that we use 'decibel' as a unit for sound.

19. When sound of all frequencies are processed through only one loudspeaker, we get 'monophonic' sound. When low and high frequencies are processed by separate loudspeakers we get a 'stereophonic' sound, which is a quality sound. This is also called a HiFi (high fidelity) sound.

20. Sound waves are longitudinal waves.

21. The graph between frequency and time period of a sound is a parabola.

22. Strength of sound is decided by its amplitude. Pitch of sound is decided by frequency. Tight membranes (of loudspeakers) produce more frequency.

23. Sound is produced by vibrations. All musical instruments vibrate.

**(*b*) Facts about light**

The T.V. receivers etc., deal with light. Few facts about light are:

1.  Light travels in a straight line.

2.  Light energy can be converted into electrical energy by video camera.

3.  The electrical energy can be converted back to light energy by a picture tube.

4.  The light waves are transversal waves.

## 1.11 MODULATION

The process of changing some characteristic (amplitude, frequency, phase, etc.) of a carrier by the signal (audio or video) is called 'modulation'.

'Modulation' means modification, variation or change. We modify the carrier according to the signal and hence the name. This is an important process of wireless communication.

## 1.12 NEED FOR MODULATION

In carrier (wireless) transmission, modulation is a necessity. This is explained below:

(*a*) The first and the foremost reason is that the original sound produced by microphone (or video camera in case of video signal) is very weak and it has a very low frequency. The energy contained by the signal is proportional to its frequency. Thus due to losses in energy, the signal will die after some distance. So, it cannot travel long distance. Therefore, the low frequency signal is **made to sit** on high frequency 'carrier'. Such an arrangement enables the signal to travel long distances before it dies out. At the receiver the signal is separated out and the carrier is grounded. The phenomenon can be illustrated by the following analogy.

Suppose a man travels 'on foot' to deliver a message. Naturally he will take a long time to reach the destination; moreover, he cannot travel long distance. But if the man is provided a horse, the message can reach longer distance in shorter time. At the destination, the 'receiver' will take out the message and will leave the horse.

Assume here, the message as a signal, horse as a carrier and the receiver as the radio or TV receiver. This explains the principle of radio transmission and reception.

($b$) The next reason describes the height of the antenna needed.

The transmitting antenna should have a height equal to the wavelength. This condition gives best results. We know that

$$V = f\lambda$$

where
$V$ = velocity of radio waves = $3 \times 10^8$ m/s

$f$ = frequency

$\lambda$ = wave length

($i$)  If the frequency of the signal is 20 kHz, the length of the antenna

$$l = \lambda = \frac{V}{f} = \frac{3 \times 10^8 \, \text{m/s}}{20 \times 10^3} = 15000 \, \text{m} = 15 \, \text{km}$$

*i.e.*, if the sound produced at mike is to be transmitted as such, we need an antenna of 15 km height, which is totally impractical.

($ii$)  If $f$ = 1 MHz

now length of the antenna $l = \lambda = \dfrac{3 \times 10^8}{1 \times 10^6} = 300$ m

*i.e.*, if the frequency of the signal is raised to 1 MHz, it can be transmitted through a 300 m high antenna. This is a practical height. Therefore we can 'modulate' the signal according to the requirement and need. In other words, the signal is superimposed on a high frequency carrier.

($c$) The last and the most important reason is that modulation permits the transmission without wire. We can receive audio/video signals from any corner of the world through wireless communication. We can witness a match being played at France, sitting in our bedroom. Imagine the length of wire needed if wireless communication were not possible.

## 1.13 TYPES OF MODULATIONS

Modulation is an important process in all wireless (carrier) communications. In this, the signal is superimposed on a high frequency carrier wave. Some characteristic (amplitude, frequency, phase; etc.) of the carrier wave is changed in accordance with the instantaneous value of the signal. A sine wave may be represented by

$$e = E_m \sin (\omega t + \varphi)$$

where
$e$ = instantaneous value of modulated wave

$E_m$ = maximum amplitude

$\omega$ = angular velocity

$\varphi$ = phase relation

Accordingly, modulation is of three types (see the above equation)

1. Amplitude modulation: By changing amplitude of the carrier.

2. Frequency modulation: By changing frequency of the carrier.

3. Phase modulation: By changing phase of the carrier.

**However, the complete classification of modulation processes are given below:**

1. Amplitude modulation (AM)

   (*a*)  Single sideband AM (SSBAM)

   (*b*)  Double sideband AM (DSBAM)

   (*c*)  Frequency division multiplexing (FDM)

In India, for sound, amplitude modulation is used

2. Frequency modulation (FM)

   In India, for television signals, frequency modulation is used.

3. Phase modulation.

**Other modulation processes are:**

(1) Pulse modulation (used in telephone and telegraphy)–these may be:

   (*a*)   Pulse amplitude modulation (PAM)

   (*b*)   Time division multiplexing (TDM)–used in long play records

   (*c*)   Pulse time modulation (PTM)

   (*d*)   Pulse division multiplexing (PDM)

   (*e*)   Pulse code modulation (PCM)

(2) Digital modulation (DM)–They may be:

   (*a*)   Differential PCM (DPCM)

   (*b*)   Adoptive PCM (ADPCM)

   (*c*)   Data modulation (DM)

   (*d*)   Adoptive data modulation (ADM)

   **Note: 1.** The modulations may also be:

   (*a*)   analog modulation

   (*b*)   Digital modulation

2. The amplitude modulation is often referred as *linear modulation*. The frequency and phase modulations are known as *non linear, angular* or *exponential* modulation. While there may be many forms of exponential modulations but only two *i.e.*, frequency and phase modulations are practical. In particular, both linear as well as non-linear modulations are continuous wave (CW) type modulations.

## 1.14 RADIO (WIRELESS) BROADCASTING,TRANSMISSION AND RECEPTION

The process of sending radio or T.V. signals by an antenna to multiple receivers which can simultaneously pick up the signal is called 'broadcasting'.

In simple words 'to radiate radio waves from a station into space' is called broadcasting or, to send signal in all directions (broad) is called broadcasting.

After the waves are thrown into the space, the transmission starts and all the receivers in 'the range' can simultaneously pick up the signal. This is called 'reception'. There is a little difference between broadcasting and transmission. However, the process of reception is quite different.

Important components of a typical network are under: See Fig. 1.8

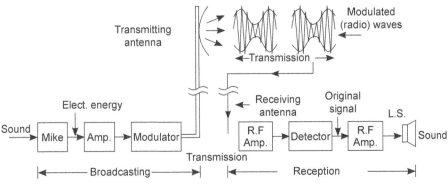

**Fig. 1.8**

(*a*) **Broadcasting**

1. **Microphone:** At the broadcasting station, the person speaks before mike. The mike is a transducer and converts sound energy into electrical energy. The speaker generates a sound of frequency between 20 Hz and 20 kHz (*i.e.*, audio frequency).

2. **Amplifier:** The electrical signal obtained from microphone (mike) is weak and the same is amplified through an amplifier(s) to the required strength.

3. **Modulator:** Here the modulation of the signal occurs. A local oscillator generates high frequency waves called 'carrier'. The signal modulates the carrier or the signal is superimposed on the carrier. The resultant waves are called *radio waves* or modulated waves.

4. **Transmitting antenna:** Through the transmitting antenna, the radio waves are propagated into the space.

(*b*) **Transmission:** After broadcasting, the transmission starts. These radio waves travel in space at a speed of $3 \times 10^8$ m/s, as they are electromagnetic waves.

(*c*) **Reception:** The picking of these radio waves by radio (or T.V.) receiver is called reception. A radio receiver has the following important parts:

1.  **Receiving antenna:** The radio waves induced an e.m.f. on the antenna.

2.  **RF amplifier:** The radio waves are of radio frequency (R.F.) range. The e.m.f. induced is amplified through R.F. amplifier(s).

3.  **Detector:** Now the original signal is detected (separated) from the carrier by the detector circuit. The signal starts its forward journey while the carrier is grounded.

4.  **A.F. amplifier:** The signal is now passed through the amplifier. Note that now the signal is of audio frequency range. It should have sufficient energy to strike the loudspeaker.

5.  **Loudspeaker (L.S.):** This is the final stage. The electric signal is again converted into the original sound signal which was produced in the broadcasting station:

**Note:**

(*i*) Here 'Radio' does not mean radio receiver. The 'radio' means wireless.

(*ii*) Radio means 'radiations for wireless transmission'. The principle of radio broadcasting, transmission and reception described above is same for radio, T.V. signals and also for all such wireless devices.

(*iii*) 'Radio' is the abbreviated form of 'radio telegraph' or 'radio telephone'.

(*iv*) Broadcasting means to 'send out' in all directions. It may be:

(*a*) **A.M. radio broadcast band:** Its range is 540–1600 kHz. The stations are assigned every 10 Hz in the above band.

(*b*) **F.M. radio broadcast band:** Its range is 88–108 MHz. The stations are assigned every 200 Hz in the above band.

(*c*) **T.V. broadcasting band:** A T.V. channel is 6 MHz wide to include picture and sound signals for each broadcast station.

(*v*) (*a*) Analog form of a broadcast signal is a continuous variation as shown in Fig. 1.9 (*a*),

(*b*) Digital form of broadcast signal is shown in Fig. 1.9 (*b*).

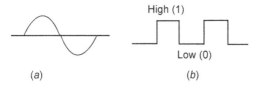

High (1)

Low (0)

(a)                              (b)

Fig. 1.9

## 1.15 REASONS OF ERRORS IN COMMUNICATION SYSTEM

When a distortion occurs in the received message, an error is said to have occurred. The distortion may be a frequency, phase or any other distortion.

There are following reasons which produce errors:

1. **Band Width:** When the bandwidth allowed to a transmission is not sufficient it causes an error. The bandwidth allowed to the AM transmission is only 10 kHz but a human ear requires a bandwidth of 15 kHz for full satisfaction. Hence AM transmission lacks fidelity. The bandwidth allowed to F.M. transmission is 200 kHz and it can reproduce a transmitted signal to our full satisfaction.

2. **Noise:** The unwanted sound is called noise. The noise is another reason that produces error. There are many types of noises: external or internal. The signal to noise ratio can be improved by changing the band width.

## 1.16 TYPES OF COMMUNICATION SYSTEMS

(*a*) **The electronic communication systems according to medium are of following types:**

1. **Wire communication:** *i.e.,* where communication is done through wires, *e.g.,* cable T.V., wire telephony, etc.

   The wires may be made of galvanised steel. See Fig. 1.10.

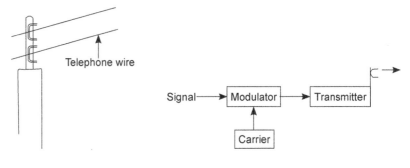

Telephone wire

Signal ⟶ Modulator ⟶ Transmitter

Carrier

Fig. 1.10                              Fig. 1.11

2. **Wireless or carrier communication (Fig. 1.11)**; *i.e.,* where communication is done without wires. In this system, a carrier

wave is used. In other words, modulation is carrier out, *e.g.* radio, T.V., radar, telephony, etc.

A carrier is a high frequency wave which 'carries' the signal, *i.e.*, the signal is 'superimposed' on the carrier.

3.  **Optical fibre communication:** An optical fibre cable (OFC) consists of an inner glass core surrounded by glass cladding and then a protective covering. (Fig. 1.12).

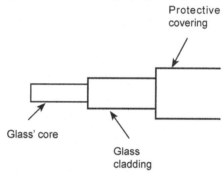

Fig. 1.12

Digital signals are transmitted in the form of intensity modulated light signals, which are trapped in the glass core of the cable. The signal is detected on the other side, using a photo device.

This communication is used above $10^{10}$ Hz frequencies.

4.  **Satellite communication:** In this, the signals are transmitted and received through a satellite positioned in the space.

5.  **Wave guide communication:** In this communication, **wave guides** are used. These are used for frequencies of $10^8$ to $10^{10}$ Hz.

6.  **Co-axial cables communications:** The co-axial cables are used for $10^5$ to $10^8$ hertz.

(*b*) The communication may be also classified as **Analog and digital communication:** In the analog communication, the signal is modulated and transmitted directly.

In digital communication, the analog signal before transmission is converted into digital signal. See Fig. 1.13.

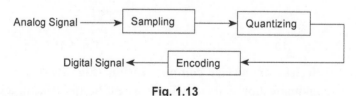

Fig. 1.13

## 1.17 SIGNALS AND CLASSIFICATION

(*a*) **Signal**

A signal may be a function of time, temperature, position, pressure, distance etc., *e.g.* speech signal, video signal. Systematically it may be defined as below:

"A function of one or more independent variables, which contains some information is called a **signal**."

The voltage and current are examples of electrical signals; which are the function of time as an independent variable.

(**b**) **The signals may be classified as**

1. **Continuous time and discrete time signals:** A signal $x$ ($t$) is a continuous time signal, if $t$ is a continuous variable (Fig. 1.14).

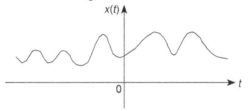

**Fig. 1.14**

In other words, a continuous time signal may be defined continuously in time domain.

If $x$ ($t$) is defined at discrete times, it is a discrete time signal. (Fig. 1.15)

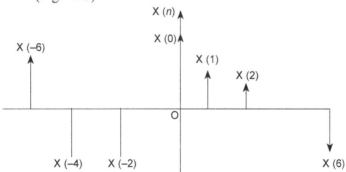

**Fig. 1.15**

This signal is therefore identified as a sequence of numbers and is denoted by $x$ ($n$), where $n$ is an integer.

2. **Real and complex signals:** A signal $x$ ($t$) is a real signal, if its value is a real number and it is a complex signal if its value is a complex number.

3. **Deterministic and non-deterministic signals:** A **deterministic signal** can be completely specified in time. The pattern of this signal is regular and its amplitude can be determined at any time.

   A **non-deterministic signal** has an irregular pattern and its occurrence in nature is always random, so some times it is also referred as **Random** signal.

   An example of such a signal is "Thermal noise".

4. **Even and odd signals:** An **even signal** is that, which shows symmetry in time domain and is identical about the origin. See Fig. 1.16 (*a*).

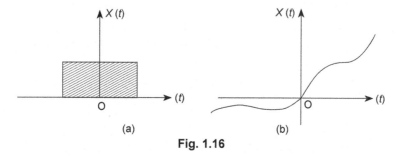

(a)                                    (b)

**Fig. 1.16**

   On the other hand an **odd signal** has no symmetry and is not identical about the original. See Fig. 1.16 (*b*).

5. **Power and energy signals:** A **power signal** is that, which has finite average power ($P$) and infinite energy ($E$) *i.e.*: $0 < P < \infty, E = \infty$.

   On the other hand, an **energy signal** has finite energy and zero average power;

   *i.e.*: $0 < E < \infty, E = \infty$.

   A signal sometimes, may neither be a power signal nor an energy signal.

6. **Periodic and non-periodic signals:** A **periodic signal** is that which has a definite pattern and repeats it after a definite time period ($T$) *e.g.*, a sinusoidal wave. [Fig. 1.17 (a)]

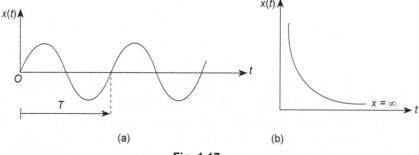

(a)                                    (b)

**Fig. 1.17**

A non periodic signal is that, which does not repeat itself *e.g.* an exponential signal. It has an infinite time period [Fig. 1.17 (*b*)]

7. **Analog and digital signals:** An **analog signal** is that which varies smoothly and continuously with time. It derives its name from the fact that it is **analogous** to the physical signal, it represents. This signal can be defined for every value of time. See Fig. 1.18 (*a*).

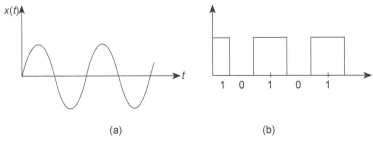

(a)                                        (b)

**Fig. 1.18**

A **digital signal** [See Fig. 1.18 (*b*)] is not a continuous representation of the original signal.

A digital signal is represented in sequence of numbers. Each number represents its magnitude at an instant of time.

The digital representation is considered as a code, which approximates the actual value.

This has also been explained earlier.

## 1.18 REPRESENTATION OF SIGNALS

An electrical signal can be represented as a **voltage source**, [Fig. 1.19 (*a*)] or a **Current source** [Fig. 1.19 (*b*)]

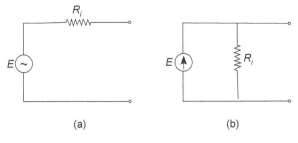

(a)                                        (b)

**Fig. 1.19**

Further, either a voltage source or a current source may be presented in two forms:

1. **Time domain representation:** In this a signal is a time varying quantity, See Fig. 1.20 (*a*).

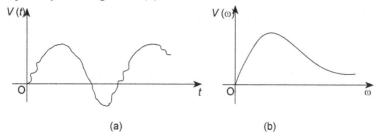

(a)　　　　　　　　　　　　　　　　(b)

**Fig. 1.20.**

2. **Frequency domain representation (Fig. 1.20 b):** In this, a signal is represented by its frequency spectrum. The representation shows the frequency content but does not necessarily indicate the shape of the waveform.

The Fig. 1.20 (*b*) shows frequency domain representation of a time domain signal, for this **Fourier transform** is used.

## 1.19 FOURIER ANALYSIS OF SIGNALS

A periodic function $x(t)$ may be analysed by using a mathematical series developed by a French physicist and mathematician "Baron Jean Fourier". This series is called **Fourier series**.

A periodic function $x(t)$ may be expressed in the form of **Trigonometric Fourier series** comprising the following sine and cosine terms:

$$x(t) = a_0 + a_1 \cos \omega_0 t + a_2 \cos 2\omega_0 t + \ldots + a_n \cos n\,\omega_0 t + \ldots + b_1 \sin \omega_0 t + b_2 \sin \omega_0 t + \ldots + b_n \sin n\,\omega_0 t + \ldots$$

$$= a_0 + \sum_{n=1}^{\infty}(a_n \cos \omega_0 t + b_n \sin n\omega_0 t)\,[t_0 \le t \le (t_0 + T)]$$

$$\ldots (i)$$

The equation is the trigonometric fourier series representation of function $x(t)$ over an interval $[t_0, (t_0 + T)]$ *i.e.*, over one period.

Here
$$T = \frac{2\pi}{\omega_0},\ \omega_o = \frac{2\pi}{t}$$

The equation (*i*) can be written as:

$$x(t) = a_0 + \sum_{n=1}^{\infty}\left(a_n \cos\frac{2\pi n}{T}\cdot t + b_n \sin\frac{2\pi n}{T}\cdot t\right) \quad \ldots(ii)$$

The $a_n$ and $b_n$ are the coefficients.

The $\omega_0$ is called **Fundamental frequency** and $2\omega_0$, $3\omega_0$, $n\omega_0$ ... are second, third and $n^{th}$ ... harmonics of the fundamental frequency ($\omega_0$).

The process of finding values of $a_0$, $a_n$, $b_n$ is called **Fourier analysis**. The $a_0$ with zero frequency is called **dc component**.

## 1.20 FOURIER TRANSFORMATION AND PROPERTIES

The conversion of a time domain function $x(t)$ into corresponding frequency domain function $X(\omega)$ and vice versa with the help of fourier series is called **Fourier transform**

$$X(\omega) = F[x(t)] = \int_{-\infty}^{\infty} x(t) \cdot e^{-j\omega t} dt$$

In the above expression, $X(\omega)$ is the frequency domain representation of time domain function $x(t)$.

Conversely, if a frequency domain function $X(\omega)$ is to be converted into corresponding time domain function, we will take "Inverse fourier transform", *i.e.*:

$$X(t) = F^{-1}[X(\omega)] = \int_{-\infty}^{\infty} x(t) \cdot e^{-j\omega t} dt$$

So,                    $X(\omega) = F[x(t)]$

and                    $x(t) = F^{-1}[x(\omega)]$

The $x(t)$ and $X(\omega)$ are called **"Fourier transform pair"**.

Symbolically, we can express it as

$$x(t) \rightarrow X(\omega)$$

(*b*) **Properties of Fourier Transform:** The various properties of fourier transform are explained below:

1. **Time scaling property:** This states that "the time compressions of a signal results in its spectral expansion and its time expansion results into its spectral compressions.

    If,                    $x(t) \leftrightarrow X(\omega)$

    Then for *a*-real constant, *a*

    $$x(at) \leftrightarrow . \frac{1}{|a|} \cdot X(\omega/a)$$

2. **Linearity property:** This states that "a fourier transform is linear".

    $$x_1(t) \leftrightarrow X_1(\omega)$$

    and                    $x_2(t) \leftrightarrow X_2(\omega)$

    Then $a_1 x_1(t) \cdot a_2 x_2(t) \leftrightarrow a_1 X_1(\omega) + a_2 X_2(\omega)$

3. **Duality or symmetry property:** According to this property:

   If                             $x(t) \leftrightarrow X(\omega)$

   then                           $x(t) \leftrightarrow 2\pi \, x(-\omega)$

4. **Time shifting property:** This property states that a shift in the time domain by an amount ($b$) is equivalent to multiplication by $e^{-j\omega h}$ in the frequency domain:

   If,                            $x(t) \leftrightarrow X(\omega)$

   Then                           $x(t - b) \leftrightarrow X(\omega) . \, e^{-j\omega h}$

5. **Frequency shifting property:** According to this property, the multiplication of a function $x(t)$ by $e^{j\omega_0 t}$ is equivalent to shifting its fourier transform $X(\omega)$ in the positive direction by $\omega_0$. In other words, the $X(\omega)$ is translated by $\omega_0$, therefore, this property is also called as **frequency translated theorem**.

   If                             $x(t) \leftrightarrow X(\omega)$

   $$e^{j\omega_0 t} . \, x(t) \leftrightarrow X(\omega - \omega_0)$$

6. **Time differentiation property:** According to this property, differentiation of a function $x(t)$ in time domain is equivalent to multiplication of its fourier transform by a factor $j\omega$.

   If                             $x(t) \leftrightarrow X(\omega)$

   $$\frac{d}{dt} x(t) \leftrightarrow j(\omega) \, X(\omega)$$

**Problem 1.1** *One of the five possible messages $p_1$ to $p_5$ having possibilities* $\dfrac{1}{2}, \dfrac{1}{4}, \dfrac{1}{8}, \dfrac{1}{16}$ *and* $\dfrac{1}{16}$ *respectively is transmitted, calculate the average information.*

**Solution:** Average information

$$H = p_1 \log_2(i/p_1) + p_2 \log_2(1/p_2) + \dots$$

$$= \frac{1}{2}\log_2(2) + \frac{1}{4}\log_2(4) + \frac{1}{8}\log(8) + \frac{1}{16}\log_2(16) + \frac{1}{16}\log_2(16)$$

$$= \frac{1}{2} + \frac{1}{2} + \frac{3}{8} + \frac{1}{4} + \frac{1}{4} = 1.875 \text{ bits/message. } \textbf{Ans.}$$

**Problem 1.2.** *Calculate capacity of a standard 4 kHz telephone channel working in the range of 300 – 3400 Hz with Signal noise (SN) ratio equal to 32 dB.*

**Solution:**

$$B = 3400 - 300 = 3100 \text{ Hz}$$

$$\frac{S}{N} = 32\text{dB} = 1585$$

$$C = B \log_2 \left(1 + \frac{S}{N}\right) = 3100 \log_2 (1 + 1585)$$

$$= 32.95 \text{ K bits/sec. } \textbf{Ans.}$$

**Problem 1.3.** *A gaussian channel is band limited to* 1 *MHz. If the signal to noise spectral density S/$\eta$* = $10^5$ *Hz, calculate channel capacity and max. information rate.*

**Solution:**

(i)
$$C = B \log_2 \left(1 + \frac{S}{N}\right) = B \log_2 \left(1 + \frac{S}{\eta}\right)$$

$$= 10^6 \log_2[1 + 10^5]$$

$$= 0.1375 \times 10_6 \text{ bits/sec.}$$

$$= 0.1375 \text{ m Bit/sec. } \textbf{Ans.}$$

(ii) Maximum information rate

$$= 1.44. \frac{S}{\eta} = 1.44 \times 10^5$$

$$= 0.144 \text{ m bits/sec. } \textbf{Ans.}$$

**Problem 1.4.** *For the wave shown (Fig. 1.21), find Fourier coefficients $a_0$, $a_3$ and $b_4$.*

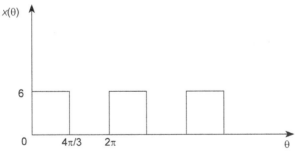

**Fig. 1.21**

**Solution:** The function $x$ ($\theta$) has a constant value for $\theta = \left(0 \text{ to } \dfrac{4\pi}{3}\right)$ radians and its value is zero for $\theta = \left(\dfrac{4\pi}{3} \text{ to } 2\pi\right)$ radians.

(i) $\qquad a_0 = \dfrac{1}{2\pi}$ (Area of one cycle) $= \dfrac{1}{2\pi}\left(6 \times \dfrac{4\pi}{3}\right) = 4$ **Ans.**

(ii) $\qquad a_3 = \dfrac{1}{\pi}\displaystyle\int_0^{2\pi} x(\theta)\cos 3\theta d\theta = \dfrac{2}{\pi}(\sin 4\pi) = 0$ **Ans.**

(iii) $\qquad b_4 = \dfrac{1}{\pi}\displaystyle\int_0^{2\pi} x(\theta)\cos 4\theta d\theta = \dfrac{-3}{2\pi}\left[\cos\dfrac{16\pi}{3} - \cos 0°\right] = \dfrac{9}{4\pi}$ **Ans.**

**Problem 1.5.** *Find Trigonometric Fourier series from a full wave rectified sine wave shown in Fig. 1.22.*

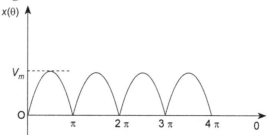

**Fig. 1.22**

**Solution:** The equation for the sinusoidal wave is given by $v = V_m \sin\theta$

or $\qquad\qquad x(\theta) = V_m \sin\theta$

$$a_0 = \dfrac{1}{2\pi}\int_0^{2\pi} x(\theta)d\theta = \dfrac{1}{2\pi}\int_0^{2\pi} V_m \sin\theta d\theta = \dfrac{V_m}{\pi}\left|-\cos\theta\right|_0^{\pi}$$

$$a_0 = \dfrac{2V_m}{\pi} \qquad\qquad\qquad\qquad ...(i)$$

$$a_n = \dfrac{1}{\pi}\int_0^{2\pi} x(\theta)\cos n\theta\, d\theta = \dfrac{2V_m}{\pi}\int_0^{\pi}\sin\theta\cos n\theta\, d\theta$$

Solving, we get $\qquad a_n = \dfrac{2V_m}{\pi}\int_0^{\pi}[\sin(1+n)\theta + \sin(1-n)\theta]d\theta$

$$= \dfrac{V_m}{\pi}\dfrac{\cos(1+n)\theta}{1+n} + \dfrac{\cos(1-n)\theta}{1-n}$$

Solving further, we get $\quad a_n = \dfrac{2V_m}{\pi}\left[\dfrac{1}{1+n} + \dfrac{1}{1-n}\right] = \dfrac{-4V_m}{\pi(n^2-1)} \qquad ...(ii)$

$\therefore \qquad\qquad x(\theta) = a_0 - \dfrac{4V_m}{\pi}\displaystyle\sum_{n=2}^{\infty}\dfrac{\cos 2\theta}{n^2-1}$

The trigonometric Fourier series:

$$x(\theta) = \dfrac{2V_m}{\pi} - \dfrac{4V_m}{\pi}\left[\dfrac{1}{3}\cos^2\theta - \dfrac{1}{15}\cos 4\theta + \dfrac{1}{35}\cos 6\theta +...\right] \text{ **Ans.**}$$

**Note:** As the given function (waveform) has even symmetry ($b_n = 0$); it will contain only cosine terms in the Fourier series.

**Problem 1.6.** *Find Trigonometric Fourier series for the square voltage waveform shown (Fig. 1.23):*

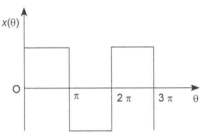

**Fig. 1.23**

**Solution.** The Fourier series for this wave will be written as:

$$x(\theta) = \sum_{n=1}^{\infty} b_n \sin n\theta t$$

where,

$$b_n = \frac{1}{\pi} \int_0^{2\pi} x(\theta) \sin n\theta. d\theta$$

$$= \frac{1}{\pi} \left[ \int_0^{\pi} V \sin n\theta. d\theta + \int_{\pi}^{2\pi} -V \sin n\theta. d\theta \right]$$

$$= \frac{V}{\pi n} \left| -\cos n\theta \right|_0^{\pi} + \frac{V}{\pi n} \left| \cos n\theta \right|_{\pi}^{2\pi}$$

$$= \frac{V}{\pi n}[(-\cos \pi n + \cos 0) - \cos \pi n] = \frac{2V}{\pi n}(1 - \cos \pi n)$$

When $n$ is odd, $1 - \cos \pi n = 2$

When $n$ is even, it is $= 0$

$\therefore$

$$b_1 = \frac{2V}{\pi.1} \times 2 = \frac{4V}{\pi} \text{ (Putting } n = 1) \qquad \qquad ...(i)$$

$$= \frac{2V}{\pi.3} \times 2 = \frac{4V}{3\pi} \text{ (Putting } n = 3) \qquad \qquad ...(ii)$$

$$= \frac{2V}{\pi.5} \times 2 = \frac{4V}{5\pi} \text{ (Putting } n = 5) \qquad \qquad ...(iii)$$

and so on.

$\therefore$ The Fourier series comes to be

$$x(\theta) = \frac{4V}{\pi} \sin \theta + \frac{4V}{3\pi} \sin 3\theta + \frac{4V}{5\pi} \sin 5\theta + ... + ...$$

$$= \frac{4V}{\pi}\left[\sin \omega_0 t + \frac{1}{3}.\sin 3\omega_0 t + \frac{1}{5}\sin 5\omega_0 t + ...\right]. \text{ Ans.}$$

**Note:** As the given function (waveform) is "odd" and it possesses half wave symmetry, the Fourier series contains only "sine terms" and "odd" harmonics.

**Problem 1.7.** *Find Fourier series for the saw tooth wave shown in Fig.* 1.24:

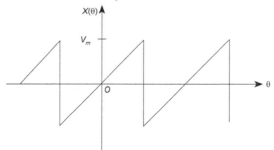

**Fig. 1.24**

**Solution:**

$$x(t) = \sum_{n=1}^{\infty} b_n \sin n\theta$$

where,
$$b_n = \frac{1}{\pi}\int_0^{2\pi} x(\theta)\sin n\theta.d\theta$$

Slope of the curve $= V_m/\pi$.

∴
$$x(\theta) = (V_m/\pi)\theta.$$

Taking limits from $-\pi$ to $+\pi$

$$b_n = \frac{1}{\pi}\int_{-\pi}^{\pi}\left(\frac{V_m}{\pi}\right)\theta\sin n\theta d\theta = -\frac{2V_m}{\pi b}\cos \pi n$$

[When $n$ is even, cos $\pi n$ will be positive and when $n$ is odd, cos $\pi n$ will be negative]

The required Fourier series will be

$$= x(\theta) = \frac{2V_m}{\pi}\left[\sin \theta - \frac{1}{2}\sin 2\theta + \frac{1}{3}\sin 3\theta - \frac{1}{4}\sin 4\theta + ...\right]$$

**Note:** As the average value of the given wave is zero over a cycle, hence $a_0 = 0$; The curve has odd symmetry, an $= 0$. The series, therefore, contains only sine terms.

**Problems 1.8.** *Find exponential Fourier series for the saw tooth wave shown in Fig.* 1.25.

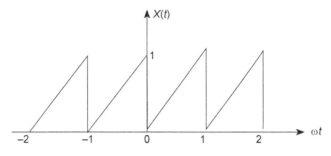

**Fig. 1.25**

**Solution:** Over one period,

$$x(t) = t \ [\text{for } 0 < t < 1]$$

$$\text{Period, } T_0 = 1$$

Fourier series coefficient, $C_n = \dfrac{1}{T_o} \displaystyle\int_0^{T_0} x(t)e^{-j2\pi nt/T_0} dt$

$$= \dfrac{1}{1}\displaystyle\int_0^1 t.e^{-j2\pi nt} dt \quad (T_0 = 1)$$

To simplify above, we will use

$$\int xe^{ax} dx = e^{ax}\left(\dfrac{x}{a} - \dfrac{1}{a^2}\right)$$

taking $a = j2\pi n$, $C_n = -j/2\pi n$ (for $n \neq 0$)

with $n = 0$, $\qquad C_0 = \dfrac{1}{T_0}\displaystyle\int_0^{T_0} x(t)dt = \displaystyle\int_0^1 t.dt = \dfrac{1}{2}$

$$|C_n| = \sqrt{0 + \dfrac{1}{(2\pi n)^2}} = \dfrac{1}{2\pi n}$$

**Note:** (*i*) For conversion of exponential fourier series into trigonometric fourier series:

$$a_0 = C_0 a_n = C_n + C_{-n}$$
$$b_n = j(C_n - C_{-n})$$

(*ii*) For conversion of trigonometric fourier series into exponential fourier series:

$$c_0 = a_0$$

$$c_n = \dfrac{1}{2}(a^n - jb_n)$$

$$c_{-n} = \dfrac{1}{2}(a_n - jb_n).$$

**Problem 1.9.** *Obtain Fourier series representation for the periodic rectangular wave form shown in Fig.* 1.26.

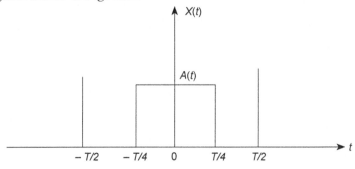

**Fig. 1.26**

**Solution:** The expression for Fourier series is given by:

$$x(t) = a_0 + \sum_{n=1}^{\infty}(a_n \cos n\omega_0 t + b_n \sin n\omega_0 t) \qquad \ldots(i)$$

Since the given waveform is symmetrical about the vertical axis, *i.e.*, $x(t) = x(-t)$, so in the fourier series representation, only cosine terms are present. This means $b_n = 0$.

As the given waveform is also symmetrical about horizontal axis, the average area is zero. Hence dc term $a_0 = 0$.

Putting values of $a_0$ and $b_n$ in the equation (*i*), we get

$$x(t) = \sum_{n=1}^{\infty}a_n \cos n\omega_0 t \qquad \ldots(ii)$$

Also from the Fig, $\quad x(t) = -A$ for $\quad -\dfrac{T}{2} < t < -\dfrac{T}{4} + A$

for $\qquad -\dfrac{T}{4} < t < +\dfrac{T}{4}$

$\qquad\qquad -A$ for $+\dfrac{T}{4} < t < +\dfrac{T}{2}$

Now, $\qquad\qquad a_n = \dfrac{2}{T}\int_{-T/2}^{T/2} x(t)\cos n\omega_0 t\, dt$

On solving above, we get $a_n = \dfrac{8A}{n\omega_0 T}\sin\left(\dfrac{n\omega_0 T}{4}\right) - \dfrac{4A}{n\omega_0 T}\sin\left(\dfrac{n\omega_0 T}{2}\right)$

$$\omega_0 = \dfrac{2\pi}{T}$$

or $\qquad\qquad \omega_0 T = 2\pi$

So for all integer values of *n*, the second term in the above expression is zero.

$$a_n = \frac{8A}{2n\pi}\sin\left(\frac{n\pi}{2}\right) = \frac{4A}{n\pi}\sin\left(\frac{n\pi}{2}\right)$$

Putting value of $a_n$ in equation $(ii)$,

$$x(t) = \sum_{n=1}^{\infty}\frac{4A}{n\pi}\sin\left(\frac{n\pi}{2}\right)\cos n\omega_0 t$$

$$x(t) = \frac{+4A}{3\pi}\sin\left(\frac{3\pi}{2}\right)\cos 3\omega_0 t + \ldots \qquad (n=1)$$

$$x(t) = \frac{+4A}{3\pi}\times(1\cos\omega_0 t) + 0 + \frac{4A}{\pi}(-1)\cos 3\omega_0 t + \ldots$$
$$(n=2)$$

$$x(t) = \frac{+4A}{\pi}\left[\cos\omega_0 t - \frac{1}{3}\cos 3\omega_0 t + \frac{1}{5}\cos 5\omega_0 t \ldots\right](n=3)$$

## 1.21 POLAR FOURIER SERIES REPRESENTATION

The polar fourier series can be derived from Trigonometric Fourier series. Here we reproduce trigonometric Fourier series

$$x(t) = a_0 \sum_{n=1}^{\infty}(a_n \cos n\omega_0 t + b_n \sin n\omega_0 t)$$

$$= a_0 \sum_{n=1}^{\infty}\sqrt{a_n^2 + b_n^2}\left[\frac{a_n}{\sqrt{a_n^2 + b_n^2}}\cos n\omega_0 t + \frac{b_n}{\sqrt{a_n^2 + b_n^2}}\cdot\sin n\omega_0 t\right]$$

Putting $\qquad \dfrac{a_n}{\sqrt{a_n^2 + b_n^2}} = \cos\phi_n$

and $\qquad \dfrac{b_n}{\sqrt{a_n^2 + b_n^2}} = \sin\phi_n$

$\therefore \qquad \tan\phi_n = \dfrac{b_n}{a_n}$

or $\qquad \phi_n = \tan^{-1}\left(\dfrac{b_n}{a_n}\right)$

$$x(t) = a_0 + \sum_{n=1}^{\infty}\sqrt{a_n^2 + b_n^2}\cdot[\cos n\omega_0 t \cos\phi_n + \sin n\omega_0 t \sin\phi_n]$$

$$= a_0 + \sum_{n=1}^{\infty}\sqrt{a_n^2 + b_n^2}\cdot\cos(n\omega_0 t - \phi_n)$$

Let $\qquad D_0 = a_0$ and $D_n = \sqrt{a_n^2 + b_n^2}$

Then $\qquad x(t) = D_0 + \sum_{n=1}^{\infty}D_n \cos(n\omega_0 t - \phi_n)$

and $\qquad \phi_n = \tan^{-1}(b_n/a_n)$.

## 1.22  COMPLEX FOURIER EXPONENTIAL SERIES

This can also be derived from trigonometric Fourier Series.

The trigonometric fourier series is given as:

$$x(t) = a_0 \sum_{n=1}^{\infty} (a_n \cos n\omega_0 t + b_n \sin n\omega_0 t) \qquad \ldots(1)$$

**From Euler Identiy:**

$$e^{j\theta} = \cos \theta + j \sin \theta \qquad \ldots(2)$$

and
$$e^{-j\theta} = \cos \theta + j \sin \theta \qquad \ldots(3)$$

Adding (2) and (3), we get: $2 \cos \theta = e^{j\theta} + e^{-j\theta}$

and
$$\cos \theta = \frac{e^{j\theta} + e^{-j\theta}}{2}$$

Subtracting (2) and (3), we get: $2j \sin \theta = e^{j\theta} + e^{-j\theta}$

$$\sin \theta = \frac{e^{j\theta} - e^{-j\theta}}{2j}$$

$\therefore$
$$\cos n\omega_0 t = \frac{e^{jn\omega_0 t} - e^{-jn\omega_0 t}}{2}$$

and
$$\sin n\omega_0 t = \frac{e^{jn\omega_0 t} - e^{jn\omega_0 t}}{2}$$

Substituting these values in equation (1), we get

$$x(t) = a_0 + \sum_{n=1}^{\infty} a_n \left[ \frac{e^{jn\omega_0 t} + e^{-jn\omega_0 t}}{2} \right] + b_n \left[ \frac{e^{jn\omega_0 t} - e^{-jn\omega_0 t}}{2} \right]$$

$$= a_0 + \sum_{n=1}^{\infty} \left[ \frac{(a_n + jb_n)e^{jn\omega_0 t}}{2} + \frac{(a_n + jb_n)e^{-jn\omega_0 t}}{2} \right] \ldots(4)$$

Let
$$C_n = \frac{1}{2}(a_n - jb_n)$$

$$C_{-n} = \frac{1}{2}(a_n + jb_n)$$

$$(C_{-n} \text{ is the complex conjugate of } C_n \therefore C_n = C_{-n})$$

and
$$C_0 = a_0$$

Putting these values in equation (4).

$$x(t) = C_0 + \sum_{n=1}^{\infty} C_n . e^{jn\omega_0 t} + \sum_{n=1}^{\infty} C_{-n} . e^{-jn\omega_0 t}$$

$$= C_0 + \sum_{n=1}^{\infty} C_n . e^{jn\omega_0 t} + \sum_{n=-\infty}^{\infty} C_{-n} . e^{-jn\omega_0 t}$$

$$= \sum_{n=-\infty}^{\infty} C_{-n} \cdot e^{-jn\omega_0 t}$$

Substituting for $a_n$ and $b_n$

$$C_n = \frac{1}{2}(a_n - jb_n) = \frac{1}{T} \int_{-x/2}^{\pi/2} x(t)[\cos n\omega_0 t - j\sin n\omega_0 t] dt$$

Hence

$$C_n = \frac{1}{T} \int_{-\pi/2}^{\pi/2} x(t) e^{-jn\omega_0 t} dt$$

and

$$C_{-n} = \frac{1}{T} \int_{-\pi/2}^{\pi/2} x(t) \cdot e^{-jn\omega_0 t} dt .$$

## 1.23 CONCEPT OF NEGATIVE FREQUENCY

The negative frequencies are present in complex exponential fourier series expression. In fact, negative frequency signals are not physical signals. They are only a mathematical tool to provide a real signal by combination of complex exponentials of positive and negative frequencies.

A real function $x$ $(t)$ cannot be represented by complex exponential functions such as $e^{jn\omega_0 t}$, (of positive frequency $n\omega_0 t$) without associating with their opposite rotating counter part $i.e.$, $e^{-jn\omega_0 t}$ (of negative frequency $-n\omega_0$).

The functions like $e^{jn\omega_0 t}$ and $e^{-jn\omega_0 t}$ have no physical meaning, except that they provide real functions like ($\cos n$ $e\omega_0 t$ and $\sin n\omega_0 t$), when combined together.

## 1.24 HARTLEY AND SHANNON HARTLEY THEOREMS

(*a*) **Hartley Theorem:** According to Hartley theorem, "The highest frequency required to pass *b* bits per second is only *b*/2 Hz."

Hence using Binary coding, the channel capacity (in bits per second) is equal to twice the bandwidth in Hz. In general, we can state that in the absence of noise,

$$C = 2B \log_2 N \qquad\qquad ...(1)$$

where

$C$ = channel capacity in Bits/sec

$B$ = channel bandwidth in Hz

$N$ = No. of coding levels.

Using binary coding system, the equation (1) reduces to

$$C = 2B$$

Hence, the Hartley law shows that the B.W. required to transmit information at a given rate is proportional to the rate of information. The law further shows that in absence of noise, greater the number of levels in the coding system, the greater is the information, which may be sent through a channel. The Hartley law can be extended to

$$H = C.t = 2B.t. \log_2 N$$

where $H =$ total information in bits sent in time $t$.

$$t = \text{time in seconds.}$$

(b) **Shannon Hartley theorem:** The Shannon Hartley theorem gives the capacity of a channel when its B.W. and noise level are known.

$$C = B \log_2 \left[ 1 + \frac{S}{N} \right]$$

where,                              $C$ = channel capacity in Bits/sec

$$B = \text{Bandwidth (Hz)}$$

$$\frac{S}{N} = \text{Single-noise ratio}$$

## Note that

(i) The Shannon Hartley theorem gives max. signalling speed in a channel in which the noise is purely random. The theorem is used to find channel capacity.

Limiting channel speed for a typical telephone channel is about 33 K. bits per second. However putting this value in equation $C = 2 \log_2 N$, we get code level $N = 39.8$, such a system will be very complex. Practically the channel speed does not exceed 11 K bits/sec.

(ii) Interpreting that by doubling the speed, the equation, $C = B \log_2 \left[ 1 + \frac{S}{N} \right]$ will give double Bandwidth will be wrong. Actually capacity gets increased only by 80% depending on S-N ratio, we can bargain B.W. for S-N. ratio. Low channel capacity does not mean that the desired amount of information cannot be sent over a given channel, it simply means that sending this amount of information will take longer time as evident from equation $H = 2 B. t. \log^2 N$.

(iii) Lastly, the Shannon Hartley theorem puts a limitation. Any attempt to exceed the limit will result in an error. The max. acceptable error is 1 in $10^5$ only.

## SUMMARY

1. The transfer of message from one place to another is called communication.

2. The communication may be oral like talking or telephone and written like through a letter or FAX.

3. The steps of communication are encoding, transmission, reception, decoding and use.

4. At the beginning, communication was carried out through pigeons and then through men. The first post box started in India was around 1930.

5. The electronic communication is carried out through telephony, telegraphy, FAX, radio, T.V., radar etc.

6. The electronic communication may be wire communication *e.g.*, telephony and *wireless communication e.g.*, T.V., mobile etc.

7. In wireless communication, the one of the parameters (amplitude, frequency, phase) of a wave (called carrier) is changed according to that of the information (called signal) before transmission. The process is called modulation.

8. Proper bandwidth is required for carrying out communications.

9. The signals may be continous/discrete and periodic/non-periodic etc.

10. The signals may be analysed by Fourier series.

11. The Hartley Shannon theorem is used to find channel capacity.

❑❑❑

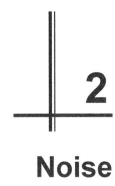

# Noise

Noise in electrical sense, is an unwanted form of signal or energy that may interfere with the reception and reproduction of wanted signal, *e.g*, in radio receivers the noise may produce hissing sound and in T.V. receivers, noise in the form of "snow" becomes superimposed on the picture.

Reception of a signal in a telecommunication system is marred by noise. In radio receivers, the noise may cause the loudspeaker to sound fuzzy. In pulse communication, the noise may obliterate the pulses, thus causing serious error.

## 2.1 SOURCES OF NOISE

The source of noise are:

(*i*) It may be due to faulty connection in an equipment.

(*ii*) It may be make or break of an electrical circuit *e.g.*, in ignition system of an automobile or at the brushes of electrical machines.

(*iii*) It may be a natural phenomenon *e.g.*, storms, lightening, solar flares or radiations in space.

(*iv*) The fundamental source within an equipment *i.e.*, due to materials used to make the equipment, *e.g.*, wires, components etc.

## 2.2 CLASSIFICATION OF NOISE

The noise can be classified as shown below:

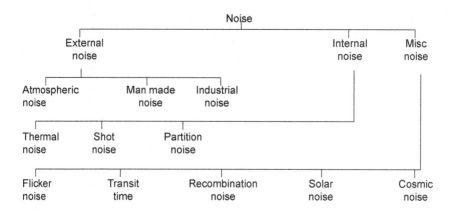

## 2.3 POWER SPECTRAL DENSITY (PSD)

The PSD is a curve, which shows the energy distribution as a continuous function of frequency. A typical PSD curve has been shown in Fig. 2.1.

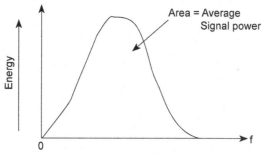

**Fig. 2.1**

The units for PSD are Watt per hertz (or Joules). The area under the curve has units of Joules; which is dimensionally equivalent to Watts ; But as Watts are the units of power, therefore total area under the curve gives average signal power ; this means the total average power is equal to the product of PSD

(Watt/hertz) and the Bandwidth (in Hz); $\left[ \dfrac{\text{Watts}}{\text{Hz}} \times \text{Hz} = \text{Watts} \right]$.

## 2.4 EXTERNAL NOISES

These are the noises created outside the receiver and are of following types:

(*a*) **Atmospheric Noise:** The atmospheric noise is caused by lightning and such other electrical disturbances in the atmosphere. These are in the form of impulses; called *static*.

The atmospheric noise is less severe at frequencies above 30 MHz, and also these are increased in level at night at both broadcast and short wave frequencies.

The field strength is inversely proportional to the frequency, thus this noise interferes more with the reception of radio than that of T.V. The impulses of this noise are non sinusoidal waves having harmonics whose amplitude falls with increase in the harmonic, *i.e.*, the fifth harmonic has less amplitude than that of the fourth harmonic and so on.

(*b*) **Man Made Noise or Industrial Noise:** The man made noise arises due to electrical devices such as fluorescent tubes, switch gears, electric motors, high tension lines etc. This is also produced due to the arc produced during automobile and aircraft ignitions.

The man made noise is difficult to analyse as it does not obey the general principles.

Between 1 to 600 MHz frequencies in industrial areas, the intensity of man made noise easily strips out the noise created by any other source: external or internal.

## 2.5 INTERNAL NOISES

These are the noise created by active or passive devices used inside the receiver. This is a random noise and the noise power is proportional to the bandwidth over which it is measured.

The internal noise may be of the following types:

(*a*) Thermal noise

(*b*) Shot noise

(*c*) Partition noise

We will study these in brief.

## 2.6 THERMAL OR RESISTANCE NOISE

The free electrons within a conductor are always in random motion due to the thermal energy. Though the average voltage resulting from this motion is zero, the average power available is not zero, because the noise power results from thermal energy.

This is known as **thermal noise** (or Johnson noise on the name of the discoverer). The another name for this noise is **agitation or white noise**.

This is also called as **resistance noise**. In transistors, the resistance noise depends upon the emitter, collector and base resistances. The base resistance plays the major role.

The average noise power ($P_n$) generated by a resistor (or a resistive component of an impedence) is proportional to the bandwidth as well as to its absolute temperature. Mathematically,

$$P_n \quad \alpha \; BT \; \text{Watts}$$

$$P_n = KBT \; \text{Watts}$$

where,         $P_n = $ noise power

$K = $ Boltzmann's constant $= 1.38 \times 10^{-23}$ J/kelvin

$B = $ Bandwidth, over which the noise is to be measured.

$T = $ absolute temperature $= 273 + °C$.

***Resistance noise generator:*** A resistance may be considered as a noise generator and there may be even a quite large voltage across it. But since it is due to random motion of the molecules of the resistance, therefore, the voltage has a definite rms (root mean square) value, but no dc component. An alternating current (ac) meter can only measure this alternating voltage. Due to the noise voltage, a noise current is also constituted.

The circuit of a resistor as noise voltage generator is shown in Fig. 2.2.

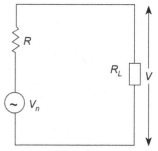

$R = $ The resistance, as noise generator

$V_n = $ noise voltage (to be calculated)

**Fig. 2.2**

$R_L = $ The load across which noise power ($P_n$) is to be developed.

For maximum transfer of power, the source impedence should be equal to the load impedence *i.e.*,

$$R = R_L$$

Power generated in         $R_L = P_n = \dfrac{V^2}{R_L} = \dfrac{V^2}{R}(R = R_L)$

$\therefore$         $P_n = \dfrac{(V_n/2)^2}{R}[V = V_n/2]$

$P_n = \dfrac{V_n^2}{4R}$

$\therefore$         $V_n^2 = 4RP_n$

$V_2^n = 4R.K.B.T \qquad\qquad [P_n = K.B.T.]$

or         $V_n = \sqrt{4RKBT}$

From the derived relation, $V_n^2 = 4RKBT$ we observe that

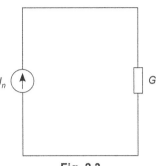

**Fig. 2.3**

(*i*) The rms noise voltage square is proportional to the absolute temperature, bandwidth and the value of the resistance.

(*ii*) The noise voltage is independent of the frequency.

(*iii*) The noise voltage is at random and evenly distributed over the entire frequency spectrum.

The equivalent circuit for thermal noise can also be drawn as shown in Fig. 2.3 and for the similar reasons, the noise current

$$I_n^2 = 4 \, GKBT$$

where,                    $G = \text{Conductance} = \dfrac{1}{R}$

If a resistance $R$ delivers thermal noise power to a load $R_L$, the $R_L$ must also deliver the same power to $R$ and in thermal equilibrium, the net exchange of power is zero. But if the load is a loudspeaker, the noise output will be that resulting both from $R_L$ and $R$.

*Noise Curve:* The curve between instantaneous thermal noise voltage and the time can be drawn as shown in Fig. 2.4.

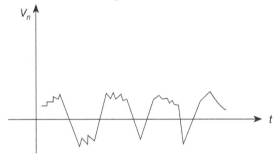

**Fig. 2.4**

*Power Spectrum Density:* For thermal noise the power spectrum density is the average noise power per hertz of its bandwidth. This is represented by $S_n$ and,

$$S_n = KT \text{ watt per Hz. (where } K = \text{Boltzmann Constant)}$$

The spectrum density graph for thermal noise has been shown in Fig. 2.5.

At room temperature of 23°C, the power/Hertz.

$$S_n = KT = 1.38 \times 10^{-23} \times (273 + 23)$$
$$= 1.38 \times 10^{-23} \times 296$$
$$= 408.48 \times 10^{-23} \text{ W/Hz.}$$

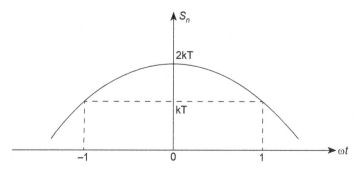

**Fig. 2.5**

The available noise power for a bandwidth ($B$) of 1 MHz.

$$p_n = S_n \cdot B$$
$$= 408.48 \times 10^{-23} \times (1 \times 10^6)$$
$$= 408.48 \times 10^{-17} \text{ W}$$

**Note:**

(*i*) We have seen that a resistor acts as a thermal noise voltage generator. If a no. of resistors are joined in series, the resultant thermal noise voltage is given by

$$V_n^2 = 4(R_1 + R_2 + R_3...) \text{ KTB}$$

(*ii*) If a no. of resistors are joined in parallel, the resultant thermal noise voltage is given by.

$$V_n^2 = \left( \frac{1}{R_1} + \frac{1}{R_2} + \frac{1}{R_3}... \right) \text{KTB}$$

and the thermal noise current, can be given as

$$I_n^2 = 4 \, (G_1 + G_2 + G_3...) \text{ KBT}$$

(*iii*) The readers should recall that input to an oscillator is noise voltage only.

(*iv*) The contribution of thermal noise is limited by the bandwidth of the circuit. The thermal noise, therefore, has a constant power density spectrum, *i.e.*, it contains all frequencies in equal amount. That is why thermal noise is also called as **white noise** as white implies all colours as frequencies.

## 2.7 SHOT NOISE

This occurs in all active devices (Vacuum diode, Vacuum Triode, PN Junction, transistor etc.). It occurs due to random variation in the arrival of no. of electrons (or holes) at the output electrode (plate in case of vacuum devices, and *collector* in case of semi conductor devices). When this sound is amplified by a loud speaker in a radio receiver, it appears like a **shower of gun** shots falling on a metal sheet hence the name as *shot noise*.

We take example of a bipolar transistor. Though the collector current is more or less constant, minute variations occur in its value as no. of electrons or holes reaching the collector is varying. The effect is known as "shot effect" and this effect is the source of shot noise.

The shot noise behaves in the same manner as the thermal noise, though both have a different source.

The shot noise is inversely proportional to the transconductance of the triode or transistor and directly proportional to the output current. The shot noise has also a uniform spectrum density like thermal noise and its root mean square (rms) noise current depends upon the direct current.

The root mean square (rms) shot noise current in case of diode is given by

$$I_n^2 = 2\, I_{dc} \cdot Q_c \cdot B$$

where $\qquad\qquad I_{dc}$ = dc component of current

$\qquad\qquad Q_e$ = charge of one electron = $1.6 \times 10^{-19}$ C

$\qquad\qquad B$ = effective bandwidth.

The most convenient method of dealing with shot noise is to find the value of an equivalent "noise resistor". In other words, the noise current is replaced by a resistor $(R_n)$ so that it is easy to add the shot noise and thermal noise. The value of equivalent value $(R_{eq})$ is quoted on the devices by the manufacturers.

The noise resistance is only a fictitious resistance, whose sole function is to make shot noise calculations easy.

Everytime a charge carrier moves from cathode to plate (or to anode), the current in vacuum tubes/transistors flows in the form of discrete pulses. Hence, though the current appears to be continuous, but it is not. The Fig. 2.6 shows shot noise current variation with time.

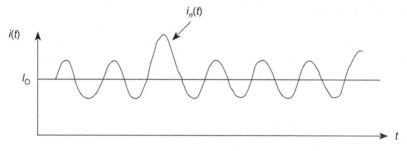

**Fig. 2.6**

The current fluctuates about a mean (average) value $I_0$. The total current is:

$$i\,(t) = I_0 + i_n\,(t) = \text{mean current} + \text{shot noise current}.$$

**Power density spectrum for shot noise:** The shot noise current is given by

$$i(t) = I_0 + i_n\,(t)$$

For all practical purposes, it may be assumed that power density of statistically independent non-interacting and random noise current $i_n\,(t)$ is expressed as: $S_n = q.\,I_0$

where, $q$ is the charge on an electron ($1.6 \times 10^{-19}$ C) and $I_0$ is the mean value of current. The variation of power density with frequency has been shown in Fig. 2.7.

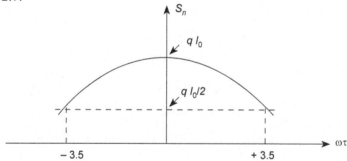

**Fig. 2.7**

## 2.8 PARTITION NOISE

This noise occurs when a current divides between two or more paths and results from the random fluctuations in this division. It can be understood that a transistor will be more noisy than a diode, since the former has more no. of paths than the latter. The noise in a three terminal device can be reduced if the third terminal does not draw any current. Recently gallium arsenide (Ga-As) field effect transistors are used in microwave applications, as these transistors draw zero gate current. **The spectrum for the partition noise is flat.**

The root mean square (rms) value of noise current in partition noise for transistors may be given as:

$$I_n^2 = 2I_C B \left[ 1 - \frac{|\alpha|^2}{\alpha_0} \right]$$

where $I_C$ = Collector current

$\alpha$ = Common base short circuited current amplification factor

$\alpha_0$ = Current amplification factor (low frequency value.)

$B$ = effective bandwidth.

Over a 10 kHz frequency band, the rms value of the partition noise current in a transistor carrying 1 mA collector current and common base amplification factor of $\alpha = 0.97$, is approximately 0.3 mA.

**Problem 2.1.** *Two resistors of 20 and 30 kilo-ohms are at room temperature. Calculate thermal noise voltage.*

(*a*) For each resistor

(*b*) For both the resistors in series

(*c*) For both resistors in parallel.

Take bandwidth as 100 kHz and $KT = 4 \times 10^{-21}$ W/Hz.

**Solution.**

(*a*) For $R = 20$ Kilo-ohms

$$V_n^2 = 4\,RBKT$$

$$= 4 \times (20 \times 10^3)\,(100 \times 10^3)\,(4 \times 10^{-21})$$

$$= 32 \times 10^{-12}\,\text{V}$$

$$V_n = 5.66\,\mu V \quad \textbf{Ans.}$$

For $\qquad R = 30$ Kilo-ohms

$$V_n = 5.66 \times \sqrt{\frac{30}{20}} = 6.93\,\mu V \quad \textbf{Ans.}$$

(*b*) The resistors in series

Total resistance $= 30 + 20 = 50$ Kilo-ohms

$$V_n = 5.66 \times \sqrt{\frac{50}{20}} = 8.94\,\mu V \quad \textbf{Ans.}$$

(*c*) The resistances is parallel

Total resistance $= \dfrac{20 \times 30}{20 + 30} = 12$

$\therefore \qquad V_n = 5.66 \sqrt{\frac{12}{20}} = 4.38\,\mu V \quad \textbf{Ans.}$

**Problem 2.2.** *An amplifier operating over the frequency range of* 18 *to* 20 *MHz has a* 10 *Kilo-ohms input resistance. Calculate the rms noise voltage of the input of the amplifier. Take ambient temperature of* 27°*C and Boltzman's constant* $K = 1.38 \times 10^{-23}$ *J/Kelvin.*

**Solution.**

Frequency bandwidth

$$B = 20 - 18 = 2 \text{ MHz} = 2 \times 10^6 \text{ Hz}$$

$$R = 10 \times 10^3 \text{ ohms.}$$

$$T = 27 + 273 = 300° \text{ Kelvin}$$

$$K = 1.38 \times 10^{-23}$$

$$V_n^2 = 4 \, RBKT$$

$$= 4 \times (10 \times 10^3) \times (2 \times 10^6) \times 1.38 \times 10^{-23} \times 300$$

$$V_n = \sqrt{4 \times 2 \times 1.38 \times 3 \times 10^{-11}}$$

$$V_n = 18.2 \, \mu V \quad \textbf{Ans.}$$

**Problem 2.3.** *The emmision current of a diode is* 10 *mA. Determine the rms value of shot noise current for a 10 MHz bandwidth.*

**Solution.** From the equation, Noise Current,

$$I_n^2 = 2 \, I_{dc} \cdot Q_e \cdot B$$

$$= 2 \times (10 \times 10^{-3}) \times 1.6 \times 10^{-19} \times (10 \times 10^{-6})$$

$$= 3.2 \times 10^{-14}$$

$$I_n = 1.78 \times 10^{-7} \text{ Amp} = 0.178 \, \mu A \quad \textbf{Ans.}$$

## 2.9 OTHER NOISES

The Other noises important in communication are described below:

1. **Flicker noise/Modulation noise/Low frequency noise:** At low audio frequencies, a poorly understood form of noise is found in transistors, this is called flicker or modulation noise. It is proportional to emitter current and inversely proportional to the frequency. It can be ignored above 500 Hz. As the **spectrum density** increases as the frequency decreases, this is also called 1/f noise.

   In vacuum tubes, the flicker noise occurs due to slow changes taking place in the oxide coated cathodes. In semi conductor devices, the flicker noise is more serious. The mean square value of noise voltage is proportional to the square of the direct current flowing.

2. **Transit Time or High Frequency Noise:** When the transit time of electrons (or holes) crossing a junction in semi conducting devices (or from cathode to the control grid in a vacuum tube) is comparable with the periodic time of the signal, some of the carriers may diffuse back to the emitter (or to the cathode), this gives rise to a noise known as transit time noise. The **spectrum density** of this noise increases with the frequency. So it is more significant at high frequencies.

   **Note:** The transist time of an electron (time of an electron to reach from cathode to anode) in a diode depends upon anode voltage $V$ and may be expressed as:

   $$\tau = 3.36 \times \frac{d}{\sqrt{V}} \, \mu \sec.$$

   where $d$ is the spacing between anode and cathode.

3. **Generation/Recombination Noise:** In semi conductor devices, whenever a random generation or recombination of carriers occur, this gives rise to a noise current, (when direct current flows through the semiconductor). The noise generated is known as Generation or recombination noise.

4. **Solar Noise:** Along with many other radiations, sun also radiates noise. Under normal conditions too, the sun radiates constantly noise radiation due to high temperature at its surface (about 6500°C). It radiates a very broad frequency spectrum, which includes the frequencies we use for communication.

5. **Cosmic Noise:** The distant stars have high temperature and they also radiate noise in the same manner as the sun does. This is called cosmic noise which is similar to thermal noise and is uniformly distributed with frequency. It is interesting to know that **we also receive noise from galaxies (milky ways).**

   The solar noise and the cosmic noise come under the heading "space or terrestrial noise". This is observed in the range from 10 MHz to 1.43 GHz (1 GHz = $10^9$ Hertz).

6. **Electrical noise or Humming:** This noise is generated due to power line frequencies (50 Hz) in the audio device. The power frequencies contain its harmonics (*i.e.,* 100 Hz, 200 Hz etc.) which causes noise. A good filter circuitry can reduce this noise.

7. **Cross Talk.** This noise enters in an audio equipment due to presence of other circuits in the vicinity, which causes inductive or capacitive coupling with the audio circuits. This noise can also be reduced by proper filtering.

8. **Generated Noise.** This noise is generated in amplifiers. When signal enters an amplifier, this noise is produced or generated due to various components (resistors, capacitors, transistors etc., of the amplifier). This noise is added to the signal (See Fig. 2.8).

**Fig. 2.8**

Generated noise further may be of the following types:

(*a*)  *Johnson Noise.* This noise is produced when a signal passes through a resistor. The atoms of the resistor remain in continuous vibration and produce noise, which is added up in the signal.

(*b*)  *White Noise.* The vibrations produced by thermal effect within a resistor cover a wide frequency range and therefore the noise generated consists of a wide spectrum of frequencies. This wide-band noise is sometimes called white noise.

(*c*)  *Pink Noise.* Pink noise is a signal, whose noise power per unit frequency interval is inversely proportional to frequency itself over a specified range. It is white noise passed through filter.

9. **Conducted Noise.** The power supply to the amplifiers may itself be a source of noise as it contains ripples, spikes etc. This noise is called conducted noise.

10. **Radiated Noise.** There may be electrical or magnetic fields in the surrounding. Unwanted signal may radiate into the audio device and produce noise. Such noise may be called Radiated noise.

The Fig. 2.9 shows sources of radiated noise.

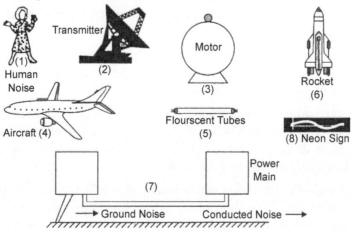

**Fig. 2.9**

11. **General noise (or Ambient noise).** The noise present all the time *e.g.*, class room, auditorium is called General noise.

## 2.10 HARMFUL EFFECTS OF NOISE AND TO REDUCE NOISE

(*a*) Working continuously in a noisy environment has the following ill effects on human beings:

    (*i*)   It produces hypertension or high B.P.

    (*ii*)   It causes mental fatigue and lowers efficiency.

    (*iii*)   It may retard mental growth.

    (*iv*)   It may cause harm to our hearing system.

    (*v*)   It may cause nervous breakdown.

(*b*)   (*i*)   The *Noise* may be reduced by using regulated power supply, use of synchronous and servo motors in drives, and by using special techniques (*e.g.*, Dolby method).

    (*ii*)   The noise may be reduced by using sound absorbers/insulators in the room. In general, room should be provided with thick curtains.

## 2.11 NOISE RESISTANCE

It is convenient to represent the noise which originates in a device by means of a *fictitious resistance* $R_n$ assumed to be equal to the generated noise at the room temperature. Now the actual device is then assumed as noiseless. The value of the noise resistance is generally mentioned by the manufacturer.

The [Fig. 2.10 (*a*)] shows an amplifier with internal resistance $R_i$ connected with a supply source $E_s$. The [Fig. 2.10 (*b*)] shows the amplifier replaced by an equivalent noise resistance $R_n$. The equivalent r.m.s. noise voltage at the input terminals.

$$V_n^2 = 4\,(R_i + R_n)\,KTB.$$

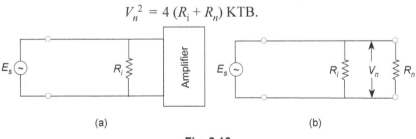

(a)                         (b)

**Fig. 2.10**

The amplifier noise is referred to the input circuit, so that the amplifier may be assumed as noiseless. The important thing to be noted is that $R_n$ is a fictitious resistance and does not affect the real resistance at the input. In Fig. 2.10(b) the input resistance remains as $R_i$ and the effective resistance from the point of view of noise generation is $(R_i + R_n)$.

**Problem 2.4.** *Calculate equivalent **noise resistance** for a triode to be assumed as noiseless. The mutual conductance of the triode is 3.6 mA/V.*

**Solution.**                              $g_m = 3.6$ mA/V $= 3.6 \times 10^{-3}$ A/V

If the triode to be assumed as noiseless, the equivalent noise resistance,

$$R_n = \frac{2.5}{g_m} = \frac{2.5}{3.6 \times 10^{-3}}$$

$$R_n = 694.44 \text{ ohm.} \quad \textbf{Ans.}$$

**Problem 2.5.** *Calculate noise voltage generated at input terminals of an amplifier for a bandwidth of 10 kHz and at a room temperature of 17°C. Assume $R_{in} = 600 \ \Omega$ and $R_n = 400 \ \Omega$.*

The source voltage is 1.0 eV and source resistance is 50 $\Omega$.

**Solution.**

Here,                                $K = 1.38 \times 10^{-23}$

(Recall that $K$ is the Boltzmann constant)

$$T = 17 + 273 = 290 \text{ Kelvin}$$

$$B = 10 \times 10^3 = 10^4 \text{ Hz}$$

Now, source resistance, $R_s = 50 \ \Omega$

Input resistance of amplifier $R_{in} = 600 \ \Omega$

The total input resistance ($R_s$ in parallel with $R_i$)

$$R_t = \frac{50 \times 600}{50 + 600} = 46.15 \ \Omega$$

∴ The noise voltage   $V_n^2 = 4 (R_t + R_n) \text{ K.T.B.}$

$$V_n^2 = 4 (46.15 + 400) \times (1.38 \times 10^{-23}) \times 290 \times 10^4$$

$$= 4 \times 446.15 \times 1.38 \times 10^{-23} \times 290 \times 10^4$$

$$= 71419.7 \times 10^{-18}$$

$$V = 2.672 \times 10^{-9} \text{ V} = 0.2672 \ \mu\text{V} \quad \textbf{Ans.}$$

**Problem 2.6.** *Calculate noise voltage at the input of an amplifier having 300 $\Omega$ equivalent noise resistance and 500 $\Omega$ input resistance. The bandwidth of the amplifier is 6 MHz and room temperature is 17°C. Assume Boltzmann constant $K = 1.38 \times 10^{-23}$.*

**Solution.**                       $R = R_i + R_n = 500 + 300 = 800 \ \Omega$

$$B = 6 \text{ MHz} = 6 \times 10^6 \text{ Hz.}$$
$$T = 17 + 273 = 290$$

Using equation, $\qquad V_n^2 = 4R \text{ K.B.T.}$

$$= 4 \times 800 \times 1.38 \times 10^{-23} \times 6 \times 10^6 \times 290$$
$$V_n^2 = 6888.96 \times 10^{-14}$$
$$V_n = 82.99 \times 10^{-7} \text{ volts}$$
$$V_n = 8.299 \ \mu\text{V} \qquad \textbf{Ans.}$$

**Problem 2.7.** *The Table shows parameters of a two stage amplifier.*

| Parameters | First Stage | Second Stage |
|---|---|---|
| Voltage gain | $A_1 = 10$ | $A_2 = 25$ |
| Input resistance | $600 \ \Omega$ | $81 \ K\Omega$ |
| Equivalent noise resistance | $1600 \ \Omega$ | $10 \ K\Omega$ |
| Output resistance | $27 \ \Omega$ | $1 \ M\Omega$ |

**Calculate the overall noise resistance of the whole cascade.**

**Solution.**

$$\text{Here, } R_1 = 600 + 1600 = 2200 \ \Omega$$
$$R_2 = \frac{27 \times 81}{27 + 81} + 10 = 30.2 \ K\Omega$$
$$R_3 = 1 \ M\Omega$$

The overall equivalent noise resistance,

$$R_n = R_1 + \frac{R_2}{A_1^2} + \frac{R_3}{A_1^2 A_2^2}$$
$$= 2200 + \frac{30.2 \times 10^3}{10^2} + \frac{1 + 10^6}{10^2 \times 20^2}$$
$$= 2200 + 302 + 16$$
$$R_n = 2518 \ \Omega \quad \textbf{Ans.}$$

## 2.12 SIGNAL-NOISE RATIO (SNR)

The signal noise ratio may be defined as the ratio of signal power to the noise power at the same point.

$$\text{SNR} = \frac{\text{Signal power}}{\text{Noise power}} = \frac{P_s}{P_n}$$

Since Power is proportional to (voltage)$^2$

$$\text{SNR} = \frac{V_s^2 / R}{V_n^2 / R} = \left[ \frac{V_s}{V_n} \right]^2$$

where $V_s$ is the signal voltage and $V_n$ is noise voltage.

The above equation applies whenever the resistance ($R$), across which signal is developed is the same across which the noise is developed.

In a telecommunication system, the comparison of signal power with the noise power is important. The S-N ratio should be higher for good results.

**Problem 2.8.** *The signal voltage and noise voltage of a system is* 0.923 µV *and* 0.267 µV *respectively. Find SNR in number and decibel.*

**Solution.**

(*a*)  SNR = $V_s^2 / V_n^2 = [V_s / V_n]^2$

$$= \left[ \frac{0.923}{0.267} \right]^2 = 11.95 \quad \textbf{Ans.}$$

(*b*)  In decibel

SNR = 10 log 11.95 = 10.77 dB   **Ans.**

## 2.13  NOISE FIGURE OR NOISE FACTOR (F or NF)

The noise figure (or factor) is the ratio of signal noise power supplied at the input to the signal-noise power obtained at the output of a system. Thus,

$$F = \frac{\text{SNR at input}}{\text{SNR at output}} = \frac{P_{si} / P_{ni}}{P_{so} / P_{no}}$$

where $P_{si}$ = signal input power, $P_{ni}$ = Noise input power, $P_{so}$ = Signal output power, $P_{no}$ = Noise output power.

In a practical receiver, the output SNR will be lower than input SNR and the noise figure will be *more than* 1. The value of noise figure for an ideal receiver will be 1, which shows that it does not produce (add) any noise of its own. It fact, the noise factor is the measure of noise added by the receiver.

The noise figure or factor is frequency dependant hence frequency must be mentioned with the value of noise factor. When the noise factor is determined at one particular frequency, it is called as "spot noise factor," but when the noise factor is measured for a range of frequencies, it is known as "average noise factor".

Now,       *Power gain (G)* $= \dfrac{\text{SNR Power at output}}{\text{SNR Power at input}} = \dfrac{P_{so}}{P_{si}}$

Now                       F $= \dfrac{1}{G} \cdot \dfrac{P_{no}}{P_{ni}}$

or                        $P_{no} = F.G. \, P_{ni}$

$$P_{no} = F.G. \ (K.B.T.) = F.G. \ KBT$$

$[P_{ni} = KBT$, where $T$ is the room temperature$]$.

This shows that output noise power is increased by a factor $F$.

In decibel,                    $F_{db} = 10 \ \log_{10}F$

Generally noise factor is expressed as a number and the noise figure in db. Thus a system having noise figure as 18 db, will have a noise factor of 63.09.

Noise figure is a means of measuring deterioration in SNR produced by amplifiers.

## 2.14 NOISE FIGURE (FACTOR) OF CASCADED AMPLIFIERS

If number of amplifiers are in cascade and their noise figure is $F_1$, $F_2$, $F_3$ .... and power gain is $G_1$, $G_2$, $G_3$,... respectively, the overall noise figure of the cascade is given as

$$F = F_1 + \frac{F_2 - 1}{G_1} + \frac{F_3 - 1}{G_1 G_2} + ...$$

**Problem 2.9.** *Calculate the value of signal voltage in a receiver with a bandwidth of 20 kHz and input circuit having a resistance of 10 K$\Omega$ at 17°C.*

**Solution.** The mean square noise voltage.

$$
\begin{aligned}
V_n^2 &= 4 \ KTRB \\
&= 4 \times (1.38 \times 10^{-23}) \times (17 + 273) \times (10 \times 10^3) \\
&\quad \times (20 \times 10^3) \\
&= 4 \times 1.38 \times 10^{-23} \times 290 \times 10^4 \times 20 \times 10^3 \\
&= 4 \times 1.38 \times 29 \times 2 \times 10^{-23} \times 10^9 \\
&= 320.16 \times 10^{-14} \\
V_n &= 17.9 \times 10^{-7} \ V = 1.79 \ \mu V \quad \textbf{Ans.}
\end{aligned}
$$

**Problem 2.10.** *A receiver connected to an antenna of resistance 50 $\Omega$ had an equivalent noise resistance of 25 $\Omega$. Find the noise figure for the receiver in number and as well as in decibel.*

**Solution.**

If $R_a$ is the resistance of antenna, $R_n$ is the equivalent noise resistance of a receiver, the expression for noise figure is given by:

$$F = 1 + \frac{R_n}{R_a}.$$

Antenna resistance          $R_a = 50 \ \Omega$

Equivalent noise resistance $R_n = 25 \ \Omega$

(*a*)  The noise figure

$$F = 1 + \frac{R_n}{R_a} = 1 + \frac{25}{50} = 1.6 \quad \textbf{Ans.}$$

(*b*)  In db          $F_{db} = 10 \log 1.5 = 10 \times 0.1761 = 1.76 \quad \textbf{Ans.}$

**Problem 2.11.** *A mixer stage has a noise figure of 30 db and it is preceeded by an amplifier, that has a noise figure of 35 db and power gain 20 db. Calculate the over all noise figure referred to the input.*

**Solution.**  See Fig. 2.11

$$G_1 = 20db, \ F_2 = 30db$$
$$F_1 = 3bdb, \ F = ?$$

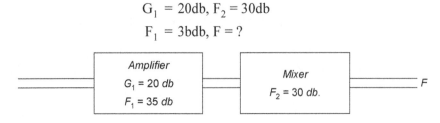

**Fig. 2.11**

Now                    $30 \ db = 10 \log F_2$

$$F_2 = \text{Antilog} \left( \frac{30}{10} \right) = 1000 \qquad \qquad \ldots(i)$$

and                    $35 \ db = 10 \log F_1$

$$F_1 = \text{Antilog} \left( \frac{35}{15} \right) = 3162.3 \qquad \qquad \ldots(ii)$$

Similarly,             $20 \ db = 10 \log G_1$

$$G_1 = \text{Antilog} \left( \frac{20}{10} \right) = 100 \qquad \qquad \ldots(iii)$$

Putting these values in the expression for the overall noise figure of the cascade:

$$F = F_1 + \frac{F_2 - 1}{G_1}$$

$$F = 3162.3 + \frac{1000 - 1}{100} = 3172.3 \quad \textbf{Ans.}$$

**Problem 2.12.**  The input resistance of the first stage of a two stage amplifier system is 500 $\Omega$. The equivalent input noise resistance of the system is 2.5 K$\Omega$.

Calculate noise figure of the system, if it is drawn by a generator, whose output resistance was found to be 60 $\Omega$. Also find its value in decibel.

**Solution.**          $R_n = (2.5 \times 1000) - 500 = 2000 \ \Omega$

(*a*) Now noise figure $F = 1 + \dfrac{2000}{60} = 34.33$   **Ans.**

(*b*) In decibel       $F = 10 \log (34.33) = 15.356 \text{ dB}$    **Ans.**

## 2.15 EQUIVALENT NOISE TEMPERATURE

The noise figure (or factor) is not convenient measure of noise specially in ultra high frequency and microwave low noise antennas or receivers. In these devices concept of equivalent "noise temperature" is employed for noise measurement. The advantages of using this concept are as follows:

(*i*) The noise temperature is additive like noise power.

(*ii*) The another advantage of using noise temperature for low noise levels is, that it shows a greater variation for any given noise level change than does the noise figure (or factor), therefore changes are easier to grasp.

As mentioned earlier, the equivalent noise temperature finds greatest use of microwave frequencies in connection with the noise at the receiver input. This noise arises from three main sources:

(*i*) The input amplifier.

(*ii*) The connections to the antenna.

(*iii*) The antenna itself.

**Note:**

(*i*) The available noise power from a resistor is directly proportional to the temperature and independent of the value of resistance, hence the concept of noise temperature is originated.

(*ii*) For devices with very low noise factor, it is very much convenient to use "noise temperature".

(*iii*) The equivalent noise temperature is denoted by $T$ and its unit is Kelvin (K°).

(*iv*) $T_{eq}$ or $T_n = T_0 (F - 1)$

where $T_{eq}$ or $T_n$ = Equivalent noise temperature, $T_0$ = room temperature $(17°C + 273 = 290° \text{ K})$ and $F$ = Noise factor.

### Derivation of Eq. Noise Temperature

We know that maximum noise power output of a resistor.

$$P_n = KBT \text{ watts}$$

The above equation can be rewritten giving total noise power from several sources:

$$P_t = P_1 + P_2 = KBT_t = KBT_1 + KBT_2$$

or
$$KBT_t = KB(T_1 + T_2)$$

$$T_t = T_1 + T_2$$

Where $P_1$ and $P_2$ are two individual noise powers, which respectively may be noise power received by the antenna and generated by antenna and $T_1$ and $T_2$ are the temperatures corresponding to $P_1$ and $P_2$ respectively and $T_t$ is the total noise temperature.

We can co-relate the noise figure $(F)$ and equivalent noise temperature $(T_{eq})$ as:

$$F = 1 + \frac{R'_{eq}}{R_a} = 1 + \frac{KT_{eq} \cdot B \cdot R'_{eq}}{KT_0 \, B \, R_a} = 1 + \frac{T_{eq}}{T_a} \qquad ...(1)$$

$R'_{eq} = R$ as assumed in the definition of $T_{eq}$.

$$T = 27°C = 27° + 273° = 300°K$$

$T_{eq}$ = Equivalent noise temperature of the receiver

or                                           amplifier.

From eq. (1),
$$T_0 F = T_0 + T_{eq}$$

$$T_{eq} = T_0 (F - 1)$$

i.e., Equivalent noise temperature can be measured, if noise figure is known.

**Problem 2.13.** *The equivalent noise temperature of an amplifier is* 30° K. *Find the noise factor. Take room temperature as* 290° K.

**Solution.** We know that:

$$T_n = 290 (F - 1)$$

$$30 = 290 (F - 1)$$

$$290 F - 290 = 30$$

$$290 F = 30 + 290 = 320$$

$$F = \frac{320}{290} = 1.103 \quad \textbf{Ans.}$$

## 2.16  ADDITION OF NOISE DUE TO SEVERAL SOURCES

(a) **Sources in series:** Let us consider several thermal noise sources *i.e.*, resistors $R_1$, $R_2$, $R_3$ etc., in series, producing noise voltage $V_{n1}$, $V_{n2}$, $V_{n3}$ etc. The rms noise voltage produced by a resistor $R$ is given as:

$$V_n = \sqrt{4RKBT} \text{ Volts}$$

Similarly,

$$V_{n1} = \sqrt{4R_1\,K\,B\,T} \ \text{Volts}$$

$$V_{n2} = \sqrt{4R_2\,K.B.T} \ \text{Volts}$$

$$V_{n3} = \sqrt{4R_3\,K\,B\,T} \ \text{Volts}$$

and so on

The resultant noise voltage will be given by the expression:

$$V_s = \sqrt{V_{n1}^2 + V_{n2}^2 + V_{n3}^2 + \dots}$$

Putting values of $V_{n1}$, $V_{n2}$, $V_{n3}$,..., we get,

$$V = 4KBT\sqrt{(R_1 + R_2 + R_3 + \dots)}$$

$$V_s = \sqrt{4\,K\,B\,T\,R_s} \ \text{Volts} \ (R_s = R_1 + R_2 + R_3 + \dots)$$

(b) **Sources in parallel:** Let us now consider several thermal noise sources *i.e.*, resistors $R_1$, $R_2$, $R_3$, ...in parallel, producing rms noise voltage $V_{n1}$, $V_{n2}$, $V_{n3}$,...

So in a similar way as explained above, The resultant noise voltage will be given as:

$$V_P = \sqrt{4\,K\,B\,T\,R_P} \ \text{Volts} \left( R_P = \frac{1}{R_1} + \frac{1}{R_2} + \frac{1}{R_3} + \dots \right).$$

## 2.17 ADDITION OF NOISE IN AMPLIFIERS IN CASCADE

The Fig. 2.12 shows a two stage amplifier. The gain of first stage is $A_1$ and gain of second stage is $A_2$. The first stage has a total input noise resistance $R_1$, the second has $R_2$. The noise voltage at the output due to output resistance $R_3$ will be given as:

$$V_{n3} = \sqrt{4R_3.K.B.T} \ \text{Volts}. \qquad \dots(i)$$

This noise voltage $V_{n3}$ can be transferred to the input point of the second stage by dividing $V_{n3}$ by $A_2$

$$\frac{V_{n3}}{A_2} = V'_{n2} = \frac{\sqrt{4R_3\,k.B.T}}{A_2}$$

$$= \sqrt{4R'_3\,K\,B\,T} \qquad \dots(ii)$$

Where $R'_3$ is the resistance, if placed at the input point of the second stage would generate a noise voltage equivalent to that, if $R'_3$ was present at the input of the second stage.

$$R'_3 = \frac{R_3}{A_2^2}$$

The eq. (*iii*) shows that if a noise resistance is transferred to the input from output stage of an amplifier, it has to be divided by the "square of the gain of that stage". The noise resistance actually present at the output of the first stage (or at the input of the second stage) is $R_2$. Thus the equivalent noise resistance at the output of the first stage is given by :

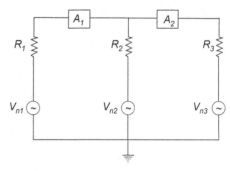

**Fig. 2.12**

$$R'_{eq} = R_2 + R'_3 = R_2 + \frac{R_3}{A_2^2}$$

Similarly, we can find $R'_2$ as

$$R'_2 = \frac{R'_{eq}}{A_1^2} = \frac{R_2 + R_3 / A_2^2}{A_1^2} = \frac{R_2}{A_1^2} + \frac{R_3}{A_1^2 A_2^2}$$

The noise resistance actually present at the input of first stage is $R_1$, therefore, total resistance of the input of first stage, $R_{eq}$ is given by

$$R_{eq} = R_1 + R'_2 = R_1 + \frac{R_2}{A_1^2} + \frac{R_3}{A_1^2 A_2^2} \qquad \ldots(iv)$$

We can extend expression (*iv*) for any no. of amplifier stages.

## 2.18 NOISE IN REACTIVE CIRCUITS

The Fig. 2.13 (*a*) shows a parallel resonant circuit. The inductor $L$ may be assumed as ideal. An ideal resonant tuned circuit does not affect the noise at the resonant frequency. The tuned circuit limits the B.W. of the noise source by not passing the noise voltage beyond its passband. The Fig. 2.13 (*b*) shows that the tuned circuit is not ideal and a resistance $R$ is associated with the inductor L. In this way, $R$ is the only source of noise voltage. Let $V_n$ is the noise voltage due to R.

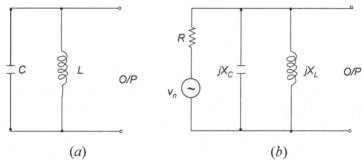

(*a*)                                                      (*b*)

**Fig. 2.13.**

The noise current $\qquad i_n = \dfrac{v_n}{Z}$ [when $Z = R + j\,(X_L - X_C)$ at resonance].

$$= v_n/R \text{ at resonance}$$

The noise voltage appearing across the capacitor $C$ (*i.e.*, across the tuned circuit) will be

$$v_c = i_n \cdot X_C = \frac{v_n \cdot X_C}{R} = \frac{v_n \cdot Q.R}{R} = Q.v_n$$

[where $X_C = Q.R$ at resonance where $Q$ is the $Q$ factor of the coil]

Further, $\qquad v_c^2 = Q^2.v_n^2 = 4Q^2.(K.B.T.R) \; [v_n = \sqrt{4\,K\,B\,T\,R}\,]$

$$= 4\,K\,B\,T\,(Q^2 R) = 4\,K\,B\,T.\,Z_p$$

$$v_c = \sqrt{4\,K\,B\,T.\,Z_p}$$

where $Z_p$ is the parallel impedance of the tuned circuit.

## 2.19 CALCULATION OF SNR AND NF

The Fig. 2.14 shows a block diagram of a 4 terminal amplifier/receiver network. This has an input resistance $R_i$, an output resistance $R_L$ and a Voltage gain of A. It is fed from a generator. In case of an amplifier/receiver, this generator is equivalent to an **antenna** of internal resistance $R_a$. The $R_a$ may or may not be equal to **$R_i$**.

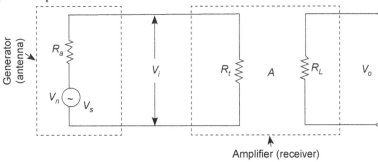

Fig. 2.14.

The calculation of noise figure ($F$) will involve the following steps:

1. **Calculation of signal input power ($P_{si}$):** From Fig. 2.14 we can obtain signal input voltage:

$$V_{si} = \frac{V_s \cdot R_i}{R_a + R_i}$$

$$P_{si} = \frac{V_{si}^2}{R_i} = \left[\frac{V_{si} \cdot R_i}{R_a + R_i}\right]^2 \div R_i \; [\text{Substituting value of } V_{si}]$$

$$P_{si} = \frac{V_s^2 R_t}{\left(R_a + R_t\right)^2} \qquad \qquad \dots(1)$$

## 2. Calculation of Noise input Power ($P_{ni}$):

Noise input voltage $\qquad V_{ni} = \sqrt{4K\,BT\,R}$

$$= \sqrt{4K\,BT\left[\frac{R_a \cdot R_i}{R_a + R_i}\right]}$$

Noise input power, $\qquad P_{ni} = \frac{V_{ni}^2}{R_t}.$

$$P_{ni} = 4K\,BT\left(\frac{R_a \cdot R_i}{R_a + R_i}\right) \div R_t \qquad \qquad \text{[Substituting the value of } V_{ni}\text{]}$$

$$P_{ni} = \frac{4K\,BT\,R_a}{R_a + R_t} \qquad \qquad \dots(2)$$

## 3. Calculation of input signal noise ratio (SNR)$_i$:

$$(\text{SNR})_i = \frac{P_{si}}{P_{ni}}$$

Putting values of $P_{si}$ and $P_{ni}$ found above:

$$(\text{SNR})_i = \frac{V_s^2 \cdot R_t}{\left(R_a + R_t\right)^2} \div \frac{4K\,BT\,R_a}{R_a + R_t}$$

$$(SNR)_i = \frac{V_s^2 \cdot R_t}{4K\,BT\,R_a\,(R_a + R_t)} \qquad \qquad \dots(3)$$

## 4. Calculation of signal output power ($P_{so}$):

$$P_{so} = \frac{V_{so}^2}{R_L} = \frac{(V_{si} \cdot A)^2}{R_L}$$

$$P_{so} = \frac{A^2 \cdot V_{si}^2}{R_L} \qquad \qquad \text{[}A\text{ is voltage gain (V.G.)]}$$

$$P_{so} = A^2 \left(\frac{V_s \cdot R_t}{R_a + R_t}\right)^2 \div R_L \quad \text{[Putting value of } V_{si} \text{ from step 1]}$$

$$P_{so} = \frac{A^2 \cdot V_s^2 \cdot R_{t^2}}{(R_a + R_t)^2} \qquad \qquad \dots(4)$$

**5. Calculation of noise output power:**

$\therefore$ Noise output power $= P_{no}$ ...(5)

**6. Calculation of output signal noise ratio $(SNR)_o$:**

$$(\text{SNR})_o = \frac{P_{so}}{P_{no}}$$

Using values of $P_{so}$ and $P_{no}$ found above:

$$(\text{SNR})_o = \frac{A^2 \cdot V_s^2 \cdot R_t^2}{(R_a + R_t)^2 \cdot R_L \cdot P_{no}} \quad \text{...(6)}$$

**7. To find the generalized form of noise figure:** The general expression of noise figure is:

$$F = \frac{(\text{SNR})_i}{(\text{SNR})_o}$$

Using the value of $(\text{SNR})_i$ and $(\text{SNR})_o$ from eqns. (3) and (6) respectively, we get

Noise Figure, $\quad F = \dfrac{V_s^2 \cdot R_t}{4KBT R_a (R_a + R_t)} \div \dfrac{A^2 V_s^2 \cdot R_t^2}{(R_a + R_t)^2 \cdot R_L \cdot P_{no}}$ ...(7)

$$F = \frac{R_L \cdot P_{no} \cdot (R_a + R_t)}{4KBT \cdot A^2 \cdot R_a \cdot R_t}$$

This is the required expression for noise figure.

## 2.20 EXPERIMENTAL MEASUREMENT OF SNR AND NF

In devices employing **Transit time effect**, the method described above cannot be used to determine "noise figure". For such cases we use experimental method.

Refer Fig. 2.15 which shows a circuit set up for measurement of **noise figure** experimentally. Here is an amplifier/receiver and vacuum diode noise generator, which is the noise power source. Recall, that in communication receivers/amplifiers, shot noise is generated by active devices (Just as vacuum diode) only.

The shot noise current is adjusted by controlling plate current of the diode. The output capacitance of the diode is resonated at the operating frequency of the receiver with the help of a variable inductance, thus it can be ignored. Thus ignoring the output capacitance of the diode, the resistance of the noise generator becomes simply as $R_a$.

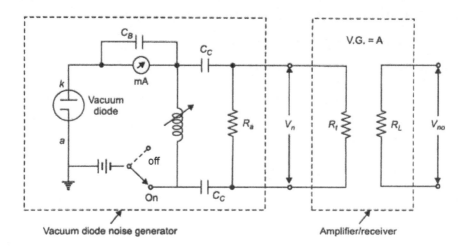

**Fig. 2.15**

Shot noise current

$$I_n = \sqrt{2q \cdot I_p \cdot B}$$

where,                                    $q$ = charge on an electron = $1.6 \times 10^{-19}$ C

$I_p$ = plate current

$B$ = effective B.W.

The noise voltage supplied by the diode to the input of the receiver is:

$$V_n = I_n Z_n$$

$$V_n = I_n \cdot Z_n = \sqrt{2q \cdot I_p \cdot B} \left[ \frac{R_a \cdot R_t}{R_a + R_t} \right] \qquad \dots (1)$$

where, $I_n$ is the shot noise current and

$$Z_n = \frac{R_a \cdot R_t}{R_a + R_t}$$

Now, the noise generator is connected to the receiver/amplifier under test and the noise output power of the receiver is measured with **zero plate current** of the diode. For this purpose, the supply voltage to the diode is switched off through selector switch adjusted at OFF position.

Now the supply voltage to the diode is switched on through selector switch adjusted at ON position, so that the plate current begins to flow. Through adjustment of the potentiometer, plate current is adjusted such that noise power developed in $R_L$ is "twice of that, which was developed when the plate current was zero". This plate current ($I_p$) is measured with the milliammeter (mA).

The additional noise power:

$$P_{no} = \frac{V_{no}^2}{R_L} = \frac{(A.V_n)^2}{R_L} = \frac{A^2 \cdot V_n^2}{R_L} \quad [A = \text{voltage gain V.G.}]$$

$$= \frac{A^2 \left[ \dfrac{R_a \cdot R_t}{R_a + R_t} \sqrt{2q \cdot I_p \cdot B} \right]^2}{R_L}$$

$$P_{no} = \frac{A^2 R_a^2 \cdot R_t^2 \cdot 2q \cdot I_p \cdot B}{R_L (R_a + R_t)^2}$$

[Substituting the value of $V_n$ from eq (1)]

The general equation for **noise figure** found is:

$$F = \frac{R_L \cdot P_{no} \cdot (R_a + R_t)}{4\, K\, B\, T \cdot A^2 \cdot R_a \cdot R_t}$$

$$= \frac{\dfrac{[A^2 R_a^2 \cdot R_t^2 \cdot 2q \cdot I_p \cdot B](R_a + R_t)}{R_L (R_a + R_t)^2}}{4\, K\, B\, T \cdot A^2 \cdot R_a \cdot R_t} \quad \text{[Putting value of } P_{no}\text{]}$$

$$F = \frac{R_a \cdot I_p \cdot q \cdot R_t}{2\, K\, T (R_a + R_t)}$$

If $R_t \gg R_a$ i.e., the system is **mismatched**, in that case (neglecting $R_t$)

$$F = \frac{R_a \cdot I_p \cdot q}{2\, K \cdot T}$$

$$= \frac{R_a I_p \times (1.6 \times 10^{-19})}{2 \times (1.38 \times 10^{-23}) \times 290} = 20\, R_a I_p$$

[Charge on electron $q = 1.6 \times 10^{-19}$ C, $K =$ Boltzmann's constant $= 1.38 \times 10^{-23}$ J/°K, $T =$ Normal Temperature $= 290°$ K]

In a practical receiver, the output SNR will be lower than input SNR and the noise figure will be *more than* 1. The value of noise figure for an ideal receiver will be 1, which shows that it does not produce (add) any noise of its own. In fact, the noise factor is the measure of noise added by the receiver.

The noise figure or factor is frequency dependant, hence frequency must be mentioned with the value of noise factor. When the noise factor is determined at one particular frequency, it is called as "spot noise factor," but when the noise factor is measured for a range of frequencies, it is known as "average noise factor".

The "noise figure" may be expressed as an actual ratio or in **decibels**. The noise figure of a practical receiver can be designed to as low as few decibels upto frequencies of the order of 1 Gigahertz (1G Hz = $10^9$ Hz), by selecting a suitable device (Transistor or Tube) to be used in the first stage. Proper circuit designs and low noise resistors can help in reducing the noise figure of a receiver. The **MASER** which is cooled by liquid nitrogen is a good example of micro-wave amplifier with low noise figure.

**Note:** The noise figure or noise factor is the "figure of merit" used to indicate how much the SN ratio deteriorates, as a signal passes through a circuit or series of circuits.

## 2.21 EQUIVALENT NOISE RESISTANCE AND NOISE FIGURE

The equivalent noise resistance of an amplifier or receiver is equal to the sum of

   (*i*)  Input resistance of the first stage.

   (*ii*)  Equivalent noise resistance of first stage.

   (*iii*)  Noise resistance of subsequent stages referred to the input of the first stage.

   All these resistances are added to get Equivalent noise resistance ($R_{eq}$), which gives the **lump resistance** corresponding to circuit whole noise. The rest of the circuit may be assumed as **"Noiseless"**.

   This implies, that all these noise resistances are added up to the parallel combination of $R_a$ and $R_t$.

   (*b*)  To co-relate the noise figure (*F*) with equivalent noise resistance, ($R_{eq}$), we define the term $R_{eq}$, as the noise resistance "without including" $R_t$ *i.e.*,

$$R_{eq}' = R_{eq} - R_t \qquad \qquad ...(1)$$

The total equivalent noise resistance of the receiver under discussion is given by:

$$R = R_{eq}' = \frac{R_a \cdot R_t}{R_a + R_t} \qquad \qquad ...(2)$$

The equivalent effective noise voltage at the input of the receiver:

$$V_{ni} = \sqrt{4KBTR}$$

After including all noise components into total equivalent noise resistance $R$, the amplifier may now be treated as **noiseless** with a voltage gain of $A$.

The output noise power is given as:

$$P_{so} = \frac{V_{so}^2}{R_L} = \frac{(A \cdot V_{ni})^2}{R_L} = \frac{A^2 \cdot (4KBTR)}{R_L} \qquad \ldots(3)$$

The expression already obtained for noise figure is :

$$F = \frac{R_L \cdot P_{no} \cdot (R_a + R_t)}{4KBT \cdot A^2 \cdot R_a \cdot R_t} \qquad \ldots(4)$$

Substituting the value of $P_{so}$ from eq. (3) in equation (4), we get

$$F = \frac{R_L (R_a + R_t)}{4KBT\, A^2 \cdot R_a R_t} \cdot \frac{A^2 \cdot 4KBTR}{R_L}$$

$$\frac{R_a + R_t}{R_a \cdot R_t} = \left[ R'_{eq} + \frac{R_a \cdot R_t}{R_a + R_t} \right] \frac{R_a + R_t}{R_a \cdot R_t}$$

$$= 1 + R'_{eq} \frac{R_a + R_t}{R_a \cdot R_t} \qquad \ldots(5)$$

This gives noise figure in terms of equivalent noise resistance.

From equation (5), we can conclude that for minimising noise figure for a given value of $(R_a)$, the ratio $\dfrac{R_a + R_t}{R_a \cdot R_t}$ should be minimum or, $R_t$ should be $>>> R_a$. The condition $R_a \neq R_t$ represents **"impedance mismatch"**, under which the transfer of power from the source (antenna) to the system (receiver) is not maximum. However, this mismatch can be utilized for reducing noise. Under the condition of $R_t$ $>>> R_a$, the ratio of $\dfrac{R_a + R_t}{R_t}$ **approaches unity** and the eq. (5) reduces to $F = 1 + \dfrac{R'_{eq}}{R_a}$.

## 2.22  NOISE BANDWIDTH

(a)  Generally we consider "Equivalent Noise bandwidth". The bandwidth of an ideal system is known as "Equivalent noise bandwidth."

The "Equivalent noise bandwidth" may be defined "as the bandwidth of an ideal bandpass system, which produces the same noise power, as produced by the actual system". With the help of the equivalent noise bandwidth, noise power can be specified at the output of a linear bandpass system.

(b)  The Fig. 2.16 shows a linear bandpass system, (consisting of noiseless elements) which is excited by an input noise voltage $V_{ni}$ and produces an output $V_{no}$. The $H(\omega)$ is the **transfer function** of the system.

**Fig. 2.16**

The Fig. 2.17($a$) shows a plot of $|H(\omega)^2|$ versus frequency for an actual system and ($b$) shows the same for an ideal system.

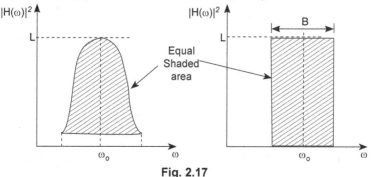

**Fig. 2.17**

Recall that power spectrum density of noise voltage of input and output are related as:

$$S_{no}(\omega) = S_{ni}(\omega)|H(\omega)|^2$$

The mean square value of the output noise signal (output noise power) can be found by integrating the output spectrum density $[S_{no}(\omega)]$ over the effective bandwidth.

$$v^2{}_{no} = P_o = \frac{1}{\pi}\int_0^\infty S_{ni}(\omega)\,|\,H(\omega)^2\,|\,d\omega \qquad [S_{no}(\omega) = S_{ni}(\omega)\,|\,H(\omega)\,|^2\,]$$

For practical purposes, the input noise power spectrum density can be taken constant (*i.e.*, independent of frequency). This constant may be taken as $K$.

$$P_o = \frac{1}{\pi}\int_0^\infty K\,|\,H(\omega)\,|^2\,d\omega = \frac{K}{\pi}\int_0^\infty |\,H(\omega)\,|^2\,d\omega$$

The integral part in the above expression is the area under the curve, which is equal to $L \times B$, where $L$ is the height of the curve and $B$ is the bandwidth [See Fig. 2.17 ($b$)]

Equating the areas of actual and ideal systems

$$L \times B = \int_0^\infty |\,H(\omega)\,|^2\,d\omega$$

or    Bandwidth $B = \dfrac{1}{L}\int_0^\infty |\,H(\omega)\,|^2\,d\omega$

= Area of the curve/maximum value of the curve.

(c) **Relation between noise bandwidth and noise power $P_o$ at the output of a system:**

$$P_o = \overline{v_{no}^2}$$

$$= \frac{C}{\pi} \times (\text{area under the curve } |H(\omega)|^2 \text{ of the actual system})$$

$$= \frac{C}{\pi} \times (\text{area under the curve } |H(\omega)|^2 \text{ of the ideal system})$$

$$= \frac{C}{\pi} \times A(B) = \overline{v_{no}^2} = \frac{C.A.B}{\pi}$$

$$P_o = \frac{C A B}{\pi}, \text{ where } B \text{ is the bandwidth.}$$

**Problem 2.14.** *Two resistors of 20 and 30 kilo-ohms are at room temperature. Calculate thermal noise voltage*

(a) *For each resistor*

(b) *For both the resistors in series*

(c) *For both resistors in parallel.*

*Take bandwidth as 100 kHz. and KT = 4 × 10$^{-21}$ W/Hz.*

**Solution:**

(a) For $R = 20$ Kilo-ohms

$$V_n^2 = 4\,R\,B\,K\,T$$
$$= 4 \times (20 \times 10^3)\,(100 \times 10^3)\,(4 \times 10^{-21}) = 32 \times 10^{-12} \text{ V.}$$
$$V_n^2 = 5.66\ \mu V \quad \textbf{Ans.}$$

For $R = 30$ Kilo-ohms

$$V_n = 5.66 \times \sqrt{\frac{30}{20}}\ 6.93\ \mu V \quad \textbf{Ans.}$$

(b) The resistors in series

Total resistance $= 30 + 20 = 50$ Kilo-ohms

$$V_n = 5.66 \times \sqrt{\frac{50}{20}} = 8.94\mu V \quad \textbf{Ans.}$$

(c) The resistance in parallel

$$\text{Total resistance} = \frac{20 \times 30}{20 + 30} = 12$$

$$V_n = 5.66\sqrt{\frac{12}{20}} = 4.38\ \mu V \quad \textbf{Ans.}$$

**Problem 2.15.** *A parallel tuned circuit is resonated at 200 MHz with a Q factor of 20. The circuit has a 10 pF capacitor. The temperature of the circuit is 17°C. Find noise voltage.*

**Solution:**

$$\text{Bandwidth, } B = \frac{f}{Q} = \frac{200}{20} = 10\,\text{MHz}$$

Also

$$Q = \frac{1}{\omega CR}$$

or

$$R = \frac{1}{Q.\omega C}$$

$$= \frac{1}{20 \times (2\pi \times 200 \times 10^{6}) \times (10 \times 10^{-12})} = 4\Omega$$

Noise voltage, $v_c = \sqrt{4.KBT.Q^2.R} =$

$$= \sqrt{4 \times (1.38 \times 10^{-23}) \times (10 \times 10^{6}) \times (273 + 17) \times (20)^2} = 16\,\pi\text{V} \qquad \textbf{Ans.}$$

**Problem 2.16.** *The Fig. 2.18 shows a passive network. Determine power density of the thermal noise voltage across terminals a and b.*

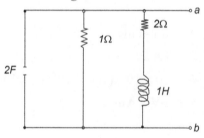

**Fig. 2.18**

**Solution.** The thevenin equivalent circuit for the given network has been shown in Fig. 2.19.

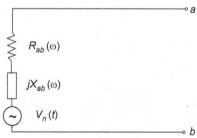

**Fig. 2.19**

The admittance of the network across terminals $a$ and $b$ can be given as:

$$Y_{a,b}(\omega) = 1 + 2j\omega + \frac{1}{1+j\omega} = \frac{2 - 2\omega^2 + 3j\omega}{1+j\omega}$$

The impedance across $a$, $b$

$$Z_{a,b}(\omega) = \frac{1}{Y_{ab}(\omega)} = \frac{1+j\omega}{2 - 2\omega^2 + 3j\omega}$$

To separate the real and imaginary parts, multiplying the above expression by conjugates, we get

$$Z_{a,b}(\omega) = \frac{1+j\omega}{2 - 2\omega^2 + 3j\omega} \times \frac{2 - 2\omega^2 - j\omega}{2 - 2\omega^2 - 3j\omega}$$

$$= \frac{(2 + \omega^2) - j\omega(1 + 2\omega^2)}{4 - 4\omega^4 + \omega^2}$$

$$= \frac{2 + \omega^2}{4 + 4\omega^4 + \omega^2} - j\frac{\omega(1 + 2\omega^2)}{4 + 4\omega^4 + \omega^2}$$

$$= R_{ab}(\omega) + jX_{ab}(\omega)$$

Hence real part is $\quad R_{ab}(\omega) = \dfrac{2 + \omega^2}{4 + \omega^2 + 4\omega^2}$

The resistance $R_{ab}(\omega)$ is the source of noise, the noise power density can be expressed as:

$$S_v(\omega) = 2\,K\,T\,R_{ab}(\omega) = \frac{2\,K\,T(2 + \omega^2)}{4 + \omega^2 + 4\omega^2} \qquad \textbf{Ans.}$$

**Problem 2.17.** Two resistors $R_1$ and $R_2$ at absolute temperature $T_1$ and $T_2$ are connected in series to form a white noise source. Find the equivalent noise temperature.

**Solution.** The equivalent resistance will be:

$$R_t = R_2 + R_2$$

The available noise power $P_n = \dfrac{V_n^2}{4R_t} = \dfrac{V_{n1}^2 + V_{n2}^2}{4R_t}$

Substituting for $V_{n1}$ and $V_{n2}$, we get

$$P_n = \frac{4KT_1BR_1 + 4KT_2BR_2}{4R_t}$$

[where $K$ is Boltzmann's constant and $B$ is bandwidth]

$$= (R_1T_1 + R_2T_2) \cdot \frac{KB}{R_t} \qquad \qquad \dots(i)$$

The equivalent noise temperature

$$T_{eq} = \frac{P_n}{KB} \qquad \qquad ...(ii)$$

Substituting (value of $P_n$) in eq. ($ii$) from eq. ($i$)

$$T_{eq} = \frac{R_1T_1 + R_2T_2 \cdot (KB/R_t)}{KB} = \frac{R_1T_1 + R_2T_2}{R_t}$$

$$= \frac{R_1T_1 + R_2T_2}{R_1 + R_2} \qquad \qquad [R_t = R_1 + R_2]$$

**Problem 2.18.** *The noise output of a resistor is amplified by a noiseless amplifier, having gain of* 60 *and B.W. of* 20 kHz. *A meter connected to the output of the amplifier reads* 1.0 mV:

   (i)  *If the resistor is operated at* 80°C, *what is the resistance?*

   (ii) *If B.W. of the amplifier is reduced to* 5 kHz, *the gain remaining constant; what the meter will read now?*

**Solution.**

($i$)                     $V_n = 10^{-3}/60$ V

Bandwidth, $B = 20$ kHz $= 20 \times 10^3$ Hz

$$T = 80° + 273° = 353° \text{ K}$$

$$V_n = \sqrt{4KBTR}$$

$$\frac{10^{-3}}{60} = \sqrt{4 \times (1.38 \times 10^{-23}) \times (20 \times 10^3) \times 353 \times R}$$

$$R = 715 \text{ k}\Omega \qquad \textbf{Ans.}$$

($ii$) Now,          B.W. $= 5$ KHz

Since B.W. is reduced to ¼ th, the noise voltage will become half *i.e.,*

$$V_n = 1 \text{ mV}/2 = 0.5 \text{ mV}$$

Now, the meter will read 0.5 mV     **Ans.**

## SUMMARY

   1. Noise is an unwanted electrical signal; which interferes with the reception and reproduction of wanted signal.

   2. **The Sources of Noise are:** Faulty connections, frequent opening of a circuit and natural phenomenon etc.

   3. **The Noise Can Broadly be classified as:** External noise and Internal noise.

4. **The External Noise May be:** Atmospheric noise and man made noise. The internal noise may be thermal noise, shot noise etc.

5. **There Are Also Other Noises Such As:** Flicker, Transit, Recombination, Solar Cosmic noise etc.

6. Thermal noise is the most important internal noise. The thermal noise power is given by $P_n = KBT$ Watts.

7. Thermal noise voltage is given by $V_n^2 = 4\,R\,KBT$ Volts.

8. The average noise power per hertz is called *power spectrum density.*

9. It is convenient to represent the noise by means of a fictitious resistance and the device is then assumed as noiseless.

10. The signal noise ratio (SNR) is the ratio of signal power to the noise power.

11. The noise figure is the ratio of SNR input to the SNR output.

12. In microwave low noise devices like antenna, "noise resistance" is a better concept used for noise measurements.

13. In devices employing transit time, experimental method is employed measure SNR and NF.

# 3

# Amplitude Modulation (AM)
## (Double Side Band with Full Carrier i.e. DSBFC)

In amplitude modulation, amplitude of the carrier wave is changed according to the amplitude of the signal. The technique is very much used in transmission of radio signals.

## 3.1 AMPLITUDE MODULATION (AM)

This may be defined as the process of modulation in which 'amplitude of the carrier is varied according to the signal'.

Figure 3.1 shows the process of amplitude modulation.

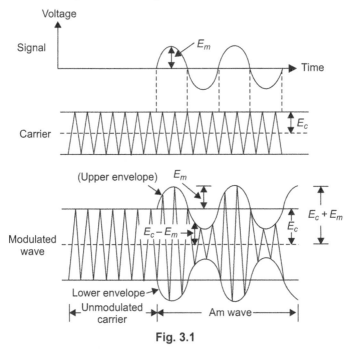

Fig. 3.1

(*i*) The signal is superimposed on a high frequency carrier and a modulated (radio) wave is obtained.

(*ii*) Only amplitude of the carrier is varied, while its frequency and phase remain unchanged.

(*iii*) When there is no signal, the amplitude of the carrier is equal to the unmodulated amplitude. When signal is present, the amplitude of the carrier changes in accordance with the instantaneous value of the signal.

(*iv*) During positive cycle of the signal, the amplitude of the carrier increases to the sum of the amplitudes of the carrier and signal ($E_c + E_m$).

(*v*) During negative cycle of the signal, the amplitude of the carrier decreases and becomes equal to the difference of the amplitudes of the carrier and the signal ($E_c - E_m$).

## 3.2  EXPRESSION FOR AMPLITUDE MODULATED WAVE

Let carrier representation be (See Fig. 3.2)

$$e_c = E_c \sin \omega_c t$$

and let signal be represented by

$$e_m = E_m \sin \omega_m t$$

Let $A$ be the amplitude of the modulated radio wave. Then

$$\begin{aligned}
A &= E_c + e_m \\
&= E_c + E_m \sin \omega_m t \\
&= E_c + mE_c \sin \omega_m t \ (E_m = mE_c) \qquad \dots(i) \\
&= E_c (1 + m \sin \omega_m t)
\end{aligned}$$

Let the voltage equation of the output modulated wave be:

$$e = A . \sin \omega_c t$$

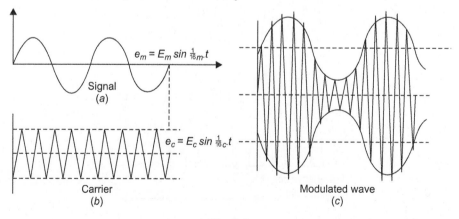

$e_m = E_m \sin \tfrac{1}{16}m·t$

Signal
(a)

$e_c = E_c \sin \tfrac{1}{16}c·t$

Carrier
(b)

Modulated wave
(c)

**Fig. 3.2**

Putting the value from Eq. (*i*)

$$e = E_c (1 + m \sin \omega_m t). \sin \omega_c t \qquad \qquad ...(ii)$$

This is the standard equation for the A.M. radio wave.

Solving further $\qquad \qquad e = E_c \sin \omega_c t + mE_c \sin \omega_c t \sin \omega_m t$

Note that $E_m = mE_c$, where $m$ is the modulating factor.

Solving again, this comes to be

$$e = E_c \sin \omega_c.t + \frac{mE_c}{2}(2\sin \omega_c.t.\sin \omega_m t)$$

[For the bracketed part, use the formulae $= 2 \sin A \sin B = \cos (A - B) - \cos (A + B)$]

Then, we get $\quad e = E_c \sin \omega_c.t + \dfrac{mE_c}{2}\cos(\omega_c - \omega_m)t - \dfrac{mE_c}{2}\cos(\omega_c + \omega_m)t$ ...(iii)

This is the expression for the equation of the modulated wave.

## 3.3 FREQUENCY SPECTRUM OF A.M. WAVE

The equation of A.M. wave is given by

$$E_c \sin \omega_c t + \frac{mE_c}{2} \cos (\omega_c - \omega_m) t - \frac{mE_c}{2} \cos (\omega_c + \omega_m) t$$

Note that the equation has three parts:

 (*i*)  First part is an unmodulated carrier wave, which remains unchanged in the process. The maximum amplitude is $E_c$.

 (*ii*)  Second part has a maximum amplitude of $mE_c/2$ and its frequency is equal to the difference of carrier and the signal frequencies. This is called **lower side band** (L.S.B.). Recall that angular velocity of the carrier $\omega_c = 2\pi f_c$, where $f_c$ is the frequency of the carrier. Similarly $\omega_m = 2\pi f_m$.

(*iii*)  Third part has also max. amplitude of $mE_c/2$ and frequency equal to the sum of carrier and signal frequencies. This is called **upper side band** (U.S.B.) (See Fig. 3.3)

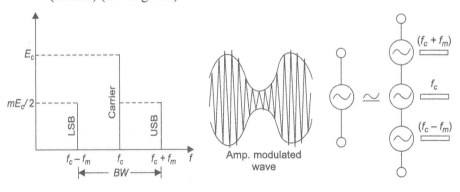

Fig. 3.3           Fig. 3.4

Figure 3.3 shows frequency spectrum of an A.M. wave which is equivalent to three sine waves in series as shown in Fig. 3.4.

Bandwidth (B.W.) of an A.M. wave:

$$\text{B.W.} = (f_c + f_m) - (f_c - f_m) = 2f_m$$

e.g., if $f_c = 100$ kHz, and $f_m = 1$ kHz,

$$\text{B.W.} = (100 + 1) - (100 - 1) = 101 - 99 = 2 \text{ kHz} \ [= 2f_m]$$

Hence, in amplitude modulation, the bandwidth is twice the signal frequency.

**Problem 3.1.** *If the modulating signal is represented by*

$$e_m = E_m \cos \omega_m t + (E_m/2) \cos 2\omega t + (E_m/3) \cos 3\omega_m t + E_m/4 \cos 4\omega_m t$$

and the carrier is represented by

$$e_c = E_c \cos \omega_c t$$

Draw the frequency spectrum of the A.M. wave.

**Solution.** Figure. 3.5 shows the frequency spectrum of the A.M. wave. Note that each modulating frequency component generates two sideband frequencies, the amplitudes of which depend upon *m*.

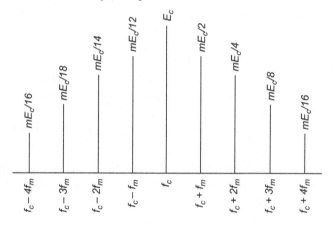

**Fig. 3.5**

**Problem 3.2.** *If one of the sidebands is removed from the modulated output, will the signal transmission be affected?*

**Solution.** Both sidebands carry intelligence (message/information) equally; therefore, if one of the sidebands is removed, the intelligence is not affected. In fact, in practice, only one of the sidebands is transmitted and the other is suppressed to save transmission power.

**Problem 3.3.** *What do you know about the following A.M. transmission systems: A3, A3J, A3H, A5C, A3B.*

**Solution.** $A3$—In this, both sidebands (SBs) are transmitted along with the carrier, *i.e.*, (S.B. $-1+$ S.B. $-2+C$), where $C$ represents carrier.

$A3J$—In this, only one of the sidebands is transmitted and the carrier is suppressed: (SB $-1$)

$A3H$—In this, one of the sidebands is transmitted along with the carrier: (SB $-1+C$)

$A5C$—In this, two sidebands are transmitted without the carrier: (S.B. $-1+$ S.B. $-2$)

$A3A$—In this, one band is transmitted along with reduced (1/3rd) carrier: (SB $-1+C/3$)

## 3.4 MODULATION FACTOR/INDEX (*m*)

The modulation factor/index ($m$) can be defined in one of the following ways:

1. It is the ratio of maximum value of the signal to the maximum value of the carrier, *i.e.*,
$$m = E_m/E_c, \text{ or } E_m = mE_c$$

2. It is the ratio of the change in the amplitude of the carrier to its original amplitude, *i.e.*,
$$m = \Delta E_c/E_c$$

3. It is the percentage change in the amplitude of the carrier, *i.e.*,
$$m = \Delta E_c/E_c \times 100$$

4. It is the ratio of minimum amplitude to the maximum amplitude of the modulated (ratio) wave.

If the modulation curve is displayed on a cathode ray oscilloscope (CRO), we get the curve as shown in Fig. 3.6.

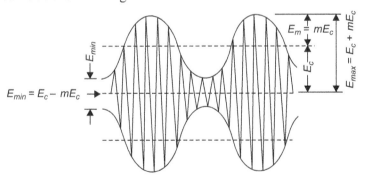

**Fig. 3.6**

The modulation index $m$ can be calculated as follow:

$$m = \frac{E_{max} - E_{min}}{E_{max} + E_{min}}$$

From Fig. 3.6            $E_{max} = E_c + mE_c$

$$E_{min} = E_c - mE_c$$

Adding,        $E_{max} + E_{min} = 2\,E_c$

or                          $E_c = \dfrac{E_{max} + E_{min}}{2}$                          ...(i)

Subtracting:  $E_{max} - E_{min} = 2\,E_m$

or                          $E_m = \dfrac{E_{max} - E_{min}}{2}$                          ...(ii)

Now:                        $m = \dfrac{E_m}{E_c}$

or                          $m = \dfrac{E_{max} - E_{min}/2}{E_{max} + E_{min}/2}$

or                          $m = \dfrac{E_{max} - E_{min}}{E_{max} + E_{min}}$

Hence $m$ can also be expressed as the ratio of minimum amplitude to the maximum amplitude of the radio wave.

**Note:**

1. The value of $m$ lies between 0 and 1.

2. The value of $m$ depends upon the amplitude of the signal as well as of the carrier.

## 3.5  SIGNIFICANCE OF $m$

The modulation factor $m$ plays a very important role in the modulation process. This will be made clear by calculating the value of $m$ for different amplitudes of signal and the carrier.

1. Let the amplitude of signal be zero (*i.e.*, signal is not present) and amplitude of carrier is $E_c$. In this case, the amplitude of modulated wave $= 0 + E_c = E_c$.

   $\therefore$ Change in the carrier amplitude $E_c - E_c = 0$.

   Modulation index $= 0/E_c = 0$ (No modulation)

2. Let the amplitude of carrier $= E_c$ and amplitude of signal $= E_c/2$.

   Then, amplitude of modulated wave $= E_c + E_c/2 = 3E_c/2$

   $\therefore$ Change in the carrier amplitude: $\dfrac{3}{2}E_c - E_c = E_c/2$

   and                          $m = \dfrac{E_c/2}{E_c} = 0.5 = 50\%$ (Under modulation)

3. Let the amplitude of signal as well as carrier be equal to $E_c$.

Amplitude of modulated wave $= E_c + E_c = 2E_c$

Change in carrier amplitude $2 E_c - E_c = E_c$

and $\hspace{3cm} m = E_c/E_c = 1 = 100\%$ $\hspace{1cm}$ (Ideal modulation)

4. Let the amplitude of the carrier be $E_c$ and that of the signal be $3/2\ E_c$

$\therefore$ Amplitude of modulated wave $= E_c + 3/2\ E_c = 5/2\ E_c$

Then, change in carrier amplitude $= 5/2\ E_c - E_c = 3/2\ E_c$

and $\hspace{3cm} m = \dfrac{3/2E_c}{E_c} = 3/2 = 7.5 = 150\%$

$\hspace{6cm}$ (Over modulation)

| Signal | Carrier | Modulated wave Signal + Carrier | Value of $m$ | Remarks |
|--------|---------|-------------------------------|--------------|---------|
| 1. No (zero) signal | $E_c$ | | $m = \dfrac{0}{E_c} = 0$ | No modulation |
| 2. $E_c/2$ | $E_c$ | | $m = \dfrac{E_c/2}{E_c} = 50\%$ | Under modulation |
| 3. $E_c$ | $E_c$ | | $m = \dfrac{E_c}{E_c} = 100\%$ | Ideal modulation |
| 4. $3/2\ E_c$ | $E_c$ | Clipping | $m = \dfrac{3/2E_c}{E_c} = 150\%$ | Over modulation (the wave is distorted ) |

Hence $m$ depends on the amplitude of both the signal and the carrier.

The value of $m$ decides the strength of the modulated wave and hence that of the signal. When $m = 1$ (100%) the signal will be strongest, perfect and clear. In the case of over modulation ($m = 150\%$), the modulated wave will be clipped off and huge distortion will occur in the reception. Hence, the ideal value of modulation is 1 or 100%.

## 3.6 POWER DISTRIBUTION IN THE A.M. WAVE

We know that the power contained in a voltage wave is proportional to the square of its amplitude ($\propto V^2$). *Note that an A.M. wave is a voltage wave.*

The total power contained in an A.M. wave will be the sum of the powers contained in the three parts of the wave.

Considering rool mean square (R.M.S.) values:

$$\left( \text{Recall that R.M.S. Value} = \frac{\text{max.value}}{\sqrt{2}} \right)$$

Power contained in the carrier:

$$P_c\alpha\left[\frac{E_c}{\sqrt{2}}\right]^2 = \frac{E_c^2}{2} \; (E_c \text{ is the maximum value of the voltage of the carrier})$$

Power contained in the lower sideband:

$$P_{SB-1}\left[\frac{mE_c}{2\sqrt{2}}\right]^2 = \frac{E_c^2 m^2}{8} = \frac{P_c m^2}{4}[P_c = E_c^2 / 2]$$

$m$ is the modulated index.

Power contained in the upper sideband:

$$P_{SB-2}\alpha\left[\frac{mE_c^2}{2\sqrt{2}}\right]^2 \alpha \frac{P_c m^2}{4}$$

The total power contained in both the bands:

$$P_{SB} = P_{SB_1} + P_{SB_2}$$

$$P_{SB} = \frac{P_c m^2}{4} + \frac{P_c m^2}{4} = \frac{P_c m^2}{4}$$

The total power in the A.M. wave

$$P_T = P_c + \frac{P_c m^2}{2} = P_c\left(1 + \frac{m^2}{2}\right) \quad \text{(where, } P_c = E_c^2/2)$$

$$\therefore \qquad P_T./P_T = 1 + \frac{m^2}{2}$$

Ratio of sideband power to the total power

$$\frac{P_{SB}}{P_T} = \frac{\dfrac{P_c m^2}{2}}{P_C\left(1 + \dfrac{m^2}{2}\right)} = \frac{m^2/2}{1 + m^2/2} = \frac{m^2}{m^2/2}$$

Note that power in both the sidebands is equal and at $m = 1$, the sidebands contain 1 /3rd (33%) power and the carrier contains 66% of the total power hence bands carry half the carrier power of the wave.

As the signal is contained only in the sidebands, useful power is contained in sidebands. This is the reason, that we are interested only in the sidebands. The power in the sidebands go on increasing with the increase in the modulating index ($m$).

## 3.7 CALCULATION FOR CURRENT

Let $\qquad I_C$ = Unmodulated current (carrier current)

$I_T$ = Modulated current of an A.M. wave (both in R.M.S. values)

and $\qquad R$ = Resistance through which current flows. Assume it to be the same in both the cases.

Now, $\dfrac{P_T}{P_C} = \dfrac{I_T^2.R}{I_C^2.R} = 1 + \dfrac{m^2}{2}$ ($R$ is same)

Note that $P_T$ is the total power of the A.M. wave, $P_C$ is the power contained in the carrier wave and power = (current)$^2$ × resistance

From above $\qquad \left(\dfrac{I_T}{I_C}\right)^2 = 1 + \dfrac{m^2}{2}$

or $\qquad \dfrac{I_T}{I_C} = \sqrt{1 + \dfrac{m^2}{2}}$

or $\qquad I_T = I_C\sqrt{1 + \dfrac{m^2}{2}}$

**Problem 3.4.** *A transmitter supplies 10 kW power to an aerial, when unmodulated. Determine the power radiated, when modulated to 30%.*

**Solution.** $\qquad$ Total power = carrier power × $\left(1 + \dfrac{m^2}{2}\right)$

$$P_T = P_C\left(1 + \dfrac{m^2}{2}\right) = 10\left(1 + \dfrac{(0.3)^2}{2}\right) (m = 30\% = 0.3)$$

$$= 10.45 \text{ kW } \textbf{Ans.}$$

**Problem 3.5.** *A wireless transmitter radiates 4 kW with an unmodulated carrier and 4.8 kW when the carrier undergoes modulation. Calculate the percentage modulation employed.*

**Solution.** $\qquad P_T = P_C\left(1 + \dfrac{m^2}{2}\right)$

$$4.8 = 4\left[1 + \dfrac{m^2}{2}\right]$$

$$m = 62\% \textbf{ Ans.}$$

**Problem 3.6.** The R.M.S. value of an aerial current is 10 A and 12 A before and after modulation. Calculate % modulation employed.

**Solution.**
$$I_T = I_C\sqrt{1+\frac{m^2}{2}}$$

$$12 = 10\sqrt{1+\frac{m^2}{2}}$$

$$m = 93.8\%$$

**Problem 3.7.** *The unmodulated carrier current to the aerial of a transmitter is 100 A. Determine increase in the currents which will result from the application of 80% modulation.*

**Solution.**
$$I_T = I_C\sqrt{1+\frac{m^2}{2}}$$

$$I_T = 100\sqrt{1+\frac{(0.8)^2}{2}} = 114.9 \text{ A}$$

Increase in the currents due to modulation

$$= 114.9 - 100 = 14.9 \text{ Amp. } \textbf{Ans.}$$

**Problem 3.8.** (*a*) *A radio transmitter using amplitude modulation has unmodulated carrier power output of 10 kW and can be modulated to a maximum depth of 90% by a sinusoidal modulating voltage. Find power of the modulated wave. Find the value of which the unmodulated carrier power may be increased, if the maximum permitted modulation is 40%.*

**Solution.**
$$P_T = P_C\left(1+\frac{m^2}{2}\right) = 10\left(1+\frac{(0.9)^2}{2}\right)(m=90\%=0.9)$$

$$= 14 \text{ kW. } \textbf{Ans.}$$

(*b*) This is the maximum power, which may be handled by the transmitter. The increased unmodulated carrier power is given by:

$$14 = P_C\left[1+\frac{(0.4)^2}{2}\right] \qquad (\text{now } m = 40\% = 0.4)$$

$$P_c = 12.96 \text{ kW } \textbf{Ans.}$$

Note that the original power of the carrier was 10 kW.

**Problem 3.9.** *A 100 V, 10 kHz carrier is modulated with the help of a 5 V, 50 Hz signal. Calculate:*

    (*a*)  Modulation factor (*m*)        (*b*)  Amplitude of each sideband
    (*c*)  Frequency of each sideband   (*d*)  Bandwidth of modulated wave.

**Solution.**

$$(a) \quad m = \frac{E_m}{E_c} = \frac{5}{100} = \frac{1}{20} = 5\% = 0.05 \text{ Ans.}$$

(b) Amplitude of each sideband $= \dfrac{mE_c}{2} = \dfrac{0.05 \times 100}{2} = 2.5$ **Ans.**

(c) Frequency of sideband

$$\text{USB} = f_c + f_m = 50 + 10000 = 10050 \text{ Hz } \textbf{Ans.}$$

$$\text{LSB} = f_c - f_m = 10000 - 50 = 9950 \text{ Hz}$$

(d) Bandwidth of the modulated wave

$$\text{B.W.} = 2f_s = 2 \times 50 = 100 \text{ Hz } \textbf{Ans.}$$

**Problem 3.10 Calculate**

(a) *Upper and lower sideband frequencies of an amplitude modulated wave when a carrier of 900 kHz is modulated by a 15 kHz signal.*

(b) *Also find the range of frequencies contained in the modulated wave.*

**Solution.** $\qquad\qquad\qquad f_c = 900 \text{ kHz}$

$$f_s = 15 \text{ kHz}$$

(a) Frequency of USB $\ = f_c + f_s = 900 + 15 = 915 \text{ kHz}$

Frequency of LSB $\ = f_c - f_s = 900 - 15 = 855 \text{ kHz}.$

(b) Range of frequencies contained in the wave

$$= \text{from } \textbf{885 kHz to 915 kHz.}$$

Bandwidth $= 915 - 885 = 30 \text{ kHz.}$

**Problem 3.11.** *An A.M. wave has peak-to-peak voltage of 600 V and valley to valley voltage of 100 V. Find the percentage depth of modulation.*

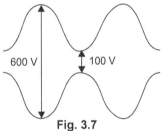

**Solution.** See Fig. 3.7

$$m = \frac{V_{max} - V_{min}}{V_{max} + V_{min}} = \frac{600 - 100}{600 + 100} = \frac{500}{700} = 70\% \textbf{ Ans.}$$

600 V    100 V

**Fig. 3.7**

**Problem 3.12.** *A 100 V. 100 kHz carrier is modulated with the help of a 10V, 1 kHz signal to the extent of 50%. Write down the equation for the A.M. wave.*

**Solution.** $\qquad\qquad\qquad m = 50\% = 0.5$

$$E_c = 100 \text{ V}; E_m = 10 \text{ V}$$

$$f_c = 100 \text{ kHz}$$

$$f_m = 1 \text{ kHZ}$$

$$\omega_c = 2\pi f_c = 2 \times 3.14 \times 100 \times 10^3 = 628000$$

$$\omega_m = 2\pi f_c = 2 \times 3.14 \times 1 \times 10^3 = 6280$$

Putting these values in the standard equation for the modulated voltage wave ;

$$e = E_c \sin \omega_c t - \frac{mE_c}{2} \cos (\omega_c - \omega_m) t - \frac{mE_c}{2} \cos (\omega_c + \omega_m) t$$

$$e = 100 \sin 628000 \, t + \frac{0.5 \times 100}{2} \cos (628000 - 6280) \, t$$

$$- \frac{0.5 \times 100}{2} \cos (628000 + 6280) \, t$$

$$e = 100 \sin 628000 \, t + 25 \cos 621720 \, t - 25 \cos$$
$$634280 \, t$$

This is the equation for the A.M. wave. **Ans.**

**Problem 3.13.** *An amplitude modulated wave is represented by the equation*

$$e = 20 \, (1 + 0.7 \sin 6280 \, t) \sin 628000 \, t$$

Determine:

   (*i*)  Modulation factor          (*ii*)  Carrier amplitude

 (*iii*)  Signal frequency           (*iv*)  Carrier frequency

   (*v*)  Maximum amplitude of A.M. wave

 (*vi*)  Minimum amplitude of A.M. wave

 (*vii*)  Bandwidth.

**Solution.** The equation of A.M. wave is given by

$$e = E_c \, (1 \, m \sin \omega_m t). \sin \omega_c t$$

Comparing with the given equation

$$e = 20 \, (1 + 0.7 \times \sin 6280 \, t). \sin 628000 \, t$$

We get

   (*i*)  Modulation factor $m = 0.7$

  (*ii*)  Carrier amplitude $E_c = 20V$

 (*iii*)  $\omega_m = 6280 \, [\omega_m = 2\pi f_m]$

    ∴ Signal frequency,      $f_m = \dfrac{\omega_m}{2\pi} = \dfrac{6280}{2\pi} = 1\text{kHz}$ **Ans.**

 (*iv*)  $\omega_c = 628,000$

    Carrier frequency,      $f_m = \dfrac{\omega_c}{2\pi} = \dfrac{628000}{2\pi} = 100 \text{ kHz}$ **Ans.**

   (*v*)  Maximum amplitude of AM wave

$$E_{max} = E + mE_c = 20 + (0.7 \times 20) = 34V \text{ **Ans.**}$$

(*vi*)  Minimum amplitude of AM wave

$$E_{min} = E_c - mE_c$$

$$E_{min} = 20 - mE_c = 20 - (0.7 \times 20) = 6V \text{ Ans.}$$

(*vii*)  Bandwidth $= 2f_m = 2 \times 1 = 2$ kHz.

**Problem 3.14.** *Calculate the percentage of the total power contained in the sidebands when m = 50%, 75%, 100%, 150% and 200%.*

**Solution.** Percentage of total power contained in the sidebands is given by

$$P_{SB} = \frac{m^2}{m^2 + 2} \times 100$$

(*i*)  If                                    $m = 50\% = 0.5$

$$P_{SB} = \frac{(0.5)^2}{(0.5)^2 + 2} = 0.11 = 11\% \text{ Ans.}$$

(*ii*)  If                                  $m = 75\% = 0.75$

$$P_{SB} = \frac{(0.75)^2}{(0.75)^2 + 2} = 0.22 = 22\% \text{ Ans.}$$

(*iii*)  If                                 $m = 100\% = 1$

$$P_{SB} = \frac{1^2}{1^2 + 2} = 0.33 = 33\% \text{ Ans.}$$

(*iv*)  If                                  $m = 150\% = 1.5$

$$P_{SB} = \frac{m^2}{m^2 + 2} = \frac{(1.5)^2}{(1.5)^2 + 2} = 0.53 = 53\% \text{ Ans.}$$

(*v*)  If                                    $m = 200\% = 2$

$$P_{SB} = \frac{2^2}{2^2 + 2} = \frac{4}{6} = 0.66 = 66\% \text{ Ans.}$$

## 3.8  LIMITATIONS OF AMPLITUDE MODULATION

The amplitude modulation suffers from the following limitations:

1. The useful power is contained in the sidebands and even at 100% modulation, the bands contain only 33% of the total power and hence the modulation efficiency is poor.

2. Due to poor efficiency, the transmitters employing amplitude modulation have very poor range.

3. The reception in this modulation is noisy. The radio receiver picks up all the surrounding noise along with the signal.

## SUMMARY

1. In amplitude modulation, amplitude of a carrier wave is changed according to the amplitude of the signal.

2. Modulation factor $m = \dfrac{\text{Change in amplitude of carrier}}{\text{Original amplitude of carrier}}$

3. The value of $m$ decides the distortion, $m = 100\%$ is ideal.

4. The AM wave has one carrier and two sidebands.

5. The sidebands contain 33% of the total power.

6. The efficiency of amplitude modulation is poor.

❑❑❑

**4**

# Various AM (SSB) Techniques

In the previous chapter, we have studied the standard amplitude modulation technique. This is called double sidebands with full carrier (DSBFC) modulation as in this, both the sidebands along with the carrier are transmitted. This has been found as an uneconomical technique, hence other advanced AM techniques have been developed, which are more economical and have many other advantages.

The two sidebands are exact "image" of each other, hence it is not necessary to transmit both the sidebands. Usually one sideband with or without the carrier is transmitted. This is called single side band amplitude modulation.

In this chapter. We shall discuss single sideband (SSB).

## 4.1 DIFFERENT FORMS OF AMPLITUDE MODULATION

The different forms of amplitude modulation are:
1. Double sideband with full carrier (DSBFC) or, A3
2. Double sideband with suppressed carrier (DSBSC) or, A 5 C
3. Single sideband (SSB) techniques.

## 4.2 DOUBLE SIDEBAND WITH FULL CARRIER (DSBFC)

Already described

## 4.3 DOUBLE SIDEBAND WITH SUPPRESSED CARRIER (DSBSC)

The carrier conveys no information as it remains constant in amplitude and frequency irrespective of the signal. Thus the carrier is superfluous. In this technique, the carrier is suppressed. The system is used in high frequency radio telephony.

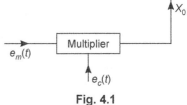

**Fig. 4.1**

If a low frequency modulating signal $e_m(t)$ is multiplied with a high frequency carrier $e_c(t)$, the result is $X_0$ i.e., the DSBSC output, as shown in Fig. 4.1.

$$X_0 = e_m(t) . e_c(t)$$

$$= E_m \sin \omega_m t . E_c \sin \omega_c t = \frac{E_m . E_c}{2} \ [2 \sin \omega_m t . \sin \omega_c t]$$

$$= \frac{E_m E_c}{2} \ [\cos (\omega_c - \omega_m) t - (\omega_c + \omega_m).t\ ]$$

$$= \frac{E_m E_c}{2} \ [\cos 2\pi \ (f_c - f_m) t - \cos 2\pi \ (f_c + f_m) t\ ]$$

## 4.4  SINGLE SIDEBAND AMPLITUDE MODULATION (SSB-AM)

In the theory of amplitude modulation (AM), we have seen that a carrier and two sidebands (SBs) are required for AM transmission. But it is not necessary to transmit all the three signals (1 carrier and 2 sidebands). The carrier and one of the sidebands may be removed (or attenuated).

The SSB modulation is the fastest spreading form of analog modulation. The greatest advantage is its ability to transmit signals by using a very narrow band width, very low power for the distances involved.

For 100% modulation ($m = 1$), only 1/3$^{rd}$ of the total power is present in one of the sidebands, while 2/3$^{rd}$ power is carried by the carrier, which contains no information. Thus if the carrier and one of the sidebands is eliminated from the signal, the transmission will need only 1/6$^{th}$ of the total power.

For comparison, the Fig. 4.2 (*a*) shows double sideband with full carrier (DSBFC) and (*b*) shows double sideband with suppressed carrier (DSBSC) and (*c*) shows single sideband transmission with suppressed carrier (SSBSC). It can be noted that (*c*) requires only half the bandwidth (BW) as required to (*a*) and (*b*).

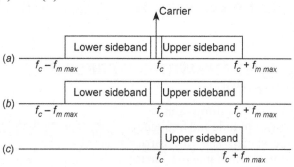

**Fig. 4.2**

## Evolution of SSBAM

The evolution of SSB amplitude modulation may be done in following steps:

(*i*) The carrier contains no power and all the power is contained in the sidebands.

(*ii*) Therefore there is no need to transmit carrier.

(*iii*) The modulated wave contains three frequencies $f_c, f_c + f_m$ and $f_c - f_m$.

(*iv*) Two sidebands are exact image of each other; since each is affected by changes in the modulating voltage via the exponent m $E_c/2$. Recall that *m* is the modulation index and $E_c$ the carrier voltage.

(*v*) Therefore all the information may be transmitted by the use of one sideband only, as the carrier is superfluous and the other sideband is redundant.

(*vi*) If the carrier is suppressed, only the two sidebands remain, power of which is equal to $= P_C \cdot m^2/4$, about 66% saving will be done. Recall that $P_c$ is the carrier power.

(*vii*) If one of the sidebands is also suppressed, the remaining power is $P_C \cdot m^2/4$, a further saving of 50% power will be achieved.

**Problem 4.1.** *Calculate the percentage power saved when the carrier and one of the sidebands is removed in an AM wave, when*

(i) $m = 1$

(ii) $m = 0.5$

**Solution.**

(*i*) When $m = 1$ total power

$$P_T = P_C \left(1 + \frac{m^2}{2}\right) = P_C \left(1 + \frac{1^2}{2}\right) = 1.5\, P_C$$

One sideband power $\quad P_{SB} = P_C \cdot \frac{m^2}{4} = P_C \cdot \frac{1^2}{4} = 0.25\, P_C$

Saving in power $\quad = \dfrac{1.5 - 0.25}{1.5} = 83.3\%$ **Ans.**

(*ii*) When $m = 0.5$ total power

$$P_T = P_C \left(1 + \frac{m^2}{2}\right) = P_C \left(1 + \frac{(0.5)^2}{2}\right) = 1.125\, P_C$$

One sideband power $\quad P_{SB} = P_C \cdot \frac{m^2}{4} = P_C \cdot \frac{0.5^2}{4} = 0.0625\, P_C$

$$\text{Saving in power} \qquad = \frac{1.125 - 0.0625}{1.125} = 94.4\% \text{ Ans.}$$

## 4.5  WAVE SHAPES

The Fig. 4.3 (*a*) shows AM (DSBFC) wave, Fig. 4.3 (*b*) shows supressed carrier (DSBSC) wave. The Fig. 4.3 (*c*) shows SSB-AM wave.

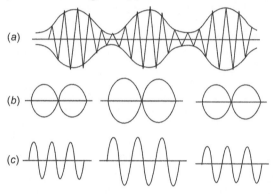

**Fig. 4.3**

## 4.6  VARIOUS SINGLE SIDEBAND (SSB AM) TECHNIQUES

The AM system, in which only one sideband is transmitted is the most popular system. The system in addition to many other advantages, needs only half the bandwidth as compared to the DSB system. The following are the various sideband techniques:

(*a*)  Single sideband with supressed carrier (SSBSC) or A 3 J

(*b*)  Single sideband with reduced carrier (SSBRC) or A 3 A

(*c*)  Single sideband with full carrier (SSBSC) or A 3 H

(*d*)  Vestigial sideband with full carrier (VSBFC)

(*e*)  Independent sideband (ISB).

Here we discuss these in brief

## 4.7  SINGLE SIDE BAND WITH SUPPRESSED CARRIER (SSBSC)

(*a*)  In SSB-SC, power is saved by eliminating the carrier component. Further increase in the efficiency of transmission is possible by eliminating one more sideband, since the two side bands are images of each other, each is affected by changes in the modulating voltage amplitude and each is equally affected by changes in modulating frequency which further changes the frequency of side band itself.

It is seen that all the information can be conveyed by the use of single side band only. The carrier is superfluous and the other side band is refundant. Suppose a lower side band is required, say :

$$e_{SSBSC}(t) = \cos(\omega_c - \omega_m) = \cos\omega_m t + \sin\omega_c t \sin\omega_m t.$$

If $\cos\omega_m t$ and $\cos\omega_c t$ are signal and carrier respectively, then the required signal can be produced by a balanced modulator, provided that both the signal and carrier are shifted in phase by $+\pi/2$. This method is called **phase shift method** of producing single side band suppressed carrier signals.

(*b*) **Analysis:** Modulating signal (as it is) lags carrier signal (by 90° phase shift) at the input of modulator $M_1$ then O/P of modulator $M_1$ will contain sum and difference frequencies (Fig. 4.4).

$$v_1 = \cos[(\omega_m t + 90°) - \omega_m t] - \cos[(\omega_c t + 90°) - \omega_m t]$$
$$= \cos(\omega_c t - \omega_m t + 90°) - \cos(\omega_c t + \omega_m t + 90°)$$

$$\qquad\qquad \text{LSB} \qquad\qquad\qquad \text{USB}$$

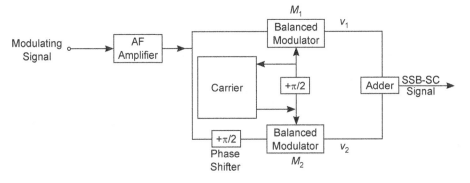

**Fig. 4.4**

Similarly, the modulator $M_2$ at its input has carrier signal (as it is ) lags modulating signal (by 90° phase shift).

Then the O/P of $M_2$,

$$v_2 = \cos[\omega_c t - (\omega_m t + 90°)] - \cos[\omega_c t (\omega_m t + 90°)]$$
$$= \cos[\omega_c t - \omega_m t - 90°) - \cos(\omega_c t + \omega_m t + 90°)$$

$$\qquad\qquad \text{LSB} \qquad\qquad\qquad \text{USB}$$

The output of the Adder is

$$V_0 = v_1 + v_2 = 2\cos(\omega_c t + \omega_m t + 90°)$$

The Fig. 4.5 shows the block diagram showing the complete process.

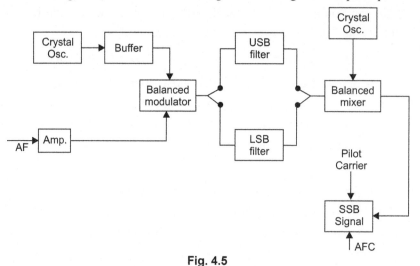

**Fig. 4.5**

The Fig. 4.6 shows the modulating signal, carrier, SSBSC output signal and frequency spectrum of the output signal.

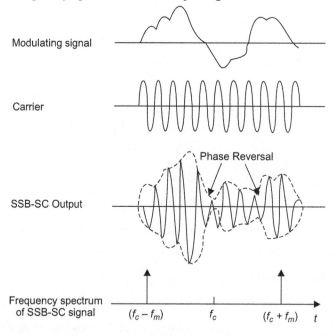

**Fig. 4.6**

(*c*) **Advantages:** The SSB-BC system has the following advantages.

1.  Bandwidth required for the system is half of that required for DSB system.

2. The effect of selective fading is minimum as only one side band exists. In long range high frequency communication, particularly in audio-range, SSB technique is employed. The quality of communication is better in this system.

3. It requires relatively low power for communication and efficiency of transmission is increased.

(*d*) **Application:** The system is used in point to point radio telephony and in marine mobile communication, specially at distress call frequencies.

(*e*) **Achieving frequency stability:** In SSBSC system, the carrier should be suppressed atleast by 45 db at the transmitter. Earlier this system was not successful because highly stable oscillators are required but with introduction of "Frequency synthesisers", this system is now improved a lot.

If a 100 Hz "frequency shift" occurs in a system through which signals of 300, 500 and 800 Hz are passed, all these signals are shifted to 200, 300 and 600 Hz just deterioting the performance of the system. So this system is not suitable for music, speech etc.

The frequency stability can be obtained by using temperature controlled crystal oscillators with the transmitter which give highest transmitting stability. As told earlier, the introduction of "Frequency sythesisers" with the receiver the frequency stability also improves a lot at the reception side.

As the SSBSC system does not transmit the carrier, it may cause a "frequency shift", if highly stable oscillators are not used at the transmitter as well at the receiver.

The frequency stability of this system is 10 PPM (parts per million) which can be said as satisfactory.

## 4.8 SINGLE SIDE BAND WITH REDUCED CARRIER (SSBRC)

This is an old system and was used before invention of frequency synthesisors. In SSBRC or "Pilot carrier" system, a pilot carrier is transmitted along with the SSB signal. Block diagram of this system (See Fig. 4.7) is just an addition of "Pilot carrier" to the SSB

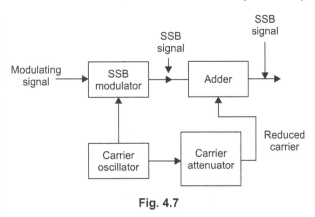

**Fig. 4.7**

system. An attenuated (reduced amplitude) carrier is added to the final SSB signal output. The inserted carrier level is of the order of 15 to 25 db below the unsuppressed carrier level. This pilot carrier is used at the receiver for demodulation and tuning.

The frequency of pilot carrier is same as that of the original carrier. This system is identical to the SSB systems studied so far. The reduced carrier and SSB signal are added in the adder to get SSBRC signal.

This system is used in transmarine point to point radio telephony and mobile communication.

## 4.9  VESTIGIAL SINGLE SIDEBAND (VSB) SYSTEM

We know that more is the information sent per second, larger is the BW required. In TV, where larger BW is required, SSB system is very important for reducing the B.W.

The BW occupied by TV video signal is atleast 4 MHz. If we use DSBFC system, minimum B W of 9 MHz will be required. If SSB system can be used, considerable BW can be saved.

Therefore a compromise between SSBSC and DSBSC has been found which is known as *Vestigial sideband system.* In VSB system, the desired sideband is allowed to pass completely but also a portion (called vestige) of the undesired sideband is also allowed to pass through. The vestige of the undesired sideband compensates for the loss of the desired sideband. Moreover, the VSB system does not need a filter.

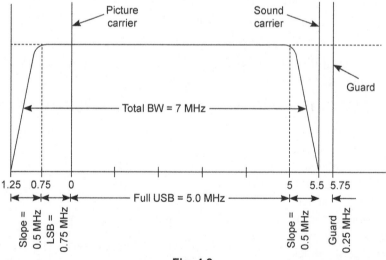

**Fig. 4.8**

In this system, 0.75 MHz of the lower sideband (along with the complete upper side band) is also transmitted to ensure that the lowest frequencies of the desired USB will not be distorted. As only 0.75 MHz of LSB is transmitted saving of 3 MHz of VHF spectrum results with every TV channel, thus making possible to allow more number of channels in the same BW. Total BW needed is 7 MHz (Fig. 4.8)

The sound occupies band near the video because it is required with the picture and it is not feasible to have a separate receiver for sound operating at some distant frequency *i.e.*, much away from the video frequency.

The VSB signals are easy to generate, whereas SSBSC signals are relatively difficult to generate. In SSB signal generation using filtering technique, the filter must have very sharp characteristics. Basically such filters must have a flat passband and extremely high attenuation outside the passband.

Suppose a frequency range of 300–3000 Hz is to be transmitted using SSBSC technique with a carrier frequency of $f_c$ and lower side band is to be suppressed. The lowest frequency, a filter must pass without attenuation is $(f_c + 300)$ Hz, where the highest frequency that must be fully attenuated is $(f_c - 300)$ Hz, as shown in Fig. 4.9 (*a*). The filters response must change from zero attenuation to full attenuation over a range of 600 Hz. If the transmitting frequency is above 10 MHz, it is quite impossible. Further, it is appreciated that the situation becomes worse, if lower modulating frequencies are used, such as 50 Hz in AM broadcasting.

To overcome this problem, a compromise between SSB and DSB is sought, what is known as vestigial sideband transmission. In this method instead of rejecting one sideband completely, a gradual cut-off of one sideband is accepted. The cut-off characteristic is such, that the partial suppression of transmitted sideband (USB) in the neighbourhood of the carrier is exactly compensated by the partial transmission of corresponding part of the suppressed sideband (LSB). Because in this arrangement, the desired signal can be recovered exactly by an appropriate detector.

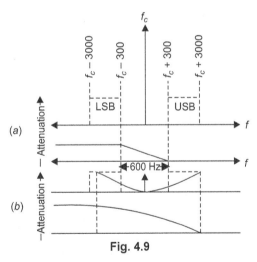

Fig. 4.9

Moreover, it requires the same BW as required for SSB. This system contains advantages of SSB as well as of DSB systems.

## 4.10 INDEPENDENT SIDEBAND (ISB) TECHNIQUE

This system is usually used for medium density traffic. It is mostly a four channel transmission system.

The system carries two independent channels simultaneously as two side bands with carrier reduced. Each sideband is independent of the other and different transmissions can be made on them. Each channel has a BW of 6 kHz and is fed to separate "Balanced modulators" along with a 100 kHz signal from a crystal oscillator. The balanced modulator suppresses the carrier by about 45 dB. The USB and LSB channels are selected by Filters and "Added" together with a carrier attenuated by 26 dB. The output of the "Adder" is mixed in a Balanced mixer with 3 MHz oscillator's output. See Fig. 4.10(*a*).

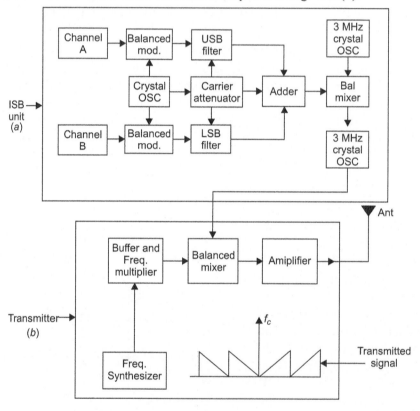

**Fig. 4.10**

The proper frequency is selected from the Balanced mixer output and amplified. (See Fig. 4.10 *b*). The signal is then given to the transmitter section, where it is again mixed in a mixer with the output of a "frequency synthesizer" and frequency multiplier to raise its frequency. The usual transmitting frequency is between 3 to 30 MHz. The resulting ISB signal is amplified to a power level of about 60 kW; and then fed to the antenna.

For high density point to point communication, multiplexing techniques are used such as frequency division multiplexing (FDM). However, for low or medium density traffic, ISB transmission is often employed.

The ISB essentially consists of two SSB channels added to form two sidebands around the reduced carrier. However, each sideband is quite independent of each other. It can simultaneously convey a totally different transmission to the extent, what the upper side band could. It is not advisable to mix telephone and telegraph channels in one sideband since "key clicks" may be heard in the voice circuit. However such hybrid arrangements are sometime unavoidable since the demand almost invariably tries to outstrip existing facilities.

## 4.11 SSB SIGNAL GENERATION OR SUPPRESSION OF UNWANTED SIDEBAND

There are 3 methods of SSB signal generation:

   (*i*)  Filter method.

  (*ii*)  Phase shift method.

 (*iii*)  Weaver (Third) method.

The Balanced modulator suppresses the carrier. To obtain SSB signal, the unwanted sideband (frequency) is to be removed. All of the above three methods have the capability of removing any of the two sidebands. These are described below:

## 4.12 FILTER METHOD

(*a*)  **This is the easiest method.** In this, the unwanted sideband is heavily attenuated using a filter as shown in the Fig. 4.11. The important blocks are "Balanced modulator" and sideband suppression "filter".

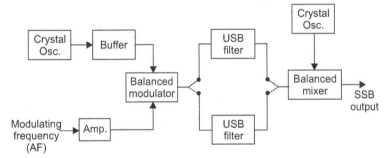

**Fig. 4.11**

In communication, the frequency range used for voice is 300 to 3400 Hz. If transmitting frequency is $f$ Hz and it is required to suppress lower sideband (LSB), then the lowest frequency, the filter should pass without attenuation is $f + 300$ and the highest frequency that must be highly attenuated is $f - 300$ Hz. In this way, the response of the filter must change from zero to full attenuation within a range of $[300 - (-300)] = 600$ Hz.

In order to obtain such a response, the tuned circuits are used with high Q (Quality factor). As the frequency increases, the value of $Q$ factor required is also higher but this has limitations. Therefore, mechanical or crystal filters are preferred.

The mechanical filters can be used upto 500 kHz and crystal filters upto 20 MHz. The mechanical filters are more preferred due to their small size, good attenuation, good band pass and high upper frequency limit. The crystal filters are preferred over 1 MHz.

As the transmitting frequencies are higher than the operating frequencies of the filters, a "Balanced Mixer" is used as shown. In this mixer, to obtain the desired transmission efficiency, the SSB signal from the filter is mixed to crystal oscillator frequency.

This has an advantage that we can achieve the desired transmission efficiency just by changing the crystal oscillator frequency. If required, two stage mixing can also be used.

(*b*) **Advantages of filter method:**

    (*i*)   It has sufficient sideband suppression of about 50 dB.

    (*ii*)  The sideband filter used also helps in suppression of the carrier to some extent.

    (*iii*)  The passband is sufficiently flat and wide except at lower frequencies when crystal filters are used.

    (*iv*)  The quality of signal generated is quite good and therefore the method is widely used for generating SSB signals.

(*c*) **Disadvantages:**

    (*i*)   The system needs heavy and large size equipment. This problem can be solved by using crystal and mechanical filters.

    (*ii*)  The system cannot generate a SSB signal at higher radio frequencies.

    (*iii*)  The filters used are very expensive and may require two filters to have the desired sideband.

## 4.13 PHASE SHIFT METHOD

(*a*) This method does not use filters and hence is free from the disadvantages of the filter method. The modulating signal is fed to an amplifier. The output of the amplifier is divided into two parts, which are fed to two "phase shifting" networks, which give them 90° phase shift. The output of the phase shifting networks are given to two separate balanced modulators, $M_1$ and $M_2$ as shown in Fig. 4.12.

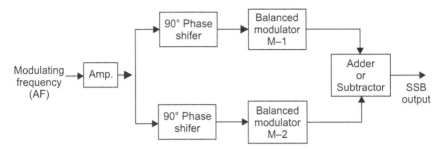

**Fig. 4.12**

The output of both the balanced modulators consists of two sidebands.

Suppression of unwanted sideband can be achieved as under:

(*i*) **Suppression of LSB:** If LSB is to be supressed, the output is given to an adder circuit.

(*ii*) **Suppression of USB:** If USB is to be suppressed, we use a subtractor circuit instead.

(*b*) **Analysis:** Let the modulating signal is $\omega_m$. $t$ and the carrier is $\omega_c t$

(*i*) **Output of the modulator $M_1$**

$$e_1 = \cos [\omega_c t - (\omega_m t + 90°)] - \cos [\omega_c t + (\omega_m t + 90°)]$$
$$= \underset{\text{LSB}}{\cos (\omega_c t - \omega_m t - 90°)} - \underset{\text{USB}}{\cos (\omega_c t + \omega_m t + 90°)}$$

(*ii*) **The Output of the modulator $M_2$**

$$e_2 = \cos [(\omega_c t + 90°) - \omega_m t] - \cos [\omega_c t + 90°) \omega_m]$$
$$= \underset{\text{LSB}}{\cos (\omega_c t - \omega_m t + 90°)} - \underset{\text{USB}}{\cos (\omega_c t + \omega_m t + 90°)}$$

(*iii*) **The Output of the adder**

$$e_0 = e_1 + e_2 = 2 \cos (\omega_c t + \omega_m t + 90°)$$

(*iv*) **If substractor is used,** instead of adder, the output of the substractor

$$e_0 = 2 \cos (\omega_c t - \omega_m t + 90°)$$

(*c*) **Advantages:**

(*i*) It can generate SSB signal at any desired frequency.

(*ii*) It is able to generate desired signal without any change.

(*d*) **Disadvantages:**

(*i*) In this method, AF phase shift network becomes very complex as it is to work for a wide frequency band. If this network fails to provide an exact phase shift of 90° to any audio frequency, this frequency does not get cancelled completely in the final output.

The Fig. 4.13 shows that the phase shift of a frequency is 88° instead of 90°. This will result an unwanted USB amplitude of $2 E \sin 2°$.

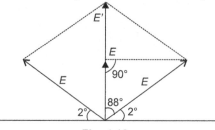

**Fig. 4.13**

(*ii*)   The relative attenuation of this signal as compared to USB will be

$$k \; = \; \frac{2E}{2E \sin 2°} = 28.66$$

In                            dB  =   20 log 28.66 = 29.14 dB

This attenuation is not sufficient for commercial systems, as this will lead to a phase error of about 0.57°.

(*iii*)   If output of the two balanced modulators is not exactly equal, the cancellation of sidebands will be incomplete and the performance will be affected.

## 4.14  THIRD (WEAVER) METHOD

(*a*)   This method is known by its inventor's name and retains the advantages of both Filter and of phase shift methods but without their disadvantages. This can generate SSB at any frequency, thus it falls in direct competition with Filter method (see Fig. 4.14).

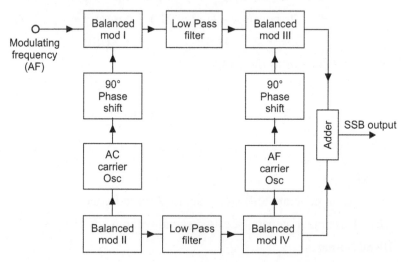

**Fig. 4.14**

It is a very complicated method as compared to filter and phase shift methods. It uses 4 balanced modulators.

Instead of phase shifting the whole range of audio frequencies, the method combines them with a fixed carrier frequency $f_c$ (say 1700 Hz). The phase shift is applied to this fixed frequency only. The resulting voltage at the output of Balanced modulators I and II is fed to low pass filters, whose cut off frequency is designed to be equal to $f_c$ to ensure that input to the last stage balanced modulators III and IV result in a proper sideband suppression.

In the final output, all LSB signals will be cancelled, regardless that the audio frequency lies below or above $f_c$. If LSB is required at the final output, the phase of the carrier voltage fed to the balanced modulator I should be shifted by 180°.

(b) **Advantages:**

    (i)   This system removes drawbacks of filter as well as phase shifter methods.

    (ii)  Low audio frequencies can be easily transmitted.

(c) **Disadvantages:**

    (i)   Though it is a superior method but it cannot superceed the filter method which is more simpler and less complicated than this method.

    (ii)  It uses four balance modulators and hence the method is costly.

**Table 4.1**  Comparison of Three SSB Generation Methods

| S. No. | Parameter | Filter method | Phase shift method | Weaver method |
|---|---|---|---|---|
| 1. | Method to cancel the unwanted side band. | By using a filter | By shifting AF and RF signals by 90° to the base modulator | Same as in phase shift method. |
| 2. | Design of 90° shifter at modulating frequency. | Not applicable | Design is critical | Design is easy, as it is to be done at a single frequency. |
| 3. | Possibility of SSB generation at any frequency. | Not possible | Possible | Possible |
| 4. | Need of up conversion. | Needed | Not needed | Not needed |
| 5. | Use of low modulating frequency. | Not possible | Possible | Possible |
| 6. | Need of linear amplifers. | Needed | Needed | Needed |

| 7. | Critical points in design. | Filter characteristics, its size, weight, cut off frequency. | Design of 90° phase shifter for modulating frequency, symmetry of Balanced modulators | Summetry of Balance modulators for proper carrier cancellation. |
|----|---|---|---|---|

**Table 4.2**   Standard AM vs SSB AM Systems

| S. No. | Standard AM system (DSBFC) | SSB–AM systems |
|--------|---|---|
| 1. | The detection or demodulation is easy and inexpensive in standard AM system. Therefore, the standard AM system is preferred for public communication in which a transmitter is associated with large no. of receivers. | The detection or demodulation is complex and expensive. The receivers need additional synchronizing circuits. |
| 2. | The standard AM system needs large and expensive power transmitters. | The suppressed carrier AM system needs low power and less expensive transmitters. The system is suitable for point to point communication, where many transmitters but a few receivers are used. |
| 3. | The standard AM signals are easy to generate. | SSB signals are generated by balance modulator and their generation is complex. |
| 4. | The transmission bandwidth needed is more in this system. | In this system the bandwidth needed is about half than the standard AM system. Therefore, the SSB systems are used for long distance communication, *e.g.* VSB is used for picture signal transmission. |
| 5. | At unity modulation index; the transmission efficiency of standard AM system is 33%. | At unity modulation index, the transmission efficiency of SSB AM systems may be achieved upto 100%. |
| 6. | While considering noise, this AM system is inferior. The Figure of merit (which is the ratio of signal noise ratio at output to input) is 1/3. | The SSB AM systems are superior than standard AM system, when noise is considered. In this case, the Figure of merit is unity. |
| 7. | The non linear distortion is maximum in standard AM system. | The non linear distortion is minimum in these systems. |

**Table 4.3**   Comparison of Various AM Systems

| S. No. | Particulars | AM systems | | | |
|--------|-------------|------------|-------|-------|-------|
| | | DSBFC | DSBSC | SSBSC | VSBSC |
| 1. | Bandwidth | $2f$ | $2f$ | $f$ | Between $f$ and $2f$ |
| 2. | Power saving (sinusoidal) | — | 66.5% | 83.25% | Between 66.5% and 83.25% |
| 3. | Power saving (Non sinusoidal) | — | — | — | Between 52% to 74% |
| 4. | Generation and Detection | Not difficult | Difficult | Difficult | Very difficult |

## SUMMARY

1. The conventional amplitude modulation (DSPFC) is uneconomical. Only single sideband (SSB) without carrier is the advance technique of communication.

2. Different forms of amplitude modulations are:

   DSBFC, DSBSC, SSBSC, SSBRC, SSBFC, VSBFC, ISB.

3. The sideband is suppressed by filter method, phase shift method and weaver method.

# AM Transmitters

Radio broadcasting employs both amplitude as well as frequency modulation. The broadcasting in medium wave (mw) and short wave (sw) frequency bands is accomplished by means of A.M. transmitters. The signal power to be transmitted in A.M. broadcast transmitters for regional coverage is of the order of kW and may be upto megawatts.

## 5.1 TRANSMITTERS

To remind the readers, a transmitter is a device in which the signal (AM/FM) is modulated, amplified and transmitted.

Radio transmitters may be classified on the following basis:

1. **Type of modulation used:** On this basis, transmitters are of the following type:

   (*a*) **A.M. transmitters:** In these transmitters, carrier is "amplitude modulated" by the signal. These are employed for:

   (*i*) Radio broadcasting—on medium and short waves

   (*ii*) Radio telephony—on short waves

   (*iii*) T.V. picture broadcasting—on very short waves.

   (*b*) **F.M. transmitters:** In these transmitters, carrier is "frequency modulated" by the signal. These are used for.

   (*i*) Radio broadcasting—in V.H.F. and U.H.F. range.

   (*ii*) T.V. sound broadcasting—in V.H.F. and U.H.F. range.

   (*iii*) Radio telephony on V.H.F. and U.H.F. range.

   (*c*) **P.M. transmitters:** In these transmitters, the carrier is "pulse modulated" and these are used in telephony and telegraphy.

2. **Type of service:** On the type of service, transmitters are of the following types:

   (*a*) **Radio broadcasting transmitters:** These transmitters are employed for transmission of sound signals for public recreation. They have low distortion and less noise. They may be A.M. or F.M. transmitters. The A.M. transmitters operate on medium and short waves and F.M. transmitters operate on very short waves.

   (*b*) **Radio telephone and telegraph transmitters:** These transmitters are employed for radio telephony, *i.e.*, to transmit telephone and telegraph signals over long distances by radio means. These transmitters are also equipped with special devices such as *volume compressors, privacy devices*, etc. Both the antennas are also specially designed. These may be A.M. and F.M. transmitters. The A.M. transmitters work on short waves, whereas F.M. transmitters work on U.H.F. range.

   (*c*) **T.V. transmitters:** Two transmitters are used for T.V. broadcasting —one for transmitting picture and other for sound. Both the transmitters operate in V.H.F./U.H.F. range. The picture transmitters are *amplitude modulated*, whereas the sound transmitters are *frequency modulated*. If Vestigial sideband transmission is used, the total BW occupied by one T.V. channel is about 6 MHz.

   (*d*) **Radar transmitters:** Radar transmitters may be pulse transmitters or a continuous wave (CW) transmitters. They are pulse or frequency modulated. They operate on microwave (3cm wavelength) frequency of 3,000 to 10,000 MHz.

3. **Frequency range:** On the basis of frequency range, the transmitters may be:

   (*a*) **Medium wave transmitters:** They operate over a range of 550 to 1650 kHz frequency.

   (*b*) **VHF/UHF transmitters:** They operate in V.H.F. (30–300 MHz) or U.H.F. (300 to 3000 MHz) range of frequencies. They are used for F.M. radio, T.V., radio telephony, etc.

   (*c*) **Microwave transmitters:** These transmitters are used on frequencies above 1000 MHz. Their application is in radars.

   (*d*) **Short wave transmitters:** They operate over a range of 3 to 30 MHz. Ionosphere propagation is employed.

## 5.2 TYPES OF A.M. TRANSMITTERS

The A.M. transmitters (on the basis of modulation) may be of the following types:

1. **Low Level A.M. Transmitter (Fig. 5.1):** This employs Low Level Modulation (LLM) scheme. If the modulated signal is generated at a power level, which is

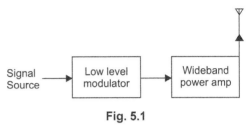

Fig. 5.1

lower than the final transmitter power required, the scheme is known as low level modulation. In this scheme, the modulated signal is amplified subsequently.

2. **High Level A.M. Transmitter:** (Fig. 5.2) This employs High Level Modulation (HLM) scheme. In this scheme, the modulation is carried out at the last stage, and no further power amplification is required.

In this case, the modulating signal should have power higher than that in low level modulation.

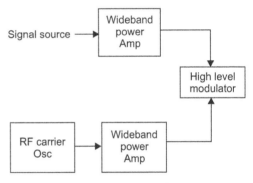

Fig. 5.2

**Table 5.1** LLM Vs. HLM

| S. No. | Low Level Modulation | High Level Modulation |
|--------|----------------------|------------------------|
| 1. | This modulation is carried out at low power level. | This modulation is carried out at high power level. |
| 2. | Needs lesser amplifier stages | Needs more amplifier stages |
| 3. | After modulation linear amplifiers can only be used. This gives lower power efficiency. | Non linear amplifiers can also be used. This leads to higher power efficiency. |
| 4. | Power loss in amplifiers is higher, the cooling problem is severe. | The power loss is less, the cooling problem is not severe. |

## 5.3 NEGATIVE FEEDBACK IN A.M. TRANSMITTERS

Generally, negative feedback is provided in A.M. transmitters in relation to the modulating signal. The radiated signal is picked up from the antenna, demodulated and combined with the modulated signal and used as negative feedback at the appropriate point. The negative feedback be used with low level modulators, as well as with high level modulators.

Advantages of using negative feedback with transmitters are same as with amplifiers. The negative feedback reduces amplitude as well as frequency distortion thus increases faithfulness of the transmitter. The negative feedback also reduces the undesirable hum and noise in the transmitter.

## 5.4 A.M. MODULATORS

In A.M. transmitters "amplitude modulation" is carried out. The A.M. modulators may be classified as

- (*i*) **Linear Modulators:** These make use of linear part of the $V - I$ curve of the transistors (or diodes). The examples of these are collector modulators and base modulators etc.

- (*ii*) **Non Linear or Square Law Modulators:** These make use of non linear part of the VI characteristic of the transistors (or diodes).

The Fig. 5.3 and 5.4 show VI characteristic of transistor and diode respectively, where OA is the linear part and beyond $A$ is non linear part of the curve.

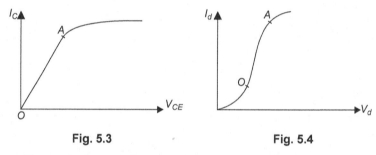

Fig. 5.3                                            Fig. 5.4

Here we discuss these modulators.

## 5.5 LINEAR MODULATORS

As mentioned above, these make use of linear part of the characteristic of transistors or diode, below we describe linear modulators using transistors, these are:

- (*i*) Collector Modulator
- (*ii*) Base Modulator

- (*i*) **Collector Modulator.** The Fig. 5.5 shows the collector modulator. The modulating signal is fed to the collector in series with $V_{CC}$ of the transistor amplifier. The carrier is fed to the base through a tuned circuit and the modulated output is obtained from the collector as shown. The modulating signal is used to vary the collector voltage. The capacitor $C$ provides a low impedance path to the carrier currents, in other words the capacitor should not bypass the modulating frequencies.

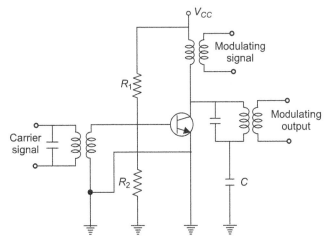

**Fig. 5.5**

The power requirement of the modulator and its load can be determined from the basic voltage and current relationship of a transistor power amplifier. To obtain 100% modulation, the maximum value of the modulating voltage $V_{max}$ must be equal to the $V_{CC}$. *i.e.,*

$$V_{max} = V_{CC}$$

In this condition the output of the modulation amplifier is zero at the negative peak of the modulating signal.

Also, the average collector current $I_C$ of the modulator amplifier reduces to zero during negative cycle of the modulating signal, therefore,

$$I_C = I_{max}$$

∴ The modulator power output

$$P = \frac{V_{max} \cdot I_{max}}{2}$$

or

$$P = \frac{V_{CC} \cdot I_C}{2} \quad [V_{max} = V_{CC}, \ I_{max} = I_C]$$

This shows that power output of the modulator is equal to one half of power supplied to the amplifier. Therefore power from the modulator provides power for sidebands also. This has been shown already that the power in sidebands is 1/3rd of the total power or one of the carrier power at 100% modulation.

Moreover, the modulator load

$$R_L = \frac{V_{max}}{I_{max}} = \frac{V_{CC}}{I_C}$$

Note that modulator may use any power amplifier circuit. **Push pull amplifiers** are mostly used for maximum power output.

(*ii*) **Base Modulator.** In base modulator, the modulating signal is fed into the base and the output is obtained from the collector of the amplifier. See Fig. 5.6

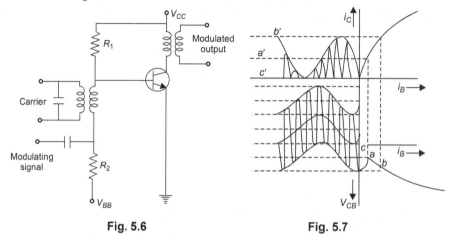

**Fig. 5.6**                                              **Fig. 5.7**

Considering the voltage current relationship (Fig. 5.7) in a base modulated amplifier, power requirement for the modulator is

$$P = \frac{i_B - V_m}{2}$$

where $i_B$ is the base current and $V_m$ is the maximum amplitude of the modulating sign.

**Note that:**

1. The power requirement for base modulator is less than the collector modulator.

2. The power output and efficiency of this modulator is comparatively low.

3. This modulator has also poor linearity and its adjustment is more critical.

The circuit is used in TV transmission as it needs little power and can meet the power requirement of larger bandwidth.

## 5.6 NON-LINEAR (SQUARE LAW) MODULATORS

Whenever a carrier and a modulating signal both are simultaneously applied to a non linear circuit, the amplitude modulation results, as the carrier modules the signal.

Note that the Square law modulators utilize the non linear part of the VI characteristic of transistor or of a diode.

Semiconductor diodes and certain other types of non linear devices have a volt-ampere $(V - I)$ characteristics like that as shown in the Fig. 5.8. When such a device is connected to a resistor, the voltage developed across the resistor can be described in the form :

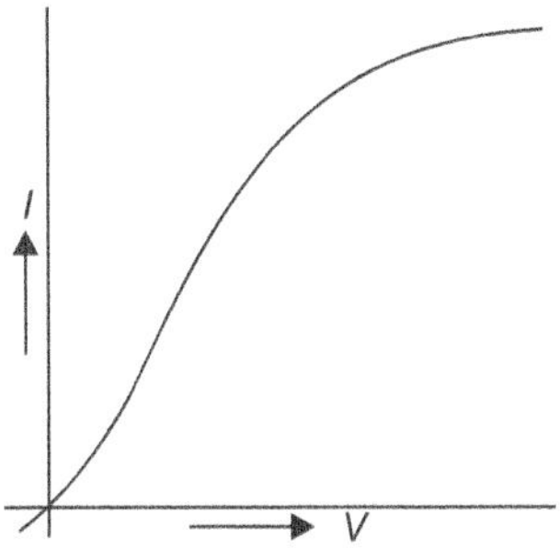

$$V_0 = \sum_{m=1}^{\infty} a_m V_i^m$$

Fig. 5.8

Where $V_0$ is taken as output voltage and $V_i$ is the voltage impressed across the device. If $|V_i|$ is small enough, we can generally approximate the transfer characteristics by retaining only the first two forms.

Then we can say that we have a square-law device. For such a device

$$V_0 = a_1 V_i + a_2 V_i^2$$

Let the input consists of sum of modulating signal plus carrier, that is
$$V_i(t) = f(t) + \cos \omega_{c.t}$$
Where 　　　　$f(t) = \cos \omega_{m.t}$
The output is then

$$V_0(t) = a_1 f(t) + a_2 f^2(t) + a_2 \cos^2 \omega_{c.t} + a_1 \left[ 1 + \frac{2a_2}{a_1} f(t) \right] \cos \omega_{c.t}$$

Which can be written as

$$V_0(t) = a_1 f(t) + a_2 f^2(t) + \frac{a_2}{2}(1 + \cos^2 2\omega'_c) + a_1 [1 + mf(t)] \cos \omega_{c.t}$$

which is an AM wave

Where 　　　　　　$m = \dfrac{2a_2}{a_1}$

**Waveform:**

The Unsealed spectrum of $V_0(t)$ is shown in the Fig. 5.9. A band pass filter of bandwidth $2\omega$ centred at $f_c$ will isolate with AM wave. The filter bandwidth can, of course, be larger, provided it does not admit the other spectra. Note that for modulation with square law device, we must have $f_c > 3\omega$ (square law modulation)

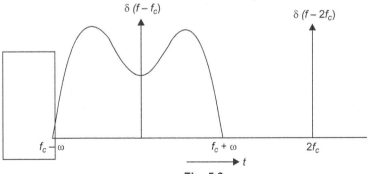

Fig. 5.9

(*c*) **A non-linear (Square law) modulator consists of the following components:**

   (*i*)   An adder circuit to add carrier and the modulating signal.

   (*ii*)  A non linear circuit element such as a transistor or a diode.

   (*iii*) A band pass filter in the range of $f_c - f_m \leq f_c \leq f_c + f_m$.

   That the relation between amplitude of the modulating signal and the resulting depth of modulation (*m*) is a square law and hence the name. Moreover, these modulators are used at low voltages.

   The Fig. 5.10 (*a*) shows block diagram and Fig. 5.10 (*b*) shows circuit for a non linear diode modulator.

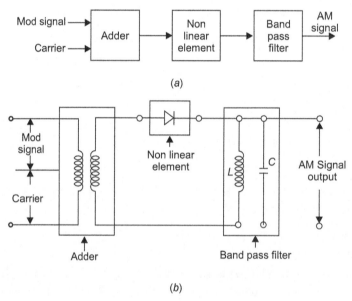

Fig. 5.10

## 5.7 TYPES OF NON LINEAR MODULATORS

Here we describe two non linear/square law modulators:

   (*i*)  Balanced modulator          (*ii*)  Ring modulator

   (*i*) **Balanced modulator:** The balanced modulator uses generally a push pull amplifier circuit.

   The push pull circuit may employ two diodes [(Fig. 5.11 (*a*)] or transistors [(Fig. 5.11 (*b*)] or FETs [Fig. 5.11 (*c*)] in push pull arrangement. The two transistors (or FETs) should have identical characteristics and the circuit should be symmetrical w.r.t. the center tapped transformer.

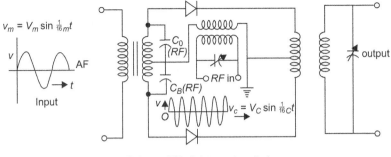

(a) Balanced Modulator using diodes

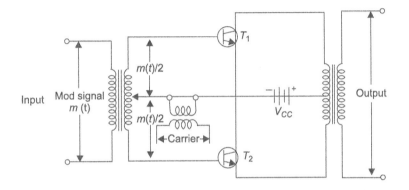

(b) Balanced Modulator using transistors

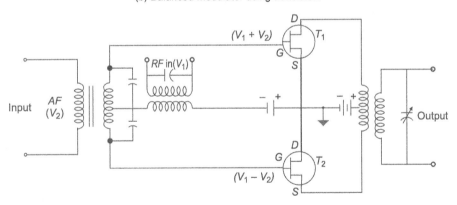

(c) Balanced Modulator using FETs.

**Fig. 5.11**

In a simple non linear circuit used as square law modulator, the **non linear terms** are eliminated by using a band pass filter, but in a balanced modulator the undesired non linear terms are *automatically balanced* out and at the output only desired terms are obtained. The band pass filter (BPF) therefore may or may not be used.

Each of three circuits shown in Fig. 5.11 utilizes non-linear principles. *Here we describe the circuit using FETS.* The AF modulating voltage $V_2$ [(Fig.5.11 (c)] is fed in push-pull, and RF carrier voltage $V_1$ in parallel, to a pair of identical FETs. In FET circuit, the carrier voltage is thus applied to two gates in phase, whereas modulating voltage appears 180° out of phase at the gates. Since these are at opposite ends of a centre-tapped transformer, the modulated output currents of the two FFTs are combined in centre-tapped primary of the output Transformer, they therefore are subtracted. If the system is made completely symmetrical, the carrier frequency will be completely cancelled. Thus the final output consists only of side bands.

(b) **Analysis.** The input voltage will be $V_1 + V_2$ at the gate $T_1$ and $V_1 - V_2$ at the gate $T_2$. If the perfect symmetry is obtained *i.e.*, devices used in balanced modulator, *i.e.*, FETs are well matched the constant of proportionality will be same for both FETs and may be called as $a$, $b$ and $c$.

The two drain currents, calculated will be

$$id_1 = a + b\,(V_1 + V_2) + c(V_1 + V_2)^2$$
$$= a + bV_1 + bV_2 + cV_1^2 + cV_2^2 + 2\,cV_1V_2$$
$$id_2 = a + b\,(V_1 - V_2) + c(V_1 - V_2)^2$$
$$= a + bV_1 - bV_2 - cV_1^2 + cV_2^2 - 2\,cV_1V_2$$

The primary current is given by the difference between the individual drain currents. Thus

$$i_1 = id_1 - id_2 = 2bV_2 + 4cV_1V_2$$

$V_1$ is given as $V_c \sin \omega_c t$ and $V_2$ as $V_m \sin \omega_c t$.

$$i_1 = 26V_m \sin w_m t + 4cV_m V_c$$
$$= 2bV_m \sin \omega_m t + 4cV_m V_c$$
$$\times \frac{1}{2}[\cos(\omega_c - \omega_m)t - \cos(\omega_c + \omega_m)t]$$

The output voltage $V_0$ is proportional to this primary current. Let the constant of proportionality be $\alpha$. Then

$$V_0 = \alpha i_1$$
$$= 2\alpha bV_m \sin \omega_m t + 2\alpha cV_m V_c$$
$$[\cos(\omega_c - \omega_m)t - \cos(\omega_c + \omega_m)t]$$

For simplicity Let $\qquad P = 2\alpha bV_m;\ Q = 2\alpha V_m V_c$

then $\qquad\qquad V_0 = P \sin \omega_m t + Q \cos(\omega_c - \omega_m)t - Q \cos(\omega_c + \omega_m)\,t$

| Modulating | Lower sideband | Upper sideband |
| frequency | (LSB) | (USB) |

Under ideally symmetrical conditions, the carrier has been cancelled out, leaving only two sidebands and modulating frequencies.

### (ii) **Ring modulator**

The ring modulator is another square law modulator. The circuit uses four diodes forming a ring and hence the name.

The four diodes are ideal and identical, moreover the transformers used have same turn ratio of $1 : 1$ (Fig. 5.12$a$). The Fig. 5.12 ($b$) shows square pulse train, Fig. 5.12 ($c$) shows modulating signal and Fig. 5.12 ($d$) shows modulated wave.

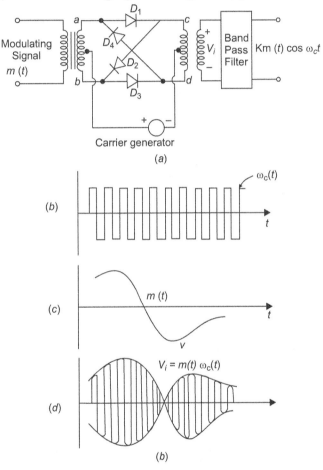

**Fig. 5.12**

In Ring Modulator during the positive half cycles of the carrier, the diodes $D_1$ and $D_3$ conduct, and $D_2$ and $D_4$ are opened. Hence, terminal '$a$' is connected to '$c$' and terminal '$b$' is connected to '$d$'. During the negative half cycles of the carrier voltage, diodes $D_2$ and $D_4$ will act while $D_1$ and $D_3$ are open, thus connecting the terminal '$a$' to '$b$' and terminal '$b$' to '$c$'. Hence the O/P is proportional to $+m(t)$ i.e., modulating signal during the positive half cycle and $-m(t)$ during the negative half cycle. The square pulse train $\omega_c(t)$, is multiplied with the signal $m(t)$ and output is shown in Fig. 5.12 ($d$). When this wave is passed through a band pass filter tuned to frequency $\omega_c$, the filter output will be the desired signal $m(t) \cos \omega_c t$. The input to the band pass filter does not contain carrier and modulating signal but only DSB (double side band).

## 5.8 EMITTER MODULATOR

An amplifier as shown in Fig. 5.13 ($a$) can be used for amplitude modulation, provided it has two inputs, one as the carrier signal and other as the modulating signal. In the absence of modulating signal $v_m$, this circuit operates as $a$ linear class A amplifier and output is simply amplified version of the carrier signal. However, when the modulating signal is also applied, the amplifier operates in non-linear region and signal multiplication occurs.

In the circuit diagram, carrier is applied to the base and modulating signal to the emitter. This circuit configuration is also known as **emitter modulator**. The modulating signal varies the gain of the amplifier at a sinusoidal equal to the frequency of modulating signal. The voltage gain for this modulator can be expressed as:

$$A_v = A_q(1 + m \sin \omega_m t) \qquad \qquad ...(i)$$

where         $A_v$ = Voltage gain of amplifier at modulation

$A_q$ = quiescent voltage gain of amplifier.

$\sin \omega_m t$ changes from a maximum value of $+1$ to a minimum value of $-1$ at $270°$, thus equation ($i$) will become as

$$A_v = A_q(1 \pm m)$$

where         $m$ = modulation index

At            $m = 1$

$$A_v \text{ (max)} = 2 A_q$$

At            $m = 0$

$$A_v \text{ (min)} = A_q$$

Waveform for modulating signal is given [(Fig.5.13 (b)]. The modulation signal is applied to emitter and carrier is directly applied to the base. The modulating signal drives the circuit into both saturation as well as in cut off. The coupling capacitor ($C_2$) at the O/P removes the modulating frequency from AM waveforms, thus, producing a symmetrical AM envelope at the output.

In emitter modulator, the amplitude of the O/P signal ([Fig. 5.13($c$)] depends on the amplitude of the input carrier and voltage gain of the amplifier. The modulation index depends entirely on the amplitude of the modulating signal.

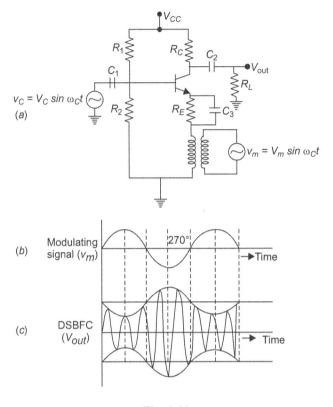

**Fig. 5.13**

## 5.9 A.M. MODULATOR (USING VACUUM TRIODE)

Widely used method of obtaining amplitude modulation for broadcasting and other high power transmission applications is by using a class $C$ amplifier shown in Fig. 5.14.

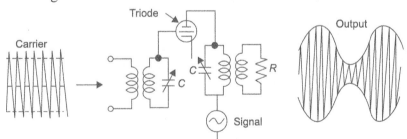

**Fig. 5.14**

In this method, the carrier is applied at the input and the signal is applied in series with the plate The voltage of class $C$ amplifier and its plate current are varied in accordance with the amplitude of the signal and the modulated output is obtained.

## 5.10 A.M. MODULATOR (USING TRANSISTOR)

Figure 5.15 shows a transistorized A.M. modulator. As can be seen it is basically a *CE* amplifier The carrier is applied at the input (between base and emitter). The signal is a part of the biasing circuit, it produces variations in the emitter circuit, this varies the gain of the amplifier. As a result, the amplitude of the carrier varies according

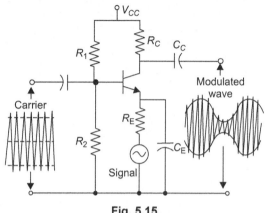

**Fig. 5.15**

to the signal and the modulated wave is obtained at the output.

Usually the carrier frequency ($f_c$) is much greater than the signal frequency ($f_m$). We need $f_c$ at least 100 times greater than $f_m$ so that the capacitors of the modulator should look like low impedance to the carrier and look like high impedance to the signal.

## 5.11 BLOCK DIAGRAM OF A.M. TRANSMITTER

The function of a transmitter is to perform process of modulation and to raise power level of the modulated signal to the desired extent for effective radiation. They may use low level or high level modulation.

The Fig. 5.16 (*a*) shows block diagram of a typical A.M. transmitter. A crystal oscillator generates the carrier frequency or its multiple. It is followed by a Buffer amplifier and a tuned driver amplifier. After this, a class *C* modulator amplifier is used which is generally a collector modulator. The audio signal is amplified by a chain of amplifiers and a power amplifier. Usually transformer coupled class B push pull power amplifier is used for power amplification.

Now, the output of the final class *C* amplifier is passed through an impedance matching network which includes the tank circuit of the final amplifier. The *Q*-factor of this circuit should be low enough so that all the sidebands of the signal are passed without any type of distortion.

The negative feedback is often used to reduce distortion in class *C* modulator system. The feedback is introduced as shown in Fig. 5.16 (*b*). A sample of RF signal sent to the antenna is extracted and demodulated to produce the feedback.

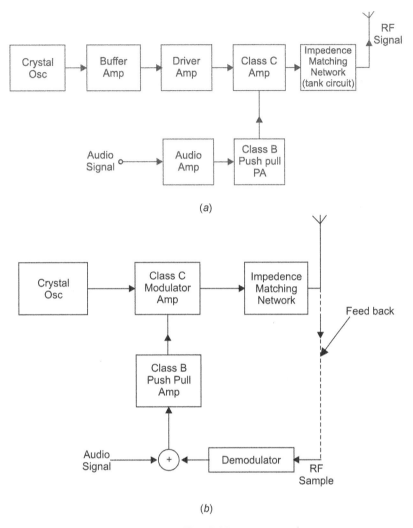

Fig. 5.16

The low power transmitters with output power upto 1 kW or so may be transistorized, but the higher power transmitters use vacuum tubes.

## 5.12 A.M. BROADCASTING TRANSMITTER

Figure 5.17 shows an A.M. broadcasting transmitter. The oscillator produces high frequency carrier waves which are fed into R.F. amplifier for amplification to the desired value. The audio signal to be transmitted is also amplified in audio frequency amplifiers. Both amplified inputs are given to an A.M. modulator and the modulated output of the modulator is further amplified and fed to the transmitting antenna.

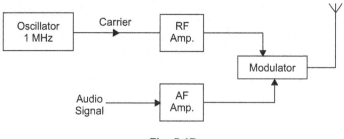

**Fig. 5.17**

## SUMMARY

1. The transmitter is the apparatus in which modulation is carried out before transmission of the signal. The apparatus which carry out the amplitude modulation and amplification are known as A.M. transmitters.

2. The A.M. transmitters may be low or high level modulation transmitters.

3. The A.M. transmitter is provided with negative feedback, which reduces noise and increases faithfulness.

4. The A.M. modulators may be linear or non linear.

5. The examples of linear modulators are Collector modulator and Base modulator. The examples of non linear modulators are Balanced modulator and Ring modulator.

❑❑❑

# AM Detectors and Receivers

The function of a receiver is to recover the original signal. There are number of signals floating in the space, the receiver should be able to select the signal of the desired frequency. It should then be able to demodulate the received signal to recover the original signal. It should be noted that the signal received is of very low level *i.e.*, of a few pecowatts, which needs to be amplified.

## 6.1 DEMODULATION OR DETECTION

The process of separating the original signal from the (AM) radio wave and grounding (rejecting) the carrier is called demodulation or detection.

The demodulation/detection is a process taking place in the receivers. After detection, the AF signal is fed to loudspeaker, which converts the electrical signal into sound. If the AF signal is not detected and the radio wave (signal + carrier) is directly given to the speaker, the diaphragm of the speaker cannot respond to such a high frequency and will not be able to perform its function. Therefore, it is necessary to detect (separate) the signal which will strike the speaker's diaphragm and not to allow the H.F. carrier to reach the speaker.

## 6.2 AM DETECTORS

The AM detectors are basically of two types.

    (*i*) Linear/diode/envelope detector

    (*ii*) Synchronous/square law detector

These are briefly discussed below:

## 6.3 LINEAR/DIODE/ENVELOPE DETECTORS

This detector is very much used in commercial receivers as it is cheap, simple and provides satisfactory performance.

A diode operating in the linear region of its characteristic can extract the modulating signal from the AM wave.

The Fig. 6.1 shows the circuit for the linear diode detector. The circuit basically consists of a diode and a *RC* net work.

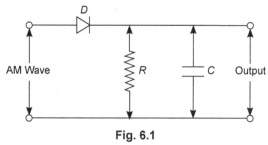

**Fig. 6.1**

## Operation (Fig. 6.2)

The A.M. wave is applied at the input terminals of the circuit. As the diode is operated in the linear region of its characteristic, during positive cycle of the A.M. wave, the output is proportional to the input signal voltage. During the negative cycle of the input, the diode does not conduct and output is theoretically zero. If the time constant (*RC*) is correctly chosen the output will follow exactly the envelope of the A.M. wave, but spikes are introduced by charging and discharging of the capacitor,

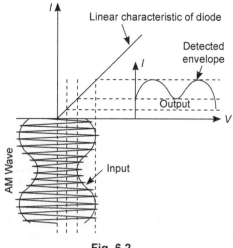

**Fig. 6.2**

which can be reduced by taking a large *RC* constant.

The Fig. 6.2 shows the linear characteristic of diode along with input and output wave shapes. The output may contain ripples which are later on filtered out.

The diode detector basically performs two functions: (Fig. 6.3)

1. The diode rectifies the A.M. wave, *i.e.*, eliminates the negative cycle of the wave. We know that average of both the cycles of an A.C. wave is zero. In such case if both the cycles of wave is fed to the speaker without rectification, it will have no impact on the speaker's diaphragm (due to its zero average value). This job is done by the diode.

   Now, the positive cycle of the A.M. wave (containing carrier + signal) starts its journey towards speaker.

2.  The positive cycle of the A.M. wave is passed through a capacitor filter, which suppresses the H.F. carrier that is grounded.

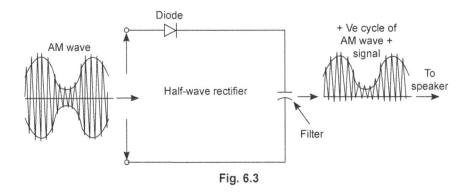

**Fig. 6.3**

## Process of Detection

The Fig. 6.4 shows complete circuit for the diode detector. The parallel combination of a capacitor $C_1$ and resistance $R$ is the load resistance, across which the rectifier output voltage is developed. The capacitor $C_1$, charges up to a voltage equal to the peak signal voltage and $V_0$ reproduces the modulating voltage accurately. The time constant $RC$ should be suitably designed to keep R.F. ripples in the output as small as possible.

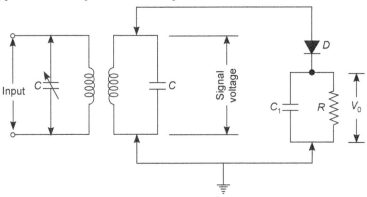

**Fig. 6.4**

The circuit suffers from the disadvantage that the output contains DC component and also R.F. ripples which are unwanted.

## Practical Diode Detector

Some modification in the previous circuit has been done. In this case, the diode (Fig. 6.5) is reverse biased. It demodulates negative envelope, which has no effect on detection. Here the resistance $R$ has been split into $R_1$ and $R_2$ to have a series path to *ground* the diode. A low pass filter circuit $R_1C_1$, removes the R.F.

ripples. The $R_2C_2$ is another low pass filter circuit to remove A.F. components. Thus, the disadvantages of the previous circuit have been removed. In the Fig. 6.5, I.F. stands for intermediate frequency and A.G.C. stands for automatic gain control.

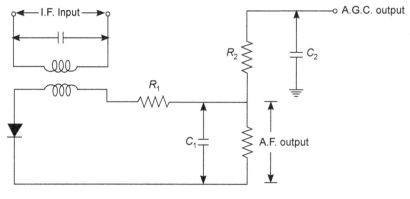

**Fig. 6.5**

## Choice of Time Constant (RC)

(*i*)  For best results, the time constant $RC$ should be such that,

$$RC = \frac{\sqrt{1-m^2}}{w_m.m}$$

where,                                        $m$ = depth of modulation

$w_m$ = angular velocity of the modulated wave.

(*ii*)  Further, $RC$ should be such that where

$$\frac{1}{f_c} \leq RC \leq \frac{1}{B}$$

where,                                        $f_c$ = Carrier frequency

$B$ = Bandwidth of the modulating signal

## 6.4  SYNCHRONOUS/SQUARE LAW DETECTOR

These detectors are used for detection of DSB-SC or SSB-SC signals. In this process, the modulating signal (Radio wave) is mixed with the carrier in a non linear device giving sum and difference frequency components as the output. The carrier may be generated by a local oscillator (LO). The output of the non linear device (which acts as a multiplicative mixer) is passed through a low pass filter (Fig. 6.6).

The synchronous detection is effective only when the locally generated carrier is properly synchronised with the transmitted carrier. Any shift in the phase or frequency of the locally generated carrier results in a distortion in the demodulated output signal.

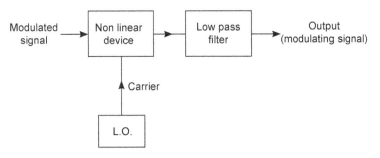

**Fig. 6.6**

The Fig. 6.7 shows circuit for Synchronous/Square law detector. The detector is similar to the square law modulator the only difference being that in place of band pass filter at the output, now we have a low pass filter of frequency range of $0 \leq f_c \leq f_m$.

Suppose at the receiver end, we obtain the modulated signal as input:

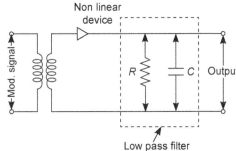

**Fig. 6.7**

$$DSB \ SC = m\,(t) \,.\, \cos \omega_c t.$$

Where $m(t)$ is the modulating signal and $\cos \omega_c t$ is the carrier. Now we have to recover the original signal *i.e.*, $m(t)$. In the non linear device, we perform a multiplying operation *i.e.*, the DSBSC signal is multiplied by $\cos \omega_c(t)$ again, which is generated in the local oscillator (LO) of the receiver.

Now we get (by multiplying):

$$m(t).\, \cos \omega_c t \,.\, \cos \omega_c t = m(t) \,.\, \cos^2 \omega_c t$$

$$= \frac{m(t)}{2}\,[1 + \cos 2\,\omega_c t] = \frac{m(t)}{2} + \frac{m(t)}{2} \,.\, \cos 2\,\omega_c t$$

The low pass filter would allow the first component to pass through and would suppress the second component. Note that the range of the filter is $0 \leq f_c \leq f_m$. The factor $\dfrac{1}{2}$ can be removed by amplification and thus we get back the original signal $m(t)$ at output.

The diode is biased positively to shift the zero signal operating point to the non linear part of the VI characteristic of the diode.

The Fig. 6.8 shows superimposition of the modulated carrier over the dynamic (VI) characteristic of the diode.

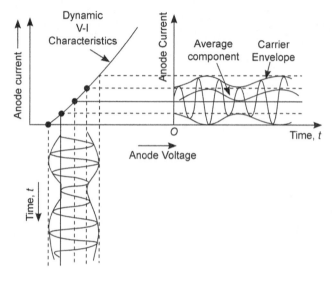

**Fig. 6.8**

## Squaring (Synchronizer) Circuit

As mentioned earlier the generated carrier in the L.O. of the receiver should be in complete synchronism with the incoming carrier. To obtain synchronism, the locally generated carrier is obtained from the incoming modulated signal. The Fig. 6.9 shows block diagram to obtain the synchronising carrier signal from the input modulated signal.

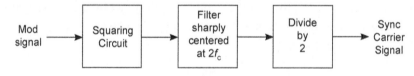

**Fig. 6.9**

The carrier signal is so obtained is used in place of locally generated carrier signal. Obviously the phase and frequency of the carrier so generated will change in the same way as the frequency and phase of the incoming carrier.

## 6.5 DISTORTIONS IN DIODE DETECTORS

Basically there are following types of distortions in diode detectors.

(*i*) **Negative clipping.** Which causes due to ac and dc load impedance. When the transmitted modulation index is small, no clipping takes place [Fig. 6.10 (*a*)], when the modulation index is large (*m* > 1) negative peak is clipped [(Fig. 6.10(*b*)].

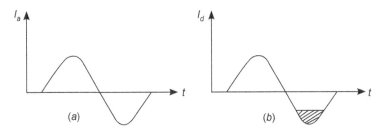

**Fig. 6.10**

**Analysis.** As diode is a current operated device, the modulation index of a *demodulated* wave in terms of currents is given as:

$$m_d = \frac{I_m}{I_c} \quad \text{[When } I_m \text{ and } I_c \text{ are peak values of currents]}$$

$$= \frac{E_m/Z_m}{E_c/R_c} = m.\frac{R_c}{Z_m} \left[ \frac{E_m}{E_c} = m \right]$$

Since, maximum value of $m_d$ is one, the max. value of *m* will be:

$$m_{max} = m_d(\text{max}) \quad . \quad \frac{Z_m}{R_c} = \frac{Z_m}{R_c} \; [m_d (\text{max.}) = 1]$$

This gives max. value of *m* to avoid negative clipping. Its value should be about 70% or 0.70.

Here,                    $m_d$ = modulation index of a demodulated wave

$m$ = modulation index of a modulated wave

$R_c$ = d.c. diode resistance

$Z_m$ = diode load impedance (assumed resistive)

other symbols have the usual meaning.

(*ii*) **Diagonal clipping.** This causes due to ac load impedance acquiring a reactive component at high frequencies. This is another type of problem that may arise with diode detectors. In the above analysis, (negative clipping) we have assumed, the diode load impedance $Z_m$ as purely resistive. At higher modulating frequencies, this may not be true and, it may have a reactive (Capacitive) component and *RC* (time constant) may be too low to follow the change.

This may result in an exponential decay of current, which will result into diagonal clipping as shown in Fig. 6.11.

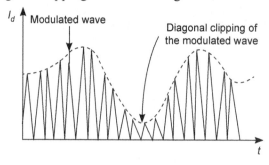

**Fig. 6.11**

The diagonal clipping does not accur when $m$ is below 0.6 (or 60%); the max. permissible value has been found as 70% or 0.7. The RC time constant of the diode detector may be designed to avoid diagonal clipping.

**Analysis:**

For best results the time constant $(RC)$ should be such that

(i)
$$RC \le \frac{\sqrt{1-m^2}}{\omega_m . m}$$

where
$$m = \text{depth of modulation}$$
$$\omega_m = \text{angular velocity of the modulated wave.}$$

(ii)   $RC$ should be such that:
$$\frac{1}{f_c} \le RC \le \frac{1}{B}$$

where
$$f_c = \text{Carrier frequency}$$
$$\text{B.W.} = \text{Bandwidth of the modulating signal.}$$

## 6.6 TYPES OF AM RECEIVERS

Recall that the AM receivers contain detector and amplifiers.

Two types of AM receivers are popular and are of commercial importance. They are:

1.  Tuned radio frequency receivers, and

2.  Superheterodyne receivers.

These are described below:

## 6.7 TUNED RADIO FREQUENCY (T.R.F.) RECEIVERS

These are also called "straight radio receivers". In these receivers, two or more R.F. amplifiers (tuned together) are employed to select and amplify the incoming frequency and rejecting all others. After this, the A.F. signal is detected and amplified suitably. See Fig. 6.12

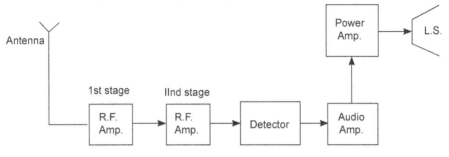

**Fig. 6.12**

The important blocks are described below:

1. **Antenna or Aerial:** The radio signals are picked up by the antenna or aerial. The signal floating in space induce small voltage (in pV) at the antenna.

2. **R.F. Amplifiers:** As mentioned above, there are one or more stages of R.F. amplifiers. They select the particular radio frequency (They are basically tuned amplifiers) and amplify to the desired level.

3. **Detector:** The diode detector detects (separates) the audio signal. As described already, this block performs two functions, *viz.*, half-wave rectification of the radio wave and detection of the AF signal.

4. **Audio Amplifier:** The detected (AF) signal is amplified. It is also done in stages— first voltage is amplified and the last stage is of power amplification.

5. **Loudspeaker:** The amplified signal is fed to the loudspeaker, which converts the electrical signal into original sound.

   The major advantage of TRF receiver is high sensitivity and low cost.

But the receiver suffers from poor selectivity, instability in gain and variation of BW over the band. The receiver works satisfactorily at low and medium frequencies but at higher frequencies it has poor reception or no reception. The receiver also suffers from a variation in $Q$ factor and BW of the tuned circuit employed in RF amplifier, at different frequencies of the band.

The TRF receivers are also available in I.C. form, one IC TRF receiver is ZN 414, which is available in TQ-18 package. (Fig. 6.13). This is a three terminal circuit, which can be used as an AM receiver by connecting an external tuning circuit.

It also offers AGC facility. The circuit (Fig. 6.13) has been designed for operation at 1210 kHz carrier. The $Q$ factor of the tank circuit should be 121 for selectivity of 10 kHz signal.

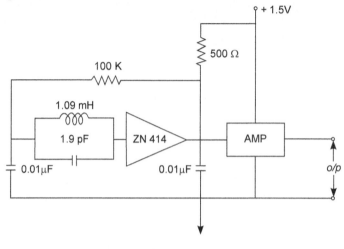

**Fig. 6.13**

## 6.8  SUPERHETERODYNE RECEIVER (SUPERHET)

All modem radio receivers are essentially the superheterodyne receivers. They are the most superior (super) circuits which utilize the principle of 'heterodyning' (mixing or beating) two frequencies.

The selected radio frequency and a high frequency (produced by a local oscillator provided in the receiver) are fed into a 'mixer' in which both frequencies are heterodyned, as a result of which lower and higher beats are produced. The mixer circuit is so designed that lower beats are accepted as 'output' and upper beats are rejected.

For example, $f_1$, is the selected radio frequency and $f_2$ is the frequency produced by the local oscillator. When both are mixed, upper beats ($f_2 + f_1$) and lower beats ($f_2 - f_1$) are produced. The lower beats are obtained as output whereas upper beats are grounded (See Fig. 6.14)

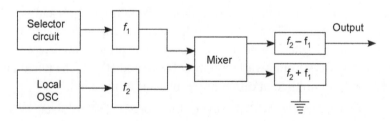

**Fig. 6.14**

Suppose the selector circuit selects a frequency of 1000 kHz and the local oscillator generates 1455 kHz. These two mixed (beated) in the mixer and $(1000 + 1455 = 2455)$ kHz and $(1455 - 1000 = 455)$ kHz are produced. The circuit will give an output as 455 kHz and the 2455 kHz will be grounded.

This output (455 kHz) is called the *Intermediate frequency* (I.F.). It is interesting to note that standard value of I.F. is 455 kHz and it is nationally (or internationally) accepted. Note that the value of I.F. is very much less than the selected frequency. Now we have to design our amplifiers, detectors and other equipment ahead at very low frequency (455 kHz) and the system becomes economical; thus the concept of I.F. is advantageous.

Whatever may be the selected frequency, the local oscillator will always produce a frequency 455 kHz more than the selected frequency. Remember that in case of T.R.F. receivers, the equipment were designed on selected frequency itself.

## 6.9 BLOCK DIAGRAM OF A SUPERHETERODYNE RECEIVER

The important blocks of a superheterodyne receiver have been shown in Fig. 6.15.

1. **R.F. amplifier:** This selects the particular radio frequency. It has parallel $LC$ circuit. By changing the value of $C$, the $LC$ tuned circuit produces a resonant frequency ($fr = 1/2\pi \sqrt{LC}$); therefore the signal of a particular frequency out of many frequencies floating in space is accepted by R.F. amplifier and all others are rejected. This signal is then amplified to the desired level.

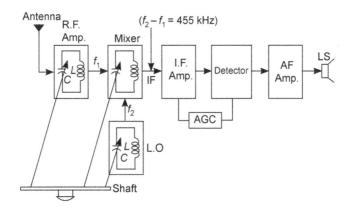

**Fig. 6.15**

2. **Local oscillator (L.O.):** The local oscillator has its own $LC$ circuit, which by varying $C$ produces a frequency equal to the selected R.F. + 455 kHz as explained above.

3. **Mixer:** The mixer is also an $LC$ circuit. It mixes the selected R.F. and the frequency generated by the local oscillator. The output of the mixer is always equal to 455 kHz (called intermediate frequency).

   Note that variable capacitors of R.F. amplifier, local oscillator and the mixer have a common 'shaft' and operated simultaneously. Whenever we 'tune' our radio receiver in fact, we vary three capacitors simultaneously to get the I.F. (intermediate frequency) of 455 kHz.

4. **I.F. amplifier:** It amplifies the I.F. output of the mixer to the desired value. The amplification is done in stages.

5. **Detector (demodulator):** The detector detects (separates) the original A.F. signal out of I.F. output which contains signal as well as the carrier. The signal starts its journey onward, whereas the carrier is grounded.

   An A.G.C. voltage is applied between I.F. amplifier and the detector. A.G.C. stands for 'Automatic gain control'.

6. **A.F. amplifier:** The original (A.F.) signal is amplified here.

7. **Loudspeaker:** It converts the signal into the original sound.

   Superheterodyne (superhets) are superior in quality than T.R.F. receivers. The superhets have superior selectivity, audio quality and are cheaper in cost and lighter in weight.

   **Note:** For its sound section, a television also employs a superheterodyne receiver.

**Problem 6.1.** *A 10 kHz signal modulates a 1100 kHz carrier. Find the frequency to be generated by the local oscillator (L.O.) of the receiver to demodulate the signal.*

**Solution.** We know I.F. = 455 kHz

Frequency of L.O. = I.F. + carrier frequency = 455 + 1100 = 1655 kHz    **Ans.**

## 6.10 DOUBLE HETERODYNE RECEIVER

It is not practical to heterodyne a 30 MHz signal down to an intermediate frequency (I.F.) of 455 kHz. The local oscillator (L.O.) and input signal are too close in frequency. In such cases the heterodyning is done in stages. A more practical answer to the problem is a double heterodyne receiver as shown in Fig. 6.16.

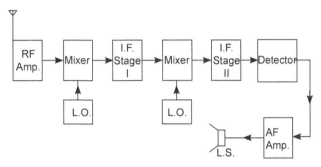

**Fig. 6.16**

# 6.11 ADVANTAGES OF SUPERHETERODYNE RECEIVERS

The superheterodyne receiver has the following advantages on TRF receivers:

- (*i*) Improved Selectivity
- (*ii*) Improved Stability
- (*iii*) Higher Gain Per Stage
- (*iv*) Uniform Band Width

These advantages make them suitable for most of the radio receiver applications such as AM, FM, SSB, communications, TV and radar receivers.

## Advantages of Using R.F. Stage

Advantages of using R.F. stage in receivers are as follows:

- (*i*) Greater gain *i.e.*, sensitivity.
- (*ii*) Improved image frequency rejection.
- (*iii*) Improved S.N.R. (Signal noise ratio).
- (*iv*) Better and improved selectivity *i.e.*, rejection of unwanted signals.
- (*v*) Better coupling of the receiver with antenna, an important factor at VHF and UHF ranges.
- (*vi*) Prevention of spurious frequencies from entering into the mixer, which may heterodyne there and may produce Interferring frequencies.

## 6.12 AM SUPERHETERODYNE RECEIVER USING I.C.

The superheterodyne receiver can be designed using discrete components. But a no. of I.C.s for superheterodyne receivers are also available in the market. One such I.C. package is L.M. 1820. This is a 14 pin DIL (dual in line) package which contains 6 major blocks. The Fig. 6.17 shows how this I.C. is connected as an AM receiver. Note that bulky components (such as tuned circuits) have to be connected externally.

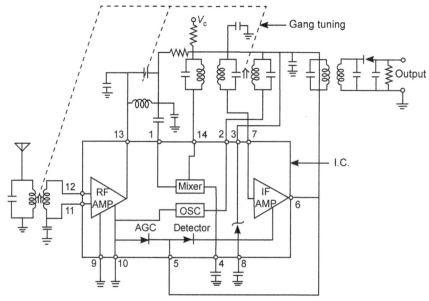

**Fig. 6.17**

The IC based AM receivers are not much popular due to their high cost and large size. This size is determined by LC tuning circuit; which is to connected externally. In place of LC tuning circuit, ceramic capacitors can be used. With these capacitors, the AM receiver can be fabricated using the IC, two external potentiometers (for volume and tuning) and an antenna.

## 6.13 AM RECEIVER USING PLL (I.C.)

It is another approach to use ICs in AM receivers. The Fig. 6.18 shows a AM receiver using a PLL (Phase Locked Loop) I.C.

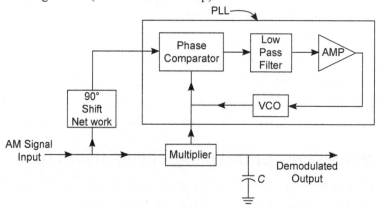

**Fig. 6.18**

The PLL is a versatile device, which is suitable for a variety of applications such as frequency selective demodulation, signal conditioning or frequency synthesis. Some of the basic applications are: AM and FM demodulation, frequency synchronization, frequency multiplications and divisions, frequency translation, AM detection etc.

When the PLL locks onto the AM signal, frequency of VCO (Voltage Controlled Oscillator) adjusts automatically equal to the AM carrier frequency. If the free running value of the VCO frequency is close to the AM carrier frequency, the VCO voltage is 90° out of phase with the AM carrier voltage.

To compensate this phase difference, the carrier is externally shifted further by 90° using a *phase shift network* as shown. Note that no external tuning circuit is required with a PLL detector, since once the VCO locks onto the incoming carrier, selectivity is automatically achieved.

## 6.14 AUTOMATIC GAIN CONTROL (AGC)

AGC is an electronic device by which gain of a radio receiver changes automatically with the changing strength of the signal so that the output remains constant. A dc bias voltage is applied to a selected number of R.F., I.F., and mixer stages. The overall result on the receiver output has been shown in the curve (See Fig. 6.19)

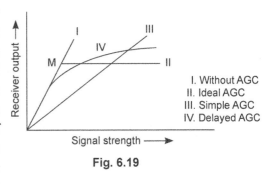

I. Without AGC
II. Ideal AGC
III. Simple AGC
IV. Delayed AGC

**Fig. 6.19**

The AGC irons out input signal amplitude variations and the gain control is not required to be adjusted everytime, the receiver is thus tuned automatically from one station to another.

**Delayed AGC:** In a simple AGC, (explained above) gain of the receiver is reduced, when strength of signal increases.

Figure 6.19 also shows two other AGC curves. The second is ideal AGC curve, here no AGC will be applied till the strength of the signal remains within limits after point M, a constant output will be obtained independent of signal strength. The fourth is a delay AGC curve. This shows that AGC in not applied till the signal strength is within limits and afterwards, the AGC is applied 'strongly'. Now, the output rises with rise in the signal strength but 'slowly'.

Figure 6.20 shows a delayed AGC circuit. There are two separate diodes, one for detector and the other for AGC. A positive bias is applied to the AGC diode to prevent its conduction till the signal strength is within limits. A '*delay control*' as shown is provided to allow manual adjustments of AGC and diode bias and hence on the signal level, at which AGC is applied. If weak stations are to be tuned, delay control setting may be quite high to keep AGC out.

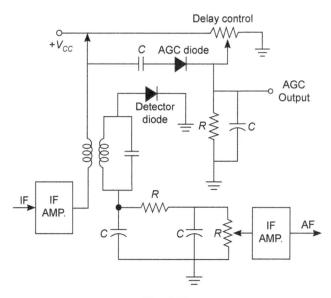

**Fig. 6.20**

**Note:** The AGC is sometimes also called as AVC (*Automatic Voltage Control*).

## 6.15 AUTOMATIC FREQUENCY CONTROL (AFC)

The heart of an AFC circuit is a frequency sensitive device such as *phase discriminator*, which produces a DC voltage whose amplitude and polarity are proportional to the amount of the local oscillator's frequency error. This DC controlled voltage is then used to vary bias on a *variable reactance* device whose output capacitance is thus changed. This variable capacitance appears across the first local oscillator and the frequency of *variable frequency* oscillator (VFO) is

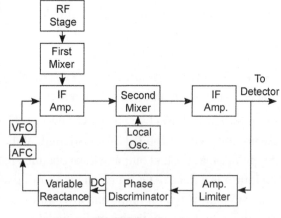

**Fig. 6.21**

automatically free from any variation with temperature, voltage, etc. (See Fig. 6.21)

It is to be noted that all receivers do not require an AFC.

## 6.16 NOISE LIMITER

Few receivers are equipped with a device called 'noise limiter', Though it is not possible to reduce 'random noise' in AM receiver. What a noise limiter does in an AM system is that it reduces noise pulses created by electrical machinery, sparks, etc., nearby. This is done by putting the receiver 'out' for that duration. In common noise limiter circuit, a diode is used along with a differentiating circuit which provides a negative bias during that duration to the diode detector which is thus cut off, and it remains cut off in that period which is in milliseconds and after that becomes ON automatically.

(*i*) The Fig. 6.22 shows a "*Series noise limiter*". It uses a limiter diode that conducts signal to the audio amplifiers as long as they represent 80% (or less) modulation. When impulses of greater amplitude come through the detector, the limiter diode stops conducting, and no signal is passed through the audio amplifier, till the amplitude falls to the limit.

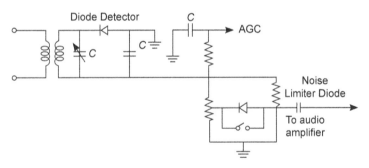

**Fig. 6.22**

(*ii*) A *shunt noise limiter* is shown in Fig. 6.23. Two zeners can be connected in series in opposite polarity across the audio amplifier. Any noise voltage above the zener voltage will cause breakdown of zeners, thus "bye passing" the impulse noise. An opposite polarity impulse breaks down the other zener.

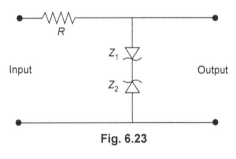

**Fig. 6.23**

(*iii*) *Two ordinary silicon diodes* in parallel and with opposite polarities as shown in Fig. 6.24 are useful as limiters across low voltage circuits such as primary or secondary of the IF transformers or across earphones. The silicon diodes require 0.7 V for conducting. When they begin to pass current they effectively short circuit the line across which they are connected for any voltage peak above 0.7 V.

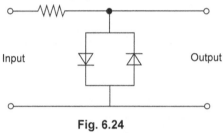

**Fig. 6.24**

## 6.17  GENERAL QUALITIES OF RECEIVERS

In general, the receiver should have the following qualities so that it can render the optimum performance.

(*i*) It should be capable to receive the weakest signal and should provide a sufficient output.

(*ii*) There may be hundreds of signals floating in space but the receiver should be capable to catch the signal which is desired and should reject all other signals.

(*iii*) The output should be an exact replica of the original modulating signal *i.e.*, the output should be free of any distortions.

(*iv*) The receivers should not pick up any noise or interference. *Accordingly, qualities of a receiver can be mentioned as below*:

    1.   Sensitivity

    2.   Selectivity

    3.   Fidelity

    4.   Signal/Noise ratio

    5.   Image frequency rejection

    6.   Double spotting.

    1.   **Sensitivity:** Sensitivity of a receiver is the measure of its ability to receive weak signals. The receiver receives the signal, which is amplified by the inbuilt amplifiers, therefore the measurement of a receiver sensitivity is actually the measurement of performance (gain) of its amplifiers. The sensitivity may be defined as:

(a)  In case of AM receiver, this is defined as the input carrier amplitude modulated to 30% with a modulating signal of 400 Hz which when applied to the antenna of the receiver produces a standard output (500 mW), when all the receiver controls are adjusted at maximum output.

(b)  *In case of FM receivers* this is defined as the input carrier frequency modulated by a 400 Hz signal so as to produce a deviation of 22.5 kHz (which is 30% of the maximum permitted deviation of 75 kHz) required to produce a standard output (500 mw) at the receiver when all the controls are adjusted at maximum output.

The sensitivity of receivers lies in the range of microvolts, (Fig. 6.25) *i.e.*, receivers can pick up pV signals. As the gain of receiver is not constant over a frequency band, the sensitivity is also different at different frequencies. The most important factor determining the sensitivity of a receiver is the gain of the IF amplifier.

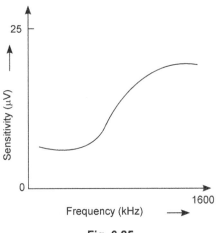

**Fig. 6.25**

2. **Selectivity:** This is the ability of a receiver to select a particular signal out of many signals floating in the space. The selection of the signal is done by resonant circuits so the selectivity of a receiver is the measure of performance of the resonant circuit in the receiver. In other words, it depends on the $Q$ factor of the circuit. A resonant circuit with high $Q$ factor has more selectivity and vice versa. The selectivity may also be expressed in terms of *sensitivity as* a ratio of

$$\frac{\text{Sensitivity of the receiver when it is mistuned}}{\text{Sensitivity of the receiver when is tuned correctly}}$$

The selectivity of receivers is determined by the characteristics of the IF systems.

3. **Fidelity:** By fidelity, we mean the ability of a receiver to reproduce correctly the different modulating frequency components present in an input signal. The fidelity curve is drawn between modulating frequencies and corresponding output of the receiver (Fig. 6.26)

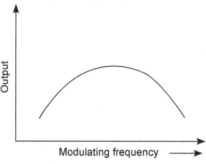

**Fig. 6.26**

The fidelity at lower modulating frequencies is determined by the low frequency characteristic of the audio/video frequency amplifiers. At higher modulating frequencies, the fidelity is determined by higher frequency characteristic of the amplifier.

4. **Signal Noise Ratio (SNR):** The Signal Noise ratio is defined as the ratio between signal power and noise power received at the output, *i.e.*:

$$\text{SNR} = \frac{\text{Signal power}}{\text{Noise power}}$$

*Thermocouple* may be used to measure the noise power. It is easy to understand that for a good receiver this ratio should be more than one.

Sometimes noise generation of a receiver is specified by a term "noise figure" (N.F.) which is the measure of the extent to which noise appearing in the receiver output in the absence of signal is greater than the noise that could be present if the receiver were a perfect one from the point of view of generating minimum possible noise.

$$\text{NF} = \frac{\text{SNR of ideal receiver output}}{\text{SNR of the receiver output under test}}$$

5. **Image frequency rejection.** Image frequency is that received undesired carrier frequency, which after mixing with the local oscillator frequency, produces a "difference frequency" equal to the Intermediate frequency (IF).

**Illustration.** Let us assume that we have tuned the receiver at 1000 kHz. The local oscillator frequency for this would be 1455 kHz, as IF = 455 kHz.

For this local oscillator frequency, a received frequency of 1910 kHz would also produce the same IF (1910 kHz - 1455 kHz = 455 kHz) and thus processed in different stages of the receiver. The 1910 kHz will be Image frequency of 1000 kHz and is highly undesireable.

In general, Image frequency corresponding to a received signal frequency of $f_s$ is $(f_s + 2f_i)$, where $f_i$ is the IF.

The rejection of image frequency is achieved in RF section, which is always tuned to the frequency intended to be received and frequency component of 910 kHz away from the desired frequency gets eliminated.

The *image frequency rejection ratio* can be defined as a ratio of the gain at the signal frequency to the gain at the image frequency. This gives the degree of image frequency rejection.

6. **Double spotting.** When a receiver picks up the same short wave station at two nearby points on the dial, it is called "double spotting." The main reason for this, is its poor "front end selectivity" *i.e.*, inadequate "image frequency rejection".

The effect of double spotting is that a weak station may be marked by the reception of a nearby strong station at the spurious point on the dial. However, double spotting may be used to determine value of I.F. of a receiver since the spurious point on the dial is precisely below the correct frequency.

## SUMMARY

1. The receiver recovers the original signal, amplfies and feeds to the loudspeaker or other output device.

2. The process of separating original signal from radio waves is known as demodulation or detection.

3. The detectors are of two types: linear detector and square detector.

4. A simple diode detector is the most basic linear detector.

5. The AM receivers are of two types—tuned radio frequency receiver and superheterodyne receiver.

6. Superheterodynes are superior and commercially used. Double hetrodyne receivers are also available

7. The IC receivers have many qualities over the conventional receivers.

8. The other components used in receivers are: noise limiter, AGC and AFC.

9. The quality of receivers are given in terms of sensitivity, selectivity, fidelity and image frequency rejection etc.

10. The ratio of signal power to noise power is called signal noise ratio.

❑❑❑

# 7

# Frequency Modulation (FM)

As described already, it is possible to convey or *transmit* an information by varying its frequency as well as angle of phase. These are known as *frequency* and phase modulation respectively and both collectively are known as "*Angle Modulation*".

The frequency and phase modulation systems have similar characteristics with minor difference. In this chapter, we will describe "Frequency modulation".

## 7.1 FREQUENCY MODULATION

The process by which the frequency of the carrier wave is changed according to the amplitude of the signal is known as "frequency modulation".

Figure 7.1 (*a*) shows the signal, (*b*) shows the carrier and (*c*) shows the frequency modulated wave.

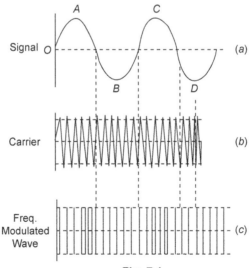

**Fig. 7.1**

Note that only the frequency of the carrier changes according to the **instantaneous** amplitude of the signal. Other characteristics (phase, amplitude) of the carrier remain unchanged. When the signal voltage is zero at $O$, (Fig. 7.1) the frequency of the carrier does not change, the modulation is zero. During positive cycle of the signal, frequency of the carrier increases and at the peak value of the positive signal ($A$ and $C$), frequency of the carrier becomes maximum as shown by **closely spaced** frequency cycles in the modulated wave. However, during negative cycle, the frequency of the carrier decreases and at the peak value of the negative cycle ($B$ and $D$), the frequency becomes minimum as shown by rarefied frequency cycles of the modulated wave.

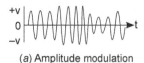

(a) Amplitude modulation

(c) Frequency modulation

**Fig. 7.2**

The Fig. 7.2 shows amplitude modulated as well as frequency modulated wave for comparison, when modulating signal is sinusoidal.

The Fig. 7.3 shows FM wave form, when intelligence (modulating signal) is in the form of rectangular pulses.

Such a situation occurs in teletypewriters.

The Fig. 7.4 shows FM waveform, when intelligence is in the form of triangular pulses.

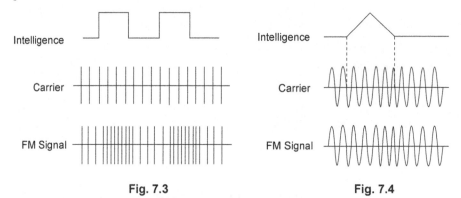

**Fig. 7.3**                                                **Fig. 7.4**

## 7.2 EXPRESSION OF FM WAVE IN TIME DOMAIN

If the modulating signal is

$$e_m = E_m \cos \omega_m t$$

and the carrier is $\qquad e_c = E_c \cos \omega_c t$ (Fig. 7.5)

From the definition of FM, instantaneous value of frequency will be

$$f = f_c (1 + kE_m \cos \omega_m t)$$

Where $k$ is the constant of proportionality and $f_c$ is unmodulated carrier frequency. Now the FM signal may be given as

$$e = E_c \sin \theta$$

where $\theta = \int \omega \, dt = f_c \, (1 + k \, E_m \cos \omega_m t) \, dt$

$$= \left[ \omega_c t + \frac{k E_m f_c \sin \omega_m t}{f_m} \right]$$

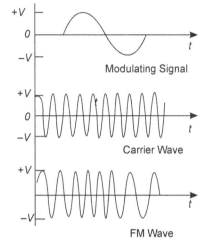

+V

0                                                                         t

−V

Modulating Signal

+V

0                                                                         t

−V

Carrier Wave

+V

−V

FM Wave

**Fig. 7.5**

If $\delta$ is the frequency deviation

$\delta$ = Max. instantaneous frequency
   – carrier frequency

$$= f_c(1 + k \, E_m) - f_c = k \, E_m f_c$$

Now the FM signal

$$e = E_c \sin \theta$$

$$= E_c \sin \left[ \omega_c t + \frac{k E_m f_c \sin \omega_m t}{f_m} \right]$$

$$= E_c \sin \left[ \omega_c t + \frac{\delta}{f_m} \sin \omega_m t \right] \qquad (\because k E_m f_c = \delta)$$

$$= E_c \sin (\omega_c t + m_f \sin \omega_m t)$$

where $m_f$ = modulation index = $\dfrac{\delta}{f_m}$.

This is the equation of FM wave.

**Note:**

1. The modulation index for FM is defined as

$$m_f = \frac{\text{Maximum frequency deviation}}{\text{Modulating frequency}} = \frac{\delta}{f_m}$$

   As the modulating frequency ($f_m$) decreases and the modulating voltage amplitude remains constant, the frequency modulation index increases. *This is the basis for distinguishing frequency modulation from phase modulation.*

2. The $m_f$, which is the ratio of two frequencies is measured in radians.

3. The max. change in the instantaneous frequency from the average frequency, $\omega_c$ is called *max frequency deviation*.

   Max frequency deviation = modulating index × modulating frequency

$$(\delta_m = m_f \times f_m)$$

The max frequency deviation is a useful parameter for determining bandwidth of FM signal.

This may be a positive deviation or a negative deviation. (Fig. 7.6)

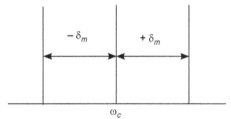

**Fig. 7.6**

4. Unlike AM, the modulation index for FM may be greater than unity.

5. The above derived equation for FM signal is single tone modulation as the modulating wave is single wave. In multitone modulation the modulating wave is usually of multitone nature *i.e.*, it consists of a group of sine waves of different frequencies, which may be harmonically unrelated.

6. Depending upon the value of modulation index $m_f$, the FM may be of two types:

   (*a*)  **Narrow band FM :** Which has small value of $f_m$.

   (*b*)  **Wide band FM:** Which has a large value of $f_m$.

The BW for a narrow band FM is closely equal to twice of the BW for standard AM (DSBFC), where as in case of wide band FM, its value is quite more.

**Table 7.1**   Narrow Band Vs Wide Band FM

| Narrow band FM | Wide band FM |
|---|---|
| 1. In this system modulating index is near unity. | 1. In this system modulating index is more than unity and ranges between 5 to 2500. |
| 2. The maximum modulating frequency is 3 kHz. | 2. The modulating frequency ranges between 20 Hz to 20 kHz and maximum deviation allowed is 75 kHz. |
| 3. In this system, the noise suppression is not good but occupies much less BW than wideband system. | 3. With large deviation, noise is better suppressed, this system occupies 15 times more bandwith than occupied by narrow band system. |
| 4. The narrow band system is more suitable for communication systems. This is also used in FM mobile services like police, ambulance etc. | 4. The wideband system is suitable for entertainment broadcasting. |

**Problem 7.1.** *In a frequency modulation system the audio frequency is 400 Hz and the audio frequency voltage is 2.5 V. The deviation is 5.0 kHz.*

*If the audio frequency voltage is 7.5V; Find the new deviation, and the modulating index.*

**Solution.** As $\delta \propto E_m$, we can write as

$$\frac{\delta}{E_m} = \frac{5.0}{2.5} = 2 \text{ kHz per volt } \textbf{Ans.}$$

(*i*)  When $E_m = 7.5$ V, the new deviation.

$$\delta = 2 \times 7.5 = 15.0 \text{ kHz.}$$

(*ii*)  Hence the modulating index

$$m_f = \frac{\delta}{f_m} = \frac{5.0 \text{ kHz}}{0.4 \text{ kHz}} = 12.5 \textbf{ Ans.} \text{ [400 Hz} = 0.4 \text{ kHz]}$$

**Problem 7.2.** *What is the frequency deviation of an FM transmitter; if its modulating index is 6 in a bandwidth of 150 kHz.*

**Solution.** According to a **thumb rule** (Carson rule) the BW of an FM wave is twice the sum of deviation and the modulating frequency.

*i.e.,*                     $\text{BW} = 2(\delta + f_m)$                          ...(*i*)

but,                     $m_f = \dfrac{\delta}{f_m} \text{ or } f_m = \dfrac{\delta}{m_f}$

In the problem,        $f_m = \dfrac{\delta}{6}$

Now putting in Eq. (*i*)

$$150 = 2(\delta + \delta/6); \; \delta = 64.2 \text{ kHz } \textbf{Ans.}$$

**Problem 7.3.** *The FM radio link having a deviation ratio of 10 is to transmit a speech band upto 5 kHz. What RF bandwidth should be used.*

**Solution.**                     $10 = \dfrac{\delta}{f_m}$

*i.e.,*                     $10 = \dfrac{\delta}{5 \text{ kHz}} \text{ or } \delta = 50 \text{ kHz.}$

Now BW to be used        $= 2(\delta + f_m) \text{ (Carson rule)}$

$$d = 2(50 + 5) = 110 \text{ kHz. } \textbf{Ans.}$$

**Problem 7.4.** *A speech signal in a telephone system occupies frequency range of 300–3400 Hz (considered on a band). In a carrier system, it is transmitted as SSB signal. Calculate saving in the bandwidth, as compared to AM transmission.*

**Solution.** Bandwidth of AM signal

$$= 2 \times 3400 = 6800 \text{ Hz}$$

Bandwidth of SSB signal $= 3400$ Hz

Saving in Bandwidth $= 6800 - 3400 = 3400$ Hz **Ans.**

**Problem 7.5.** *An FM signal is represented by*

$$e = 10 \sin (10^8 t + 15 \sin 2000\, t) \qquad\qquad ..(i)$$

*Find parameters of the FM wave.*

**Solution.** The standard equation of FM wave is:

$$e = E_c \sin (\omega_c t + m_f \sin \omega_m t) \qquad\qquad ..(ii)$$

Comparing the given equation (*i*) with the standard equation (*ii*), we have

(*a*)  The carrier amplitude     $E_c = 10$ V **Ans.**

(*b*)  The Carrier frequency    $f_c = \dfrac{\omega_c}{2\pi} = \dfrac{10^8}{2\pi} = 16$ MHz **Ans.**

(*c*)  The modulating index     $m_f = 15$ Ans.

(*d*)  The modulating frequency

$$f_m = \frac{2000}{2\pi} = 318 \text{ Hz  Ans.}$$

(*e*)  The max. frequency deviation

$$\delta = m_f f_m = 15 \times 318 = 4.7 \text{ kHz  Ans.}$$

**Problem 7.6.** *An FM wave is represented by the voltage equation.*

$$v = 16 \sin [6 \times 10^8\, t + 5 \sin 1200\, t]$$

*where t is the time in second.*

(*a*)  Find

(*i*)    carrier voltage        (*ii*)  modulating frequency

(*iii*)   modulation index      (*iv*)  max. deviation of the FM wave

(*b*)  What power will the FM wave dissipate in a 20 $\Omega$ resistor.

**Solution.** Comparing the given equation with the following standard equation of the FM wave.

$$e = E_c \sin [\omega_c t / + m_f \sin \omega_m t]$$

Now,

(*i*)  carrier voltage $= 12$ V **Ans.**

(*ii*)  modulating frequency    $f_m = \dfrac{1250}{20} = 199$ Hz **Ans.**

(*iii*)  modulation index        $m_f = 5$ **Ans.**

(*iv*)  deviation                $\delta = f_m \cdot m_f = 199 \times 5 = 995$ Hz **Ans.**

(*b*)  The power dissipated by FM wave in a 20 $\Omega$ resistor

$$P = \frac{(V_{rms})^2}{R} = \frac{(12\sqrt{2})^2}{20} = 3.6W$$

$$\text{Ans. } [V_{rms} = V_{max} / \sqrt{2}\,]$$

## 7.3 FREQUENCY SPECTRUM OF FM WAVE : BESSEL FUNCTION

The analysis of FM is more complex than of AM (Note that equation of a FM wave is *the sine of a sine*) hence only the results will be used here.

The frequency spectrum of FM wave consists of carrier component and side frequencies (harmonics) of the modulating frequency.

The amplitude of the various spectral components are given by a mathematical function called **BESSEL's FUNCTION** of first kind denoted by $J_n$ $(m_f)$, where $J$ is the coefficient, $n$ is the order of side frequency and $m_f$ is the modulation index. The Bessel functions are available both in graphical form (Fig. 7.7) as well as in tabular form (Table 7.1). Note that $J_o$ $(m_f)$ gives the carrier component.

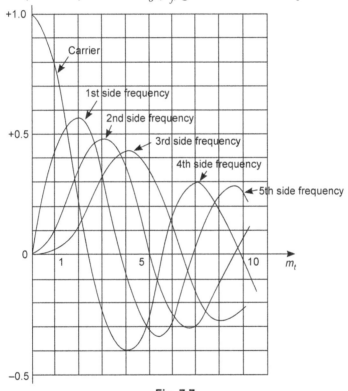

**Fig. 7.7**

**Refer Table 7.2 :** For $m_f = 1.0$, spectral components of the FM wave are:

(*i*) Carrier $(f_c) = J_n$ $(m_f) = J_0$ $(1.0) = 0.77$

(*ii*) First order side frequencies

$$(f_c \pm f_m) = J_1 \ (m_f) = J_1(1.0) = 0.44$$

(*iii*) Second order side frequencies

$$(f_c \pm 2f_m) = J_2 \ (m_f) = J_2(1.0) = 0.11$$

**Table 7.2** Bessel Functions

| Modulation Index $m_f$ | Carrier $J_0$ | Side Frequencies | | | | | | | | | | | |
|---|---|---|---|---|---|---|---|---|---|---|---|---|---|
| | | 1st $J_1$ | 2nd $J_2$ | 3rd $J_3$ | 4th $J_4$ | 5th $J_5$ | 6th $J_6$ | 7th $J_7$ | 8th $J_8$ | 9th $J_9$ | 10th $J_{10}$ | 11th $J_{11}$ | 12th $J_{12}$ |
| 0.25 | 0.98 | 0.12 | 0.01 | | | | | | | | | | |
| 0.5 | 0.94 | 0.24 | 0.03 | | | | | | | | | | |
| 1.0 | 0.77 | 0.44 | 0.11 | 0.02 | | | | | | | | | |
| 1.5 | 0.51 | 0.56 | 0.23 | 0.06 | 0.01 | | | | | | | | |
| 2.0 | 0.22 | 0.58 | 0.35 | 0.13 | 0.03 | 0.01 | | | | | | | |
| 2.4 | 0 | 0.52 | 0.43 | 0.20 | 0.06 | | | | | | | | |
| 3.0 | −0.26 | 0.34 | 0.49 | 0.31 | 0.13 | 0.04 | 0.01 | | | | | | |
| 4.0 | −0.40 | −0.07 | 0.36 | 0.43 | 0.28 | 0.13 | 0.01 | | | | | | |
| 5.0 | −0.18 | −0.33 | 0.05 | 0.36 | 0.39 | 0.26 | | | 0.02 | 0.01 | | | |
| 5.5 | 0 | −0.34 | −0.12 | 0.26 | 0.40 | 0.36 | 0.19 | 0.09 | 0.03 | 0.01 | | | |
| 6.0 | 0.15 | −0.28 | −0.24 | 0.11 | 0.36 | 0.36 | 0.25 | 0.13 | 0.06 | 0.02 | 0.01 | | |
| 7.0 | 0.30 | 0 | −0.30 | −0.17 | 0.16 | 0.35 | 0.34 | | 0.13 | 0.06 | 0.02 | 0.01 | |
| 8.0 | 0.17 | 0.23 | −0.11 | −0.29 | −0.10 | 0.19 | 0.34 | 0.02 | 0.22 | 0.13 | 0.06 | 0.03 | 0.01 |
| 8.65 | 0 | 0.27 | 0.06 | −0.24 | −0.23 | 0.03 | 0.26 | 0.34 | 0.28 | 0.18 | 0.10 | 0.05 | 0.02 |
| 9.0 | −0.09 | 0.24 | 0.14 | −0.18 | −0.27 | −0.06 | 0.20 | | 0.30 | 0.21 | 0.12 | 0.06 | 0.03 |
| 10.0 | −0.25 | 0.04 | 0.25 | 0.06 | −0.22 | −0.23 | −0.01 | | 0.31 | 0.29 | 0.20 | 0.12 | 0.06 |
| 12.0 | 0.05 | −0.22 | −0.08 | 0.20 | 0.18 | −0.07 | −0.02 | | 0.05 | 0.23 | 0.30 | 0.27 | 0.20 |
| 15.0 | −0.01 | 0.21 | 0.04 | −0.19 | −0.12 | 0.13 | 0.20 | 0.03 | −0.17 | −0.22 | −0.09 | 0.10 | 0.24 |

**Note :**

1. The amplitudes can be negative in some cases, which may not be shown on spectrum graph.

2. For certain value of $m_f$, the carrier amplitude may be zero. The value of carrier and side frequencies may even be negative.

3. In AM, there are only 3 frequencies (carrier and two side bands) but in FM there are infinite no. of side bands as well as the carrier. The modulation index determines significant no. of side bands.

4. The side bands at equal distance from $f_c$ have equal amplitude.

5. When $m_f$ increases, value of $j$ also increases. Note that, $m_f$ (modulation index) is inversely proportional of $f_m$ (modulating frequency).

6. In AM, increased value of modulation index increases the side band power and thus increases the total transmitted power. In FM, the total power transmitted remains constant but as $m_f$ increases, required B.W. increases.

7. In AM, the carrier remains constant but in FM this is not the case. Its coefficient is $J_0$, which is a function of $m_f$, so with $m_f$ the amplitude of carrier varies accordingly.

8. For certain values of $m_f$, the carrier component of FM disappears completely. These values of $m_f$ are 2.4, 5.5, 8.6, 11.8 etc.

## 7.4 POWER OF FM WAVE

Though the frequency of FM wave varies with time, the carrier amplitude remains constant. Therefore, it can be shown that average power of a FM wave remains always equal to the carrier power. When modulation applied, the total power of the carrier is redistributed among all the components of the spectrum. At certain values of $m_f$, when the carrier component becomes zero, all the power is carried by the side frequencies.

## 7.5 TRANSMISSION BW OF FM WAVE

Theoretically an FM wave contains an infinite no. of side frequencies, the BW required to transmit such a signal shall be also infinite. But in practice, an FM has a finite no. of side frequencies so we can specify BW required for transmission of an FM wave.

The number of significant sidebands produced in a FM wave may be obtained from the Bessel function $j_n.m_f$. If $n > m_f$ the values of $J_n. m_f$ are negligible. At $m_f \gg 1$, the no. of significant sidebands:

$$n \simeq m_f$$

and the BW of FM wave is given by

$$\begin{aligned} BW &= 2\,n.f_m \\ &= 2\,m_f f_m && \text{(As } n \simeq m_f) \\ &= \frac{2.\delta}{f_m}.f_m = 2\delta && [\text{As } m_f = \delta/f_m] \end{aligned}$$

Note that $m_f$ is frequency modulation factor, $f_m$ is the modulating frequency, and $\delta$ is frequency deviation. Thus we can say that BW of a FM is twice the frequency deviation, when $m_f \gg 1$.

## 7.6 CALCULATION OF BW (CARSON RULE)

The Carson's rule (a thumb rule) states that the BW required to pass an FM wave is twice the sum of deviation and highest modulating frequency, *i.e.*,

$$BW = 2\,(\delta + f_m)$$

Note that this gives an approximate value only but gives sufficient good results if modulation index is more than 6.

Further,                                  $BW = 2\,(\delta + f_m) = 2\delta + 2f_m$

$$= 2\delta\left(1 + \frac{f_m}{\delta}\right)$$

$$= 2\delta\left(1 + \frac{1}{m_f}\right) \qquad\qquad [\text{As } m_f = \delta/f_m]$$

$$= 2\delta\,(f_m.m_f + f_m) \qquad\qquad [\text{multiplying by } f_m.m_f]$$

$$= 2\delta f_m(1 + m_f).$$

**Note:**

(*i*) **When $\delta \ll f_m$ (Narrow band FM):** *i.e.,* $m_f \ll 1$

$BW = 2\delta\,[1/m_f] = 2f_m$ which is equal to AM

(*ii*) **When $\delta \gg f_m$ (wide band FM):** *i.e.,* $m_f \gg 1$

$BW = 2\delta$ [Neglecting $1/m_f$]

The Table 7.3 shows the no. of significant side frequencies for different modulating indices.

## 7.7 PLOTTING FREQUENCY SPECTRA

Using Table 7.2, we can find the size of carrier and each side band for a specific modulation index and also frequency spectrum for the FM wave can be plotted (Fig. 7.8)

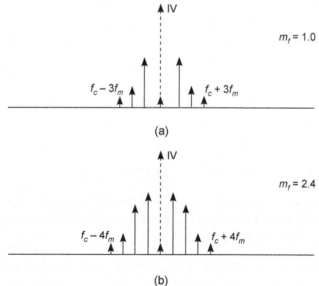

(a)

(b)

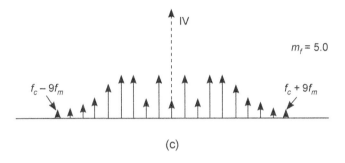

(c)

**Fig. 7.8**

Note that as modulation depth ($m_f$) increases, the bandwidth also increases and also a reduction in the modulating frequency ($f_m$) increases the number of sidebands (though not the bandwidth). Also note that though the no. of sidebands are theoretically infinite, in practice a lot of higher side bands have no significance and can be ignored.

**Notes:**

1. The unmodulated carrier frequency in an FM signal is known as "centre frequency".

2. In the modulated FM signal, the intelligence contained depends on the frequency variations and also the rate at which these variations take place.

3. The frequencies, which are having amplitude equal to or greater than 1% of the unmodulated carrier are considered to be "significant" and other as an "Insignificant". The range of all significant side band frequencies is called the **Bandwidth**. The number of significant frequencies depends upon the maximum frequency deviation.

4. Unlike AM system, where the power or the amplitude of the carrier remains same after modulation, in FM, the carrier amplitude falls after modulation because the sidebands derive their power from the carrier. So this is possible to make carrier amplitude zero and inject whole of the power into sidebands. This is highly desirable as carrier does not have any intelligence.

5. More the number of significant sidebands, more is the bandwidth, also if the modulating frequency is higher, the sidebands are widely spaced and the band width is increased.

6. The broad band FM has less distortion, but the adjacent channel interference is more, in other words, less number of FM signals can

be transmitted on a particular radio spectrum simultaneously. The FM signal having band width comparable with an AM signal is called narrow band FM. A narrow band FM has heavy distortion, but adjacent channel interference is less.

7. The carrier frequencies allotted to FM broadcast are higher than the AM broadcast, as the FM has comparatively larger bandwidth, thus to accommodate more FM signals on a particular frequency spectrum without any adjacent channel interference, higher carrier frequencies are must.

**Table 7.3**   Significant side Frequencies

| Modulating Index ($m_f$) | No. of significant side Frequencies |
|:---:|:---:|
| 0.1 | 2 |
| 0.2 | 2 |
| 0.3 | 4 |
| 0.5 | 4 |
| 1.0 | 6 |
| 2.0 | 8 |
| 5.0 | 16 |
| 10.0 | 28 |
| 20.0 | 50 |
| 30.0 | 70 |

## 7.8 FM AND THE NOISE

An FM signal is less susceptible to noise than an AM signal, because in AM, the intelligence is in the form of amplitude variations of the carrier and any noise that modulates the carrier **rides along** the AM signal and appears at the output. But in FM signal, the intelligence is in the form of frequency variations and the noise which modulates the signal does not harm the intelligence.

However, the frequency of FM signal may be minutely affected by noise, but if the signal's modulation index ($m_f$) is greater than the noise modulation index, the variations are eliminated.

The FM is more immune to noise than the AM, to establish this fact, it is necessary to study the effect of noise on the FM carrier.

## Effect of Noise on the FM Carrier : Noise Triangle

A single noise frequency will affect the output of a receiver only if it falls within its pass band, the carrier voltage and the noise voltage will then mix and if the difference is audible, it will naturally interfere with reception of the signal.

If such a single noise voltage is represented vectorially, it is seen that the noise voltage is superimposed on the carrier rotating about it with a relative angular velocity $\omega_n - \omega_c$. See Fig. 7.9 and the maximum deviation in amplitude will be $\phi = \sin^{-1}(V_n/V_c)$. Note that $V_n$ is the noise voltage, $V_c$ the carrier voltage and $\omega_n$ and $\omega_c$ are their angular velocities respectively.

## Analysis (Fig. 7.9)

Let noise voltage $\qquad V_n = \dfrac{1}{4} \times V_c \,(\text{Carrier voltage}) = 0.25 \; V_c$

In case of amplitude modulation by noise, Modulation Index

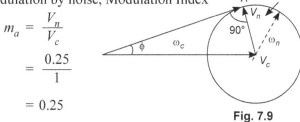

$$m_a = \frac{V_n}{V_c}$$

$$= \frac{0.25}{1}$$

$$= 0.25$$

**Fig. 7.9**

Max. Phase deviation $\qquad \phi = \sin^{-1}\left(\dfrac{0.25}{1}\right)$

$$= 14.5°$$

For voice communication, an AM receiver will not be effected by phase change but affected by amplitude change, whereas the FM receiver will not be bothered by amplitude change which can be reduced by amplitude limiter.

### Let at 15 kHz Modulating frequency :

(i)  For AM, Noise to signal ratio $= \dfrac{N}{S} = \dfrac{0.25}{1} \, (m_a = 1)$

(ii)  For FM, Noise to signal ratio $\dfrac{N}{S} = \dfrac{14.5°}{1 \text{ rad}} = \dfrac{14.5°}{57.3°} = 0.253 = 25.3\%.$
$\qquad\qquad\qquad\qquad\qquad\qquad\qquad\qquad\qquad (m_f = 1)$

When $\qquad\qquad f_m = 15 \text{ kHz}, \dfrac{N}{S} = 0.253$

When $\qquad\qquad f_m = 30 \text{ Hz}, \dfrac{N}{S} = 0.253 \times \dfrac{30}{15 \times 10^3} = 0.000505$

a, reduction from 25.3% at 15 kHz to 0.05% at 30 Hz.

For FM VHF broadcast, maximum frequency deviation is limited to 75 kHz for highest Modulating frequency of 15 kHz.

$$m_f = \frac{\delta}{f_m} = \frac{75 \text{ kHz}}{15 \text{ kHz}} = 5$$

For lower modulating frequency, the modulation Index is correspondingly higher. For modulating frequency of 1 kHz, $m_f = 75$,

Signal to noise improvement is 5 : 1 in voltage and 25 : 1 in power. Such improvement is not possible in AM.

So it is possible to reduce noise by increasing the deviation *i.e.*, bandwidth in FM.

Assuming that noise frequencies are evenly spread across the band pass of the receiver, it can be seen that noise output from the receiver decreases uniformly with noise sideband frequency for FM, For AM, however, it remains constant. This is shown in Fig. 7.10. The triangular noise distribution in FM is called **noise triangle**. The corresponding noise distribution in AM is rectangular. From the figure it can be observed that voltage improvement of FM would be twice. Such an observation can be made by considering the audio frequency, at which FM noise appears as half the size of AM noise.

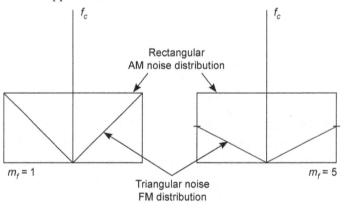

**Fig. 7.10**

Let the noise voltage is one fourth of the carrier voltage. The modulating index of amplitude modulation of the carrier by the noise will be equal to $V_n/V_c$ = 0.25/1 = 0.25. For voice communication, the FM receiver will not bother by the amplitude change which can be removed by an amplitude limiter.

The maximum permissible modulating index for AM is 1, whereas there is no such limit in FM. But in FM the maximum frequency deviation is limited to 75 kHz. Therefore at every 15 kHz audio frequency, the modulating index in FM is only 5 (= 75/15) and when audio frequency is 1 kHz, it is 75. If a given ratio of **signal to noise voltage** exists at the output of an **FM amplitude limiter**, the ratio will be reduced in proportional to the increase in the modulating index, *e.g,* if the modulating index = 3, the ratio of signal-noise voltages at the limiter output of the receiver will be tripled and so on.

This ratio therefore is proportional to the modulating index, thus **signal to noise power** ratio in the output of an FM receiver is proportional to the square of the modulating index (power $\propto$ voltage$^2$) *i.e.,* when modulating index = 5, there will be 25 : 1 ($5^2 = 25$) improvement for FM, but no such improvement is possible for AM.

## 7.9 AMPLITUDE LIMITER IN FM

The function of amplitude limiter is to remove the amplitude variations caused by noise signal. This improves signal-noise ratio. This is installed before the demodulator. Ideally, the limiters should be able to function at all levels of the input carrier and at all rates of variation. The limiters use two electrical effects to provide a relatively constant output, these are; leak type bias and drain saturation.

The Fig. 7.11 shows the circuit of an amplitude limiter used in FM. The circuit provides a leak type bias. The FET amplifier has resistance $R_d$ for drain saturation.

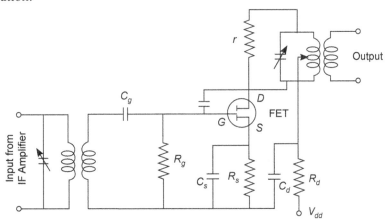

**Fig. 7.11**

(*i*) The *leak type bias* provides the limiting action. This provides the limiting action at larger inputs. As the input of IF amplifier increases, the current through capacitor $C_g$ increases. The bias on the FET will also increase and its gain will decrease thereby keeping the output constant.

(*ii*) To get limiting action at lower level of inputs, *drain saturation* action is used. This is achieved by using a low drain supply voltage $V_{dd}$ by employing a resistance $R_d$.

## Frequency Response Characteristic of Amplitude Limiter (Fig. 7.12)

The limiting action takes place only for a certain range of input voltages, beyond which output varies with the input. As the input increases from point 1 to 2, the output current also increases, thus no limiting action takes place: At point 2, limiting action starts hence, this is called **threshold of limiting**. Between points 2 to 4 there is limiting range. When the input voltage increases sufficiently, such as at point 5, the output current is reduced, this is known as **upper end of the limiting**.

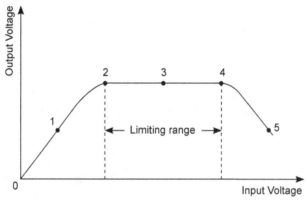

**Fig. 7.12**

## 7.10 VARIOUS FM (AMPLITUDE) LIMITER CIRCUITS

(*a*) An amplitude limiter is basically a clipping circuit, whose output remains constant despite any change in the input signal.

The frequency demodulator is preceeded by an amplitude limiter so that any amplitude change in the signal fed to the demodulator may be limited.

In ideal case, the limiter must remove from its output all variations in amplitude occurring in the input carrier voltage. The requirement of an ideal limiter are :

(*i*)  It should function at all levels of input carrier.

(*ii*)  It should function at all rates of variation in carrier voltage.

(b) **Types of limiter circuits:**

1. **Diode limiters:** The ordinary diode is the most popular circuit element used for amplitude limiting. The Fig. 7.13 shows limiter circuit comprising of ordinary diodes.

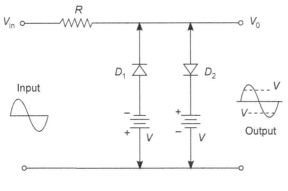

**Fig. 7.13**

As can be seen that two diodes placed back to back with individual bias (V). Forward conduction of diode begins when the applied potential exceeds 0.3V for Germanium diode (or 0.7 V for silicon diode).

The output obtained is clipped according to the voltage of the battery.

2. **Zener diode limiter (Fig. 7.14):** Two zener diodes are placed in series such that they conduct in reverse direction. The circuit provides a greater output without any bias, as zener Breakdown voltage $(V_z)$ is used for limiting action. $V_{z_1} = V_{z_2}$

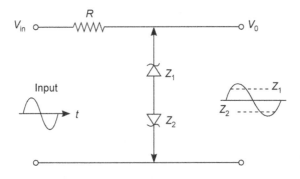

**Fig. 7.14**

3.  **FET limiter:** The Fig. 7.15 shows the amplitude limiter circuit using a field effect transistor.

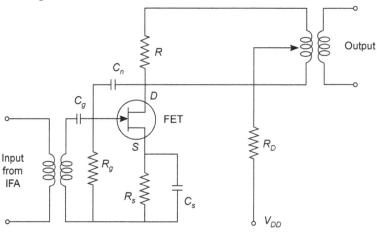

**Fig. 7.15**

In the circuit, leak type bias is provided at the gate by resistance $(R_g)$ and capacitance $(C_g)$. The capacitor $C_n$ has been connected for neutralization.

## 7.11  LIMITER/LIMITING ACTION

The following principles are used to obtain limiter (limiting) action.

(*a*) **Gate limiting action : Leak type bais** provides limiting action. As the input signal voltage rises, the current flow through capacitor $C_g$ increases and hence the negative voltage developed across $C_g$ increases. In this way, gain of the FET increases and output voltage tends to remain constant. This provides the limiting action. For small input signal, no limiting action takes place. The limiting action begins only, when signal is large enough to produce a leak bias of "0.6 times cut off bias".

(*b*) **Drain limiting action :** To obtain limiting action at low voltages of input signal, drain limiting action is utilized. This is achieved by the use of a low drain supply voltage $V_{DD}$, which results in saturation of the output current.

The amplitude limiter works on the principle of passing the stronger signal and to eliminate the weaker. This was the reason for mentioning that in FM, noise reduction is obtained only when the signal is at least twice the noise peak amplitude. A relatively weak interfering signal from another transmitter will also be attenuated in this manner as much as any other form of interference. This applies even if the other transmitter operates on the same frequency as the desired transmitter.

## 7.12 CAPTURE EFFECT/CO-CHANNEL INTERFERENCE IN F.M. LIMITERS

The "Amplitude limiter" in FM receiver passes the "stronger" signal and rejects the "weaker" signal. For an efficient working of the limiter, the "signal amplitude" should be atleast twice of the "noise amplitude". If this condition is achieved, a relatively weak interfering signal from the another transmitter is attenuated.

The phenomenon of a stronger FM signal suppressing the weaker signal is called **"Capture effect" (Fig. 7.16)**. This effect is more significant in FM mobile receivers. So long the signal from the second transmitter remains less than HALF of the signal from the first transmitter, the second transmitter remains "inaudible". After this, the transmitter towards which the receiver is moving becomes audible and becomes predominate finally excluding the first transmitter. It is said that the "receiver has been captured" by the second transmitter. If the receiver is placed at the centre of the two transmitters, the receiver is "Captured" alternately by both transmitters, which leads to a very awkward situation.

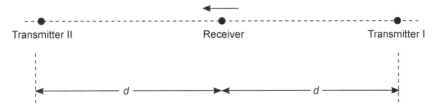

**Fig. 7.16**

## 7.13 PRE-EMPHASIS AND DE-EMPHASIS

The **"Pre emphasis"** is the process of emphasing or increasing the amplitude of high frequency components of modulating signals before modulation. This is done to increase the signal-noise ratio of the high frequency components. This makes the reception noise free.

By changing the amplitude of high frequency components some distortion is introduced in the signal. This is avoided by doing **de-emphasis** (*i.e.*, decreasing the amplitude) during demodulation.

From the **noise triangle**, it can be seen that the noise has a greater effect on higher modulating frequencies than on the low modulating frequencies. Therefore, if higher modulating frequencies are amplified at the transmitter and accordingly cut at the receiver, an improvement in the noise immunity can be obtained. The process carried out in the transmitter is known as **"Pre-emphasis"** and the process carried out in the receiver is called **de-emphasis**.

If two modulating signals have same initial amplitude and if the high frequency is amplified thrice of its amplitude and the low frequency remains unaffected, the receiver has to *de-emphasis* the first signal to one third so that both the signals have the same amplitude at the output of the receiver. When a signal is de-emphasised, any noise sideband voltages along with the signal are also de-emphasised and thus the effect of the noise is reduced.

The Fig. 7.17 (*a*) shows a circuit for pre-emphasis, and (*b*) a circuit for de-emphasis. In pre-emphasis, *RL* circuit is connected at the collector of transistor amplifier, the values of *R* and *L* are such that the time constant

$$\frac{L}{R} = \frac{0.75\ H}{10\ K} = \frac{0.75}{10 \times 10^3\ \text{ohm}} \times 10^6 = 75\mu s.$$

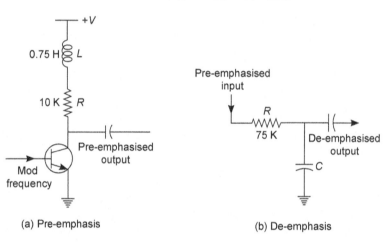

(a) Pre-emphasis                                (b) De-emphasis

**Fig. 7.17**

Similarly in de-emphasis, there is a *RC* circuit. The capacitor C is of 1 nano farad such that the time constant.

$$RC = 75K \times 1\ nf = 75 \times 10^3\ \Omega \times 1 \times 10^{-9}f \qquad [1\ nf = 10^{-9}f]$$
$$= 75 \times 10^{-6}\ s = 75\ \mu s.$$

Therefore these will be called as 75 µs pre-emphasis and 75 µs de-emphasis which is the standard for these processes.

The Fig. 7.18 shows pre-emphasis and de-emphasis curves for 75 µs. Note that de-emphasis curve corresponds to the curve which is 3 dB down at the frequency whose *RC* constant is 75 µs.

This frequency is given by:

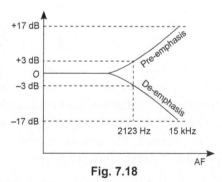

**Fig. 7.18**

$$f = \frac{1}{2\pi RC}$$

$$= \frac{1}{2\pi \times 75 \ \mu s}$$

$$= \frac{1}{2\pi \times 75 \times 10^{-6}} \ Hz = 2123 \ Hz.$$

## 7.14 FM VERSUS AM

1. In AM, there are only three frequencies (one carrier and two sidebands), but in FM there may be infinite number of carriers as well as the sidebands—separated from the carrier by $f_m$, $2f_m$, $3f_m$ and so on.

2. In AM, the modulating index ($m$) decides the power of the modulated wave but in FM, total transmitted power remains constant however, with higher value of $m$, 'bandwidth' increases.

3. In AM, the amplitude of the carrier remains constant. But in FM the amplitude of the carrier does not remain constant. Sometime in FM the carrier may disappear completely.

(*a*) **Advantages of FM over AM**

1. The amplitude of FM signal is constant. In FM the total power remains constant but increased depth and modulation increases the total BW required for the transmission.

   In AM increased depth of modulation increases the sideband power, *i.e.*, total transmitted power. In other words in AM transmission, low level modulation may be used thereby giving higher efficiency.

2. The most important and unique feature of FM is that it is noiseless. The reason being that there happens to be less noise at frequencies at which FM is used, further the FM receivers can be fitted with **Amplitude limiters**, which eliminate the amplitude variations caused by noise. Moreover, it is also possible to reduce noise by increasing deviation. In AM, the noise is also carried by the modulated wave.

3. The adjacent channel interference is less in FM as compared in AM.

(*b*) **Disadvantages**

1. In FM, a B.W. of 10 times as in AM is required.

2. The area of reception for FM is much smaller than in A.M.

3. The equipment required in FM are more complex and costly than in A.M.

**Table 7.4**   FM Versus AM

| S. No. | FM | AM |
|--------|-----|-----|
| 1. | The amplitude of FM wave is constant. It is independent of the modulating index. | The amplitude of AM wave is changing with the modulating voltage. |
| 2. | The transmitted power remains constant. It is independent of modulating index. | The transmitted power does not remain constant and is dependant on the modulating index. |
| 3. | All the transmitted power is useful. | Power of carrier and of one sideband are useless. |
| 4. | The BW depends on modulating index. The BW is quite large. | The BW is not dependant on the modulating index. The BW is much less than that of FM. |
| 5. | The FM receivers are immune to noise. | The AM receivers are noisy. |
| 6. | The FM transmitters and receivers are complex. | The AM transmitters and receivers are simpler comparatively. |
| 7. | The number of sidebands are many and depend upon the modulating index. | The number of sideband is fix *i.e.*, only two. |
| 8. | The space wave is used for propagation, so the radius of transmission is limited to the line of sight. | Ground and sky wave propagation is used, so a larger area is covered than in FM. |
| 9. | It is possible to operate several channels on the same frequency. | It is not possible to operate more channels (transmitters) on the same frequency. |
| 10. | It is used in Radio, TV broadcasting, police wireless and in *point to point* communication. | It is used in Radio and TV broadcasting |

## SUMMARY

1. When frequency of the carrier is changed according to the signal and other parameters remain constant, the process is known as frequency modulation.

2. Equation for FM wave is

$$e = E_c \sin (\omega_c t + m_f \sin \omega_m t)$$

3. The frequency spectrum of FM wave is given by Bessel functions.

4. The B.W. required to pass an FM wave is twice the sum of deviation and highest modulating frequency. This is known as Curson thumb rule.

5. The FM is much immune to noise. Amplitude limiter removes the amplitude variations caused by noise signal.

6. Higher modulating frequencies are amplified at the transmitter. This is known as pre-emphasis. The reverse process is done in receiver, known as de-emphasis.

7. In narrow band FM, the modulating index is nearly unity and in wideband FM, its value may be 2500.

8. The FM requires very large bandwidth than AM.

❑❑❑

# 8

# FM Generation, Modulators and Transmitters

The FM transmitters have a very large bandwidth as compared to AM transmitters. In FM transmission, ground wave and sky wave propagation is not possible, and signals in VHF and UHF bands are propagated by *line of sight propagation*, which restricts the range upto 50 km.

## 8.1 FM GENERATION

We can generate FM signals by using various FM modulators. The basic requirement of FM modulator (FM signal generator) is to provide a variable output frequency with varying proportions to the instantaneous amplitude of the modulating voltage. The FM signals can be generated by two methods:

1. **Direct Methods:** The FM signals can be generated directly *by frequency modulating* the carrier.

2. **Indirect Methods:** The FM signals can be generated indirectly by *integrating* the modulating signal and then allowing it to phase modulate the carrier.

3. Or by using the modulating signal first to produce a narrow band FM signal and then by frequency multiplication, the frequency deviation may be increased to the desired level.

**Table 8.1**   Comparison of AM and FM broadcasting.

|   | AM Broadcasting | FM Broadcasting |
|---|---|---|
| 1. | It requires smaller transmission bandwidth. | It requires larger bandwidth. |
| 2. | It can be operated in low, medium and high frequency bands. | It needs to be operated in very high and ultra high frequency bands. |
| 3. | It has wider coverage. | Its range is restricted to 50 km. |
| 4. | The demodulation is simple. | The process of demodulation is complex. |

| 5. | The stereophonic transmission is not possible. | In this, stereophonic transmission is possible. |
| 6. | The system has poor noise performance. | It has an improved noise performance. |
| 7. | The AM signal reception does not have any threshold in the useful range of signal noise ratio (SNR). | The FM signal reception exhibits a threshold in the useful range of signal noise ratio (SNR). The SNR value should be higher than the threshold. |

## 8.2  DIRECT METHODS OF FM WAVE GENERATION

As mentioned above, in these methods, the modulating signal varies the carrier frequency directly. In general, the oscillators/multivibrators are used to generate the carrier signal. An oscillator has a tuned LC circuit, which determines the frequency of the carrier signal. If any of the two reactive elements (inductance or capacitance) is changed in accordance to the modulating voltage, the carrier frequency is changed accordingly. An oscillator, whose frequency is changed (modulated) by the modulating voltage is called a *voltage controlled oscillator* (VCO).

A voltage variable reactance device is placed across the tuned or tank circuit of an oscillator which is tuned to the carrier frequency (in absence of modulation). Now if the modulating voltage is increased, the $L$ or $C$ will change accordingly. Larger is the departure of the modulating voltage from its normal value, larger will be the reactance variation and hence the variation in the frequency.

There are number of devices, whose reactance can be varied by the application of voltage. The 3 terminal devices are bipolar transistor (BPT), field effect transistor (FET) and vacuum tube. The most suitable 2 terminal device is a varactor diode.

The following modulators using direct methods of FM generation are discussed here:

- (*a*)  Reactance modulator
- (*b*)  Varactor diode modulator
- (*c*)  VCO modulator
- (*d*)  Stablized reactance modulator

## 8.3  REACTANCE MODULATOR

The modulator makes use of a BJT (Fig. 8.1) or FET, which exhibits a variable reactance with change in the modulating signal. The device is connected across the tank or tuned circuit of the oscillator to be frequency modulated. The reactance may be made inductive or capacitive by some change in the component. The value of this reactance is proportional to the transconductance ($g_m$) of the BJT (or of FET), which can be made to depend on the biasing of the base (or gate) of the device.

**Operation:** The operation can be understood from Fig. 8.1; Note that (*i*) the resistance $R$ should be much smaller than the reactance $X_C$ of the capacitor $C$ and (*ii*) the base current of the transistor should be less than the current through $R$ and $C$.

The above two conditions will make impedance $Z$ as pure reactive (not resistive) and the bias network large enough to be neglected. In practice, $X_C/R$ is kept 5, the circuit is therefore capacitive.

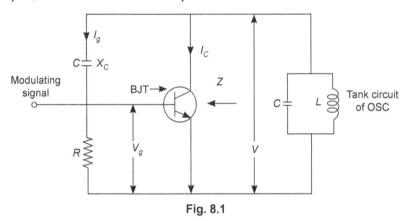

**Fig. 8.1**

From Fig. 8.1

$$V_g = I_0 R = \frac{V}{R - j X_C} \cdot R$$

The output current

$$I_c = g_m V_g$$

where $g_m$ is the trans-conductance of the device and is $= \dfrac{I_c}{V_g}$

$$= g_{m'} \frac{V \cdot R}{R - j X_C}$$

Impedence $Z = \dfrac{V}{I_C} = \dfrac{R - j X_C}{g_{m'} R}$. Putting value $I_C$ from above

If $X_C >> R$. we have

hence,         $Z = j \dfrac{X_C}{g_{m'} R}$ $R$ may be neglected.

Thus $Z$ is capacitive reactance and we can write equivalent reactance

$$X_{eq} = \frac{X_C}{g_{m'} R} = \frac{1/\omega\, C}{g_{m'} R} = \frac{1/2\, \pi f C}{g_{m'} R} = \frac{1}{2\pi f.\, g_m\, RC}$$

$$= \frac{1}{2 p_f\, C_{eg}}$$

Where         $C_{eq} =$ equivalent capacitance

$$= g_{m'}\, RC \qquad\qquad\qquad ...(i)$$

Now, Let $X_C = n. R$ at carrier frequency.

Note that $X_C$ is made $5 - 10$ times larger than $R$ ($n$ is between 5 and 10)

then,
$$X_C = nR = 1/\omega c, \; nR = \frac{1}{2nf\,C}$$

or,
$$C = \frac{1}{2nf.\,nR}$$

Putting this value of $C$ in *eq.* (*i*) equivalent capacitance
$$C_{eq} = g_m\,RC = g_m.\,R\frac{1}{2\pi f.\,nR} = \frac{g_m}{2\pi f\,n}$$

This equation is very useful for practical calculations. Refer Fig. 8.2, which shows a FET reactance modulator. Note that gate to drain impedence ($Z_{gd}$) is much greater than the gate to source impedence ($Z_{gs}$).

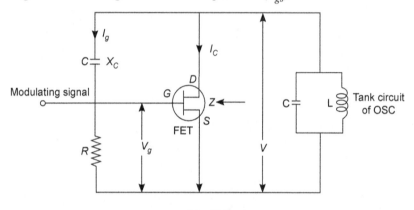

**Fig. 8.2**

The Table 8.2 shows four different arrangements of reactance modulators. The two give a capacitive reactance and the other two, give an inductive reactance.

**Table 8.2   Reactance Modulators**

| S.No. | Type of the modulator | $Z_{gd}$ | $Z_{gs}$ | Condition | Reactance formulae |
|-------|-----------------------|----------|----------|-----------|---------------------|
| 1. | RC Capacitive | C | R | $X_C \gg R$ | $C_{eq} = g_m\,RC$ |
| 2. | RL Capacitive | R | L | $R \gg X_L$ | $C_{eq} = g_m\,L/R$ |
| 3. | RC Inductive | R | C | $R \gg X_C$ | $L_{eq} = RC/g_m$ |
| 4. | RL Inductive | L | R | $X_L \gg R$ | $L_{eq} = L/g_m\,.R$ |

In the Table, $R$ = Resistive, $L$ = Inductive and $C$ = Capacitive.

**Note:** Any reactance modulator can be connected across the tank or tuned circuit of any $LC$ oscillator provided that oscillator does not require two tuned circuits for its operation. The Hartley and Colpitt oscillators are most commonly used.

**Problem 8.1.** *Determine the value of the capacitive reactance of the FET modulator, whose transconductance is 10 millisiemens. The gate to source impedence is one-tenth of the gate to drain impedence. Assume frequency as 5 MHz.*

**Solution.** $\qquad X_C = 10\,R$

*i.e.,* $\qquad\qquad\qquad n = 10$

and $\qquad\qquad\quad g_m = 10\text{ mS} = 10 \times 10^{-3}\text{ S}$

The $\qquad\qquad\qquad C = \dfrac{g_m}{2\pi f n}$

and capacitive reactance

$$X_{C_{eq}} = \frac{1}{2\pi f.C_{eq}} = \frac{1}{2\pi f} \times \frac{2\pi f n}{g_m} = \frac{n}{g_m}$$

$$X_{C_{eq}} = \frac{n}{g_m} = \frac{10}{10 \times 10^{-3}} = 1000 \text{ ohm } \textbf{Ans.}$$

**Problem 8.2.** *Determine the capacitive reactance which can be obtained from a FET reactance modulator, whose* $g_m = 12$ *mS. Assume* $Z_{gd} = 8\,Z_{gs}$. *The frequency may be taken as 3 MHz.*

**Solution.** $\qquad\qquad n = 8, g_m = 12\text{ mS} = 12 \times 10^{-3}\text{ S}$

$$X_{C_{eq}} = \frac{n}{g_m} = \frac{8}{12 \times 10^{-3}} = \frac{8 \times 1000}{12}$$

$$= 667 \text{ ohms } \quad\textbf{Ans.}$$

## 8.4 VARACTOR DIODE MODULATOR

The varactor (or capacitor) diode is a two terminal device whose capacitance varies with applied voltage. Here applied voltage is a combination of bias voltage ($V_0$) and the modulating voltage $m$ ($t$). The diode capacitance forms a part of the tuning capacitance which determines the frequency of oscillations. The capacitance varies with the change in the modulating signal and so does the frequency.

The Fig. 8.3 shows a varactor diode modulator. The voltage $V_0$ is the reverse bias voltage across the varactor diode. The value of $C$ is kept smaller than

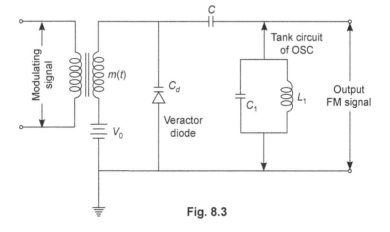

**Fig. 8.3**

the diode capacitance $C_d$ to keep the $RF$ voltage from oscillator across the diode smaller than $V_0$. The capacitance $C_d$ comes parallel to the $LC$ circuit of the oscillator, thus total capacitance $C_t$ is the sum of tank capacitance and the diode capacitance.

$$C_1 = C_1 + C_d$$

Hence the frequency of oscillation is given by

$$f = \frac{1}{2\pi\sqrt{L_1(C_1 + C_d)}}$$

Since frequency of oscillations depends on modulating signal, therefore frequency modulated signal is generated.

Though this is the simplest modulator, its applications are limited to automatic frequency control (AFC) and remote tuning.

## 8.5  VOLTAGE CONTROLLED OSCILLATOR (VCO) MODULATOR

The voltage controlled oscillator is a sinusoidal oscillator. Its frequency is controlled by an external voltage, thus it is a kind of frequency modulator. It is also available in the form of an 1C chip.

The Fig. 8.4 shows a block diagram of the modulator showing voltage controlled oscillator, VCO followed by a series of frequency multiplier and mixer. The characteristics of this modulator are good stability and a wide band FM output.

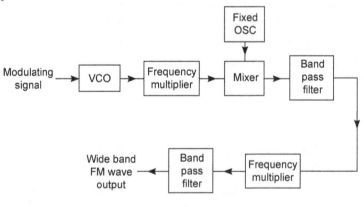

**Fig. 8.4**

## 8.6 STABILIZED REACTANCE MODULATOR

In a commercial transmitter, the oscillator on which a reactance modulator operates can not be a crystal oscillator because the latter provides stable but fixed frequency. Therefore it is used to perform the frequency stabilisation of the reactance modulator. The process is similar to the automatic frequency control (AFC).

The Fig. 8.5 shows the block diagram. The reactance modulator operates on the tank circuit of a $LC$ master oscillator. It is isolated by a **buffer**, whose output goes to the **amplitude limiter.** A fraction of the limiter's output ($f_s$) is fed to a mixer which also receives a signal ($f_o$) from a crystal oscillator. The output from the mixer which is the difference of the frequencies ($f_s - f_0$) and is about 1/20th of the master oscillator's frequency is amplified and fed to a **phase discriminator**. The output of the discriminator is given to the **reactance** modulator which provides a dc voltage to correct automatically any drift in the average frequency of the master oscillator.

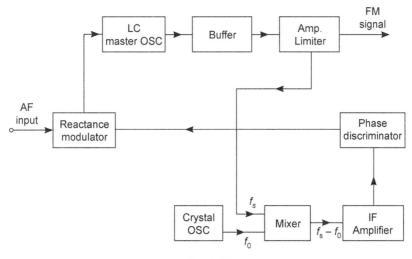

**Fig. 8.5**

**Operation:** The discriminator should be connected to give a positive output if the input frequency is higher than the discriminator's tuned frequency, and it should give a negative output if input frequency is lower than the discriminator's frequency.

We consider the case when the frequency of the master oscillator drifts higher. A high frequency ($f_s$) therefore will be fed to the mixer. The mixer output ($f_s - f_0$) will be bit higher which will be given to the discriminator. As explained above, the output of the discriminator will be a positive dc output.

This positive dc output will be fed to the Reactance modulator, which will increase the transconductance ($g_m$) of modulator. As a result the equivalent capacitance of the reactance modulator will also increase as $C_{eq} = g_m RC$. This will lower the oscillator's central frequency as $f_0 = 1/2\pi\sqrt{L.C_{eq}}$. Thus the frequency rise of the master oscillator is compensated.

Similarly, when the master oscillator drifts lower, a negative correcting voltage obtained from the discriminator will be used to increase the oscillator's frequency.

## 8.7  LIMITATIONS OF DIRECT METHODS

The direct methods of FM generation suffer from the following limitations:

(*i*) In direct methods of FM generation, it is difficult to obtain a high order of stability in carrier frequency. This is because the modulating signal directly controls the tank circuit which is generating the carrier. The crystal oscillator cannot be used as it provides a stable but fixed frequency.

(*ii*) The non linearity produces a frequency variation due to harmonics of the modulating signal hence there are distortions in the output FM signal.

## 8.8  INDIRECT METHODS OF FM GENERATION

The indirect method removes the limitations of direct method.

The indirect method employs the phase modulation to obtain FM signal. It is only necessary to integrate the modulating signal prior to applying it to the phase modulator. The general block diagram for such a method is shown in Fig. 8.6. This is used in VHF and UHF radio telephony.

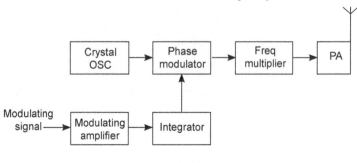

**Fig. 8.6**

Here Armstrong and phase shift indirect methods will be discussed.

## 8.9  ARMSTRONG METHOD : PRINCIPLE

This is the widely used indirect method for FM generation. This is known by the name of its inventor.

The Fig. 8.7 explains the principle of Armstrong method. The modulating signal $m(t)$ is first integrated and then the integrated output is used to *phase modulate* a carrier obtained from a crystal oscillator. The modulating index is kept small and a narrow band FM wave is generated. The narrow band signal is then multiplied in the frequency multiplier and a wide band FM wave in obtained.

Note that frequency multiplier is a non linear device, which multiplies the frequency of the input signal. In general a $n^{th}$ law frequency multiplier can multiply the frequency and the modulating index by '$n$'.

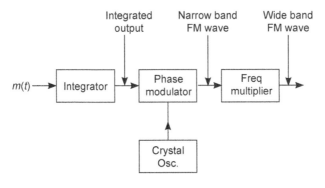

**Fig. 8.7**

The narrow band FM output of the phase modulator for a sinusoidal modulating signal is given by:

$$e_0(t) = E_c \cos [2\pi f_c t + \beta_1 \sin (2\pi f_m t)]$$

Where $f_c$ is the frequency of the carrier obtained from crystal oscillator, $\beta_1$ is the modulating index and $f_m$ is the frequency of the modulating signal.

The wide band FM wave is obtained after it is multiplied by $n$ in the $n^{th}$ law frequency multiplier and can be given as,

$$e(t) = E_c \cos [2\pi n f_c t \times \beta_2 \sin 2\pi f_m . t]$$

where          $\beta_2 = n. \beta_1$

## 8.10 FREQUENCY STABILIZED ARMSTRONG FM MODULATOR/TRANSMITTER

The crystal oscillator generates a carrier frequency of 200 kHz. The output of this crystal oscillator and the modulating signal are fed to a balanced modulator. The crystal oscillator output is also fed to phase shifting net work to produce a 90° phase shift. Both the outputs are now fed to a combining net work and FM output is obtained.

The frequency deviation obtained in this method is very small (less than 50 Hz) and therefore tremendous frequency multiplication is desired which is carried in two sections to obtain the standard deviation of 75 kHz.

The Fig. 8.8 shows a frequency stabilized indirect Armstrong FM generation in which a crystal oscillator generates 200 kHz carrier signal with a deviation of 24.4 Hz. The output is multiplied in first section by 64 (2 × 2 × 2 × 2 × 2) *i.e.*, by *frequency doubler* which raises the FM signal to 200 kHz × 64= 12.8 MHz and deviation to 24.4 Hz × 64 = 1.56 kHz. Now another crystal oscillator generates 10.925 kHz carrier which is fed to mixer, which reduces the FM signal to 12.800 – 10.925 = 1.875 MHz, the deviation remains unaltered *i.e.*, 1.56 kHz. Now this output of the mixer is passed through the second deviation multiplier to multiply it by 48 (3 × 2 × 2 × 2 × 2) *i.e.*, one triplet and four doublers to get a standard FM output = 1.875 × 48 = 90 MHz and of deviation = 1.56 kHz × 48 = 75 kHz.

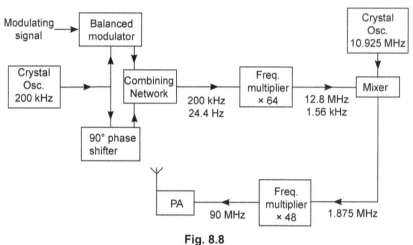

**Fig. 8.8**

**Phasor diagrams.** The Fig. 8.9 shows phasor diagram, which illustrates how Phase modulation is generated using Armstrong method. The Fig. 8.9 (*a*) shows an amplitude modulated signal. See that the resultant of two sideband phasors are always in phase with the unmodulated carrier phasor, so that there is only amplitude variation and phase or frequency variation is absent. **But for frequency variation**, an amplitude modulated signal is added to an unmodulated carrier of same frequency and both these signals are kept 90° out of phase. The resultant of these two is a complex form of phase modulation. [(See Fig 8.9 (*b*)].

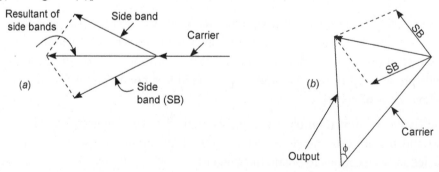

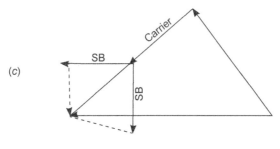

**Fig. 8.9**

If the carrier of the AM wave is suppressed, so that only two sidebands are left which are added to an unmodulated carrier and **phase modulation** is obtained [Fig. 8.9 (c)]. The carrier is supressed by a "Balanced modulator." The addition of the signal is carried out in a 'combining net work'.

Now we have obtained PM signal and FM signal can be obtained by "Bass boosting" of the modulating signal. For this a simple RL network (called RL Equalizer) is used as shown in Fig. 8.10.

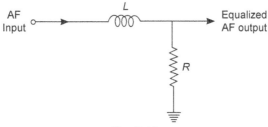

**Fig. 8.10**

For FM broadcasting $\omega L$ should be = $R$ (at 30 Hz). With increase in frequency, the output of the equalizer falls at the rate of 6 dB/octave.

The method possesses a better frequency stability but suffers from excess noise due to tremendous multiplication. The output also suffers from distortion.

## 8.11 RC PHASE SHIFT METHOD

An RC phase shift modulator is shown in Fig. 8.11. The transistor is used as an amplitude sensitive resistor. The maximum phase modulation which can be obtained is 90°.

Generally, a phase modulator is operational only for a phase change of 10°. A modulating signal, which creates a phase change results a change in the carrier amplitude, which is very small and tolerable.

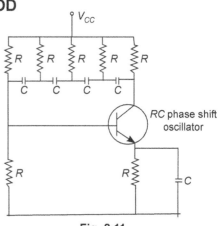

**Fig. 8.11**

The output of the modulator is narrow band FM, which is converted into wideband FM by frequency multipliers.

## 8.12 FM TRANSMITTERS

Depending upon the method of FM generation employed, the FM transmitters may be classified as:

1. Direct FM transmitters
2. Indirect FM transmitters.

## 8.13  DIRECT FM TRANSMITTERS (EMPLOYING REACTANCE METHOD)

These generally produce sufficient frequency deviation and need little frequency multiplication, but they have poor frequency stability. The frequency stability in reactance modulators may be achieved by using automatic frequency control (AFC). This system is called a **cross by system**.

The Fig. 8.12 shows a block diagram for a **cross-by** direct FM transmitter for 96 MHz which is the standard frequency for broadcasting. The modulating signal after amplification is passed through a pre-emphasis to reduce the noise effect at higher audio frequencies. The reactance modulator can produce a maximum frequency deviation of 5 kHz. The carrier oscillator generates a carrier frequency of 4 MHz and which is raised to 96 MHz by using frequency multipliers.

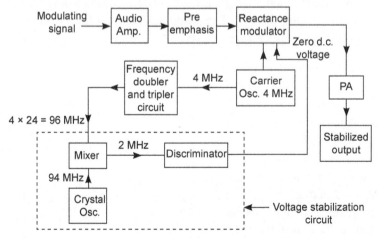

**Fig. 8.12**

Here a frequency multiplication of 24 (96/4 = 24) is required which can be done by using three "frequency doublers" and one "frequency tripler" (24 = 2 × 2 × 2 × 3). The initial deviation is selected in such a way that the frequency deviation attains the value of 75 kHz which is the prescribed frequency deviation for FM transmission.

The maximum deviation of the reactance modulation is therefore kept equal to 75/24 = 3.12 kHz.

The generated FM signal is fed to the frequency stabilization circuit. The function of this circuit is to provide a controlled dc voltage to the reactance modulator, whenever it drifts from the desired value of 4 MHz.

The generated FM signal of 96 MHz and 94 MHz output of a crystal oscillator are fed to a mixer. The output of the mixer is the difference of the two inputs *i.e.*, 96 – 94 = 2 MHz, which is given to the discriminator circuit. The discriminator provides zero dc voltage when its input is exactly 2 MHz. This is possible when the transmitter operates exactly at 96 MHz. The discriminator output will not be zero if the transmitter frequency differs than 96 Hz. Thus the output of the discriminator is utilized to adjust the frequency of transmitter output. The stabilized FM signal finally goes to power amplifiers. (PA).

## 8.14 INDIRECT FM TRANSMITTER

It is also called Armstrong FM transmitter after the name of its originator, and is based on the indirect method of FM generation. The advantage of this method is that a crystal oscillator can be used more efficiently.

The crystal oscillator generates a carrier frequency of 200 kHz. The output of this crystal oscillator and the modulating signal are fed to a balanced modulator. The crystal oscillator output is also fed to a phase shifting net work to produce a 90° phase shift. Both the outputs are now fed to a combining network and FM output is obtained.

The frequency deviation obtained in this method is very small (less than 50 Hz) and therefore tremendous frequency multiplication is desired which is carried in two sections to obtain the standard deviation of 75 kHz.

The Fig. 8.13 shows a typical method of indirect FM generation in which a crystal oscillator generates 200 kHz carrier signal with a deviation of 24.4 Hz.

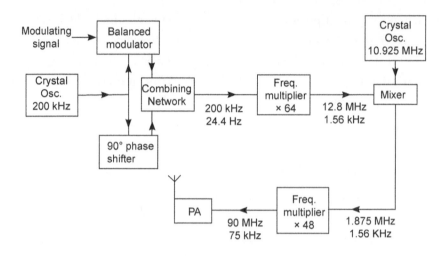

**Fig. 8.13**

The output is multiplied in first section by 64 ($2 \times 2 \times 2 \times 2 \times 2 \times 2$) *i.e.*, by *frequency doublers* which raises the FM signal to 200 kHz $\times$ 64 = 12.8 MHz and deviation to 24.4 Hz $\times$ 64 = 156 kHz. Now another crystal oscillator generates 10.925 MHz carrier which is fed to mixer, which reduces the FM signal to 12.800 – 10.925 = 1.875 MHz, the deviation remains unaltered is *i.e.*, 1.56 kHz. Now this output of the mixer is passed through the second section of multipliers to multiply it by 48 (($3 \times 2 \times 2 \times 2 \times 2$) *i.e.*, one tripler and four doublers) to get the standard FM output = 1.875 $\times$ 48 = 90 MHz and of deviation = 1.56 kHz $\times$ 48 = 75 kHz.

The method possesses a better frequency stability but suffers from excess noise due to tremendous multiplication. The output also suffers from distortions.

## SUMMARY

1. The FM transmitters have a very large bandwidth as compared to AM transmitters.

2. The FM signals can be generated by direct and indirect methods.

3. The direct method is used in reactance modulators. The indirect method is used in phase shift modulator.

4. Accordingly, the transmitters may be of two types direct transmitters and indirect transmitters.

      ❑❑❑

# 9

# FM Discriminators and Receivers

The FM receiver also uses the superheterodyne principle and is quite similar to AM receiver. The basic differences, FM receiver has: much higher operating range, need for limiting action, need for de-emphasis, methods of detection, methods of obtaining AGC etc.

The FM receivers are used (in the range of 40 MHz to 1000 MHz) for sound broadcasting, TV, police, radio, and military systems. A low noise mixer is used, where as colpitt configuration is used as local oscillator.

## 9.1 DEMODULATION (DETECTION) OF FM WAVES

The process of extracting the original signal from a FM wave is known as frequency demodulation and the circuits that perform this job are called **demodulators or detectors**.

The basic function of demodulators is to produce a *transfer characteristic*, which is inverse of that of the frequency modulation. A frequency demodulator produces an output voltage, whose instantaneous amplitude is directly proportional to the instantaneous frequency of the FM wave.

The FM detection or demodulation takes place in two stages.

(*i*) The conversion of the FM wave into the corresponding AM wave by using frequency dependent circuits *i.e.*, the circuits whose output voltage depends on input frequency (called discriminators).

(*ii*) Feeding this AM wave to a linear diode detector to recover the original modulating signal.

## 9.2 FREQUENCY DISCRIMINATORS/DETECTORS

As mentioned, the Frequency *discriminators* convert the FM signal into the corresponding AM signal to be fed to a diode *detector* to get the modulating signal as output.

These can be classified as

(*i*) **Slope Discriminator/Detector:** The operation of this *discriminator* depends upon the slope of the frequency response curve of the frequency selector circuit. The FM discriminator using this principle are

    (*a*)   Single tuned discriminator or simple slope detector.

    (*b*)   Staggered tuned discriminator or balanced slope detector.

(*ii*) **Phase Difference Discriminator/Detector:** These detector's operation depend on measurement of the phase difference. These are:

    (*a*)   Foster seely discriminator/detector.

    (*b*)   Ratio detector.

## 9.3  SLOPE DISCRIMINATOR/DETECTOR

We shall describe two such descriminators.

(*a*)  Single tuned

(*b*)  Staggered tuned.

(*a*)  **Single tuned discriminator/single slope detector (Fig. 9.1)**

It consists of a resonant circuit which is tuned to a frequency slightly above the carrier frequency. This circuit converts the FM signal into the corresponding AM signal. This AM signal is then fed to the diode detector for detection.

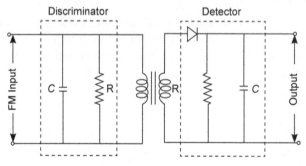

**Fig. 9.1**

By tuning the circuit to receive the signal on the slope of the frequency response curve (Fig. 9.2), the carrier amplitude is made to vary the frequency. The circuit is tuned so that its resonant frequency ($f_c + \Delta f$) is higher than the carrier frequency ($f_c$).

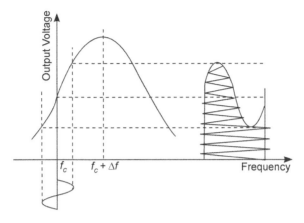

**Fig. 9.2**

When the signal frequency increases above $f_c$, the amplitude of the carrier voltage rises and when the signal frequency decreases below $f_c$ the carrier voltage falls. If the circuit is detuned so that the carrier frequency lies on the positive slope of the curve, amplitude of the resulting voltage across the tuned circuit will vary with the frequency of the input circuit.

**Advantages:**

(*i*) It is simple.

(*ii*) It is economical and hence suitable, when the cost is more important.

**Disadvantages:**

(*i*) The characteristic of the circuit is non linear.

(*ii*) To reduce the distortions, the frequency deviation has to be less.

(*iii*) The amplitude variation may rise due to noise and other factors.

(*b*) **Staggered Tuned Discriminator/Balance Slope Detector (Fig. 9.3)**

The Non linearity of the simple slope detector has been removed in this detector circuit. This detector uses two slope detectors connected **back to back** to the opposite ends of a centre tapped transformer, so that they are 180° out of phase.

In this detector, the output is the difference of two outputs $V_1$ and $V_2$. The resonant circuit No. 1 is tuned to a frequency $f_1, = (f_c + \Delta f)$ slightly higher than the carrier frequency and the resonant circuit No. 2 is tuned to a frequency $= f_2$ ($f_c - \Delta f$) slightly lower than the carrier frequency. Then as the output is the difference of $V_1$ and $V_2$, the output will be zero as $V_1$ and $V_2$ are equal at central frequency. Thus output

$$V_0 = |V_1| - |V_2|$$

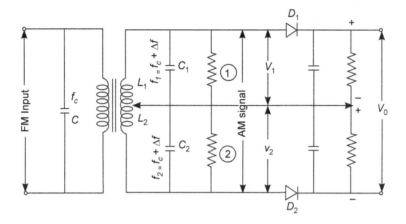

**Fig. 9.3**

The output curve will have a shape of the English alphabet '*S*' (Fig. 9.4). The output is balanced to zero hence, the name as **balanced detector**. When the carrier deviates towards $f_1$, the $|V_1|$ increases while $|V_2|$ decreases and the output $V_0$ goes positive. When the carrier deviates towards $f_2$, the $|V_1|$ decreases while $|V_2|$ increases and output $V_0$ goes negative.

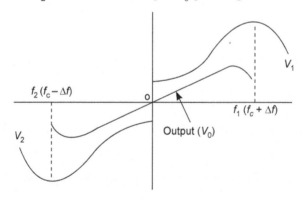

**Fig. 9.4**

The output of the device is almost linear so this is superior than the simple slope detector but this detector suffers from a disadvantage that the output is sensitive to the amplitude variation of the input signal.

**Note:** As can be seen (Fig. 9.3) the secondary is a **back to back** connection of two tuned circuits. One tuned to one side of the FM signal center frequency (intermediate frequency) and other tuned to other side of it. In commercial FM broadcast, the FM signal center frequency (IF) is 10.7 MHz and the transformer primary will be tuned to this frequency. Maximum frequency deviation ($\Delta f$) in FM broadcast is 75 kHz and the two secondaries are off tuned by 100 kHz *i.e.*, the center frequencies for these off tuned circuits would be 10.6 MHz and 10.8 MHz. When the instantaneous frequency equals to 10.7 MHz, the two tuned circuits produce equal amplitude outputs and thus cancel each other. When the instantaneous frequency equals $f_c + \Delta f = 10.8$ MHz (Fig. 9.5) in the present case, the output is positive maximum and it is negative maximum when the instantaneous frequency of the input signal is $f_c - \Delta f = 10.6$ MHz. For all other frequencies between 10.6 MHz and 10.8 MHz, the output amplitude lies between negative and positive maximum, thus giving rise to standard shaped pattern of FM detectors. (See Fig. 9.5).

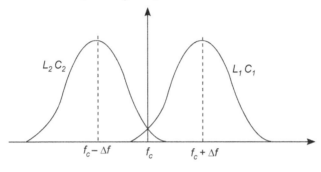

**Fig. 9.5**

## 9.4 PHASE DIFFERENCE DISCRIMINATORS/DETECTORS

As mention earlier, these depend for their operation on the phase difference. Here, we shall discuss the following two discriminators:

(*a*)  Foster seeley discriminator/center tapped detector.

(*b*)  Ratio detector.

The Foster seeley and ratio detectors both belong to the category of **"Quadrature Detectors"**. The principle of operation of a Quadrature detector can be briefly stated as below:

The FM signal is fed to a tuned circuit whose center frequency is same as the unmodulated carrier frequency of the FM signal. The FM signal is shifted 90° in phase and vectorially added to the signal appearing at the output of the tuned circuit. The amplitude of the resultant change as the instantaneous frequency deviates from the center frequency due to change in the phase difference between the two signals being added vectorially. The resultant amplitude increases or decreases, whether the magnitude of the phase difference becomes more or less than 90°.

### (a) Foster Seely Discriminator/Center Tapped Detector (Fig. 9.6)

The circuit consists of two tuned circuits $L_1$ $C_1$ and $L_2$ $C_2$ which are **inductively coupled** and both the circuits are tuned to the carrier frequency. The center of secondary inductance $L_1$, is connected with the primary tuned circuit through a capacitor $C_1$ as shown. It acts as a coupling capacitor, it couples the signal frequency from primary to the center of $L_2$. In other words, this capacitor blocks dc from primary to secondary circuit and the entire voltage applied across primary appears across the inductance $L_3$ which acts as RFC (Radio Frequency Choke). Each half of the secondary tuned circuit has envelope detector and both the detectors ($D_1$ and $D_2$) are identical.

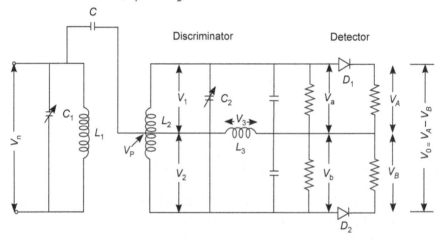

**Fig. 9.6**

The center tapping of $L_2$ has equal and opposite voltages $V_1$ and $V_2$. The voltages applied to the detectors are $V_a = V_3 + V_1$ and $V_b = V_3 - V_2$. The voltage $V_a$ and $V_b$ depend upon phase relations among $V_1$, $V_2$ and $V_3$. The voltages $V_1$ and $V_2$ are always equal and are in phase opposition, but the phase position of $V_1$, $V_2$ and $V_3$ depend on the input frequency.

Thus, if relative magnitudes of $V_a$ and $V_b$ are made to depend upon the instantaneous frequency of the input signal, the output voltage will also vary as per the frequency variation of the input signal.

Three conditions may exist.

(i) **When $V_a = V_b$ :** When input voltages applied to the detectors are equal, the rectified currents will be equal and output $V_A - V_B = 0$.

(ii) **When $V_a > V_b$ :** When input to upper detector is more than to the lower detector, the output $V_A - V_B$ will be positive, and $V_A > V_B$.

(iii) **When $V_a < V_b$ :** When the input to the lower detector is more, then the output $V_A - V_B$ will be negative and $V_B > V_A$.

**Phasor Diagrams (Fig. 9.7):** In Foster Seeley circuit if the phase difference ($\phi$) between $V_1$ and $V_P$ is $+90°$ it would be $-90°$ between $V_2$ and $V_P$. The vector addition of $V_1$ and $V_P$ gives $V_a$ and is fed to the diode $D_1$, whereas vector addition of $V_2$ and $V_p$ gives $V_b$ which is fed to diode $D_2$.

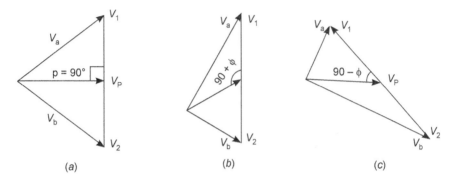

**Fig. 9.7**

When the instantaneous frequency equals the center frequency, both $V_a$ and $V_b$ are of equal amplitude with the result that final output is zero [Fig. 9.7 (a)].

When the instantaneous frequency becomes more than the center frequency, $V_a$ becomes larger than $V_b$ to give an overall positive output [Fig. 9.7 (b)].

When the instantaneous frequency becomes less than the center frequency, $V_a$ becomes smaller than $V_b$ to give on overall negative output [Fig. 9.7 (c)].

**Characteristic (Frequency response) (Fig. 9.8):** The variation of output voltage $V_0 = (V_A - V_B)$ to the instantaneous frequency is called as **characteristic**. It is zero at resonance, positive above resonance and negative below resonance. The characteristic is linear for the region between the peaks of $V_a$ and $V_b$. This range is called **Peak Saturation Range**. Note that the discriminator works satisfactorily over the peak saturation range (MN). This range should be twice the frequency deviation ($\Delta f$), Where $f$ is input frequency.

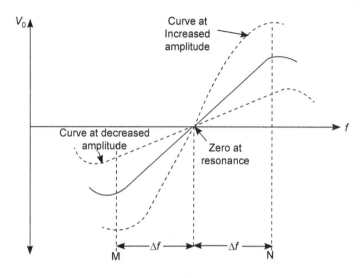

**Fig. 9.8**

**Advantages :** The Foster seeley discriminator provides almost a linear characteristic.

**Disadvantages :**

(*i*) Any variation in the input FM signal due to noise or otherwise changes the characteristic and the output gets distorted.

(*ii*) To reduce the distortion, the receiver with this discriminator requires an "amplitude limiter".

(*b*) **Ratio Detector (Fig. 9.9)**

The one important disadvantage of the Foster seeley discriminator is that it needs an amplitude limiter. This limitation has been removed in the improved version called ratio detector. Here the *ratio* of voltage change depends on the signal frequency, hence the name as *ratio detector*.

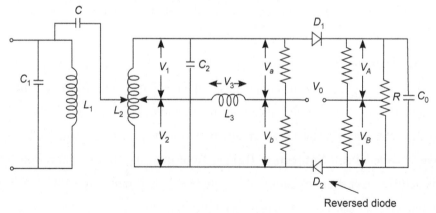

**Fig. 9.9**

The following improvements have been made to obtain the ratio discriminator:

(*i*) One of the diodes have been reversed.

(*ii*) An additional capacitor $C_0$ is connected at the output to limit the variations of amplitude. This is the reason that a ratio detector does not need an amplitude limiter.

(*iii*) The place of taking output is changed. Now, the output is taken from the center tap of a resistor $R$. The output of the ratio detector is half of that of the Foster seeley discriminator *i.e.*,

$$V_0 = \frac{V_A - V_B}{2}$$

The operation of ratio detector is almost same as that foster seeley detector except that the output is reduced to half. The output voltage varies with the input signal exactly in the same way. At the center frequency both diodes conduct equally (one diode conducts in reverse direction due to its reversed connections). But when the input frequency shifts to either side, one of the diodes conduct heavily, as a result, $V_A$ or $V_B$ increases but their sum $V_A + V_B$ remains constant. The ratio of voltages therefore changes depending upon the signal frequency. Now current flows in the circuit and charges the capacitor $C_0$ to the peak value of voltage across $L_2$.

The amplitude variations due to noise or other factors have little effect on the charge on the capacitor $C_0$ and the output remains constant. Since the output voltage is not affected by amplitude variations, amplitude limiter is not required with a ratio detector.

**Table 9.1** Comparison of FM Detectors/Discriminators.

| S. No. | Particulars | Balanced slope detector | Phase discriminator | Ratio detector |
|--------|-------------|-------------------------|---------------------|----------------|
| 1. | Alignment/ Tuning. | Critical as three circuits are to be tuned at different frequencies. | Not critical | Not critical |
| 2. | Amplitude limiting | Not provided | Not provided | Provided |
| 3. | Output characteristics | Depends upon primary and secondary frequency relationship | Depends upon primary and secondary phase relationship | Depends upon primary and secondary phase relationship |
| 4. | Linearity of output characteristics | Poor | Very good | Good |
| 5. | Application | Not used | Used in FM Radio, satellite receiver | Used in Narrow band FM receivers, in sound section of TV receiver. |

## 9.5 OTHER FM DETECTORS

Here two detectors are discussed:

($a$)  PLL detector

($b$)  Zero crossing detector.

($a$)  **Phase Locked Loop (PLL) Detector**

A phase locked loop (PLL) is a negative feedback system. In this, an **error signal** is generated which depends on phase difference between input and the feedback signals. The function of PLL is to **lock or synchronise** the frequency of a voltage controlled oscillator (VCO) with the incoming signal hence the name. The PLL technique is used for detection of signals and their processing.

The Fig. 9.10 shows the block diagram of a PLL demodulator. It has four components—phase detector, low pass filter, amplifier and a voltage controlled oscillator.

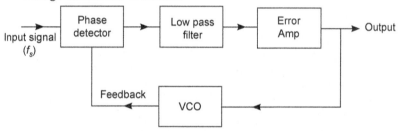

**Fig. 9.10**

With no input signal fed to PLL the output from phase detector is zero. The control signal (error signal) to the VCO is also zero, and the VCO operates at the free running frequency known as its **center frequency** ($f_0$). Now if an input signal ($f_s$) is applied, detector produces sum and difference components *i.e.*, ($f_s + f_0$) and ($fs - f_0$).

If there is a slight difference between $f_s$ and $f_0$, both the frequencies do not fall into the pass band of low pass filter and are attenuated, as a result, frequency of VCO does not change and the loop does not acquire a lock. When the input frequency $f_s$ is such that $f_s - f_0$ lies within the pass band of the filter, this is amplified and applied as a control signal to the voltage controlled oscillator (VCO). This reduces the difference between $f_s$ and $f_0$ and if their difference is not significant, the feedback action of the loop causes synchronising or locking of VCO with the incoming signal. Once in lock, the difference $f_s - f_0$ becomes proportional to the phase difference between input and the VCO signals. If the input signal frequency is varied through the 'lock range' the total variation possible for the phase angle is between 0° and 180° and its value is 90° when input frequency is equal to center frequency, (*i.e.*, $f_s = f_0$).

Below are given few terms related to PLL (Fig. 9.11)

(*i*) The range of frequency over which a PLL can maintain lock with the input signal is called as **lock in range**.

(*ii*) The range of frequencies over which a PLL aquires lock is called as **capture range**. Generally this is lesser (or at the most equal) to the **lock in range**. More over the capture of the input signal after its application takes a finite time, called as **Pull in Time**.

(*iii*) The VCO output frequency when it is not locked is called free running frequency. The frequency deviation on either side of this **frequency**, over which the VCO can track the input once in lock is called tracking range.

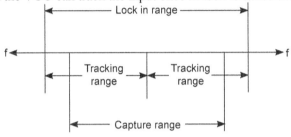

**Fig. 9.11**

## (a) PLL (I.C.) Demodulator / Detector

This system is also available in IC form.

The PLL system is particularly suitable for demodulating narrow band FM signals. The block diagram of such IC package (NE 565) is shown in Fig. 9.12. This can work for frequencies in the range of 0.5 Megahertz.

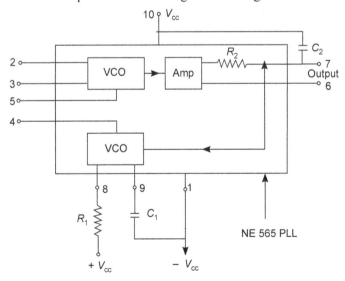

**Fig. 9.12**

**Table 9.2**  Commercially Available PLLs.

| PLL No. | Max. Operating Frequency (MHz) | Lock Range % of Center Frequency | Package | Supply Voltage (Volts) | Typical Application |
|---|---|---|---|---|---|
| NE 560 | 30 | 40% | 16 pin DIP | + 16 to + 26 | Modems, tracking filters, FSK receivers, signal generators, FM demodulators. |
| NE 561 | 30 | 40% | 16 pin DIP | + 16 to + 26 | Modems, tracking filters, FSK receivers, synchronous detectors. AM receivers, signal generators, telemetry decoders. |
| NE 562 | 30 | 40% | 16 pin DIP | + 16 to + 30 | Frequency synthesisers, data synchronisers, tracking filters, modems, tone decoders, FSK receivers, signal conditioning, FM detectors. |
| NE 564 | 50 | 40% | 16 pin DIP | + 4.5 to + 12 | High speed modems, frequency synthesisers, signal generators, FSK transmitters and receivers. |
| NE 565 | 0.5 | 120% | 10 pin DIP | ± 5.0 to ± 12 | Modems, frequency shift keying, tone decoders, wideband FM discriminators. |
| SE 565 | 0.5 | 120% | Same as NE 565 | ± 5.0 to ± 12 | Same as for NE565 |
| NE 567 | 0.5 | 14% | 14 pin DIP | + 5.0 to + 9 | Touch tone decoding, intercom, ultrasonic control (remote TV etc.). Communication paging, precision oscillator, frequency monitoring and control |
| SE 567 | 0.5 | 14% | Same as for NE 567 | + 5 to + 9 | Same as for NE 567 |

| PLL No. | Max. Operating Frequency (MHz) | Lock Range % of Center Frequency | Package | Supply Voltage (Volts) | Typical Application |
|---|---|---|---|---|---|
| NE 566 | 0.5 | | Same as for NE 567 | + 10 to + 26 | FM modulation tone generators, signal generators, clock generators, function generators, frequency shift keying (FSK). |
| SE 566 | 0.5 | | Same as for NE 567 | + 10 to + 26 | Same as for NE 566 |
| CD4046 | 1.0 | Full VCO range | 16 pin DIP | + 2 to + 20. | FM modulator and demodulator frequency synthesis and multiplication. Frequency discriminator, data synchronisers, FSK modems, tone decoding, signal conditioning. |

DIL = Dual in package

## (*a*) **Zero Crossing Detector**

The instantaneous frequency of an FM wave is given by $f = \dfrac{1}{2\Delta t}$, where $\Delta t$ is

the time difference between the two adjacent zero crossover points of the FM wave.

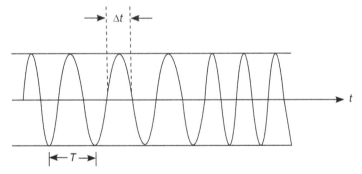

**Fig. 9.13**

Take a time interval $T$, such that:

  (*i*)   The time $T$ is small as compared to 1/B.W. of the signal.

  (*ii*)   The time $T$ is large as compared to 1 /carrier frequency.

If number of zero crossings during time $T$ is $n_0$ the

$$\Delta t = \frac{T}{n_0}$$

The instantaneous frequency $f = \dfrac{1}{2\Delta t} = \dfrac{n_0}{2T}$

As there is a linear relation between instantaneous frequency and the signal, the later can be recovered, if $n_0$ is known.

This is the principle of operation of a zero crossing detector.

### Circuit and Operation

The Fig. 9.14 (*a*) gives the block diagram of the circuit. In this method, we measure instantaneous frequency by the number of zero crossings. The limiter converts the FM input into a square wave. The output of the limiter goes to zero crossing detector. Each time the output goes positive, It triggers the monostable multibrator which generates rectangular pulse of a fixed width. The rate of zero crossing is equal to the instantaneous frequency of the input signal. The output of multivibrator is fed to an averaging circuit, which gives the input frequency. The Fig. 9.14 (*b*) shows the associated waveforms.

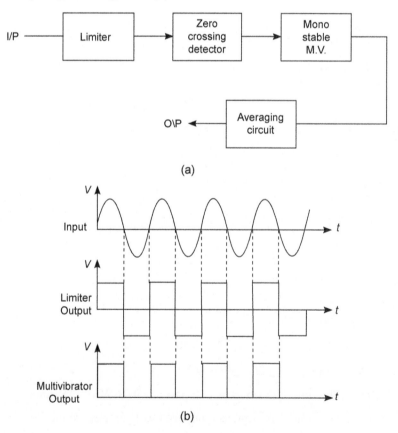

(a)

(b)

**Fig. 9.14**

## 9.6 FM RECEIVERS

The FM receiver receives the FM signal coming from FM transmitter and then recovers the original modulating signal. These are superheterodyne type receivers with **double frequency conversion**. Their RF, IF and mixer stages are also similar to AM receivers. Since the FM receivers are to operate on UHF/VHF range, their circuits are designed accordingly, moreover these receivers do not require AGC circuits and their RF and IF stages are to be designed to have adequate bandwidth (150 kHz) to accommodate the FM signals.

The Fig. 9.15 shows the block diagram of an FM receiver. The RF amplifier receives and amplifies the FM signal. The IF amplifiers amplify the intermediate frequency signals. Note that IF for FM broadcasting is 1 07 MHz (as compared to 455 kHz for AM receivers). The IF amplification is done in stages. Note that local oscillators and mixers are also to be designed at UHF/VHF ranges.

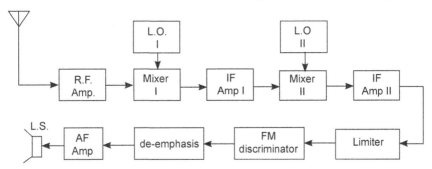

**Fig. 9.15**

The amplified IF signals are now passed through limiter stages. The limiter keeps the output voltage constant and removes all amplitude fluctuations produced due to noise etc. This is necessary, the reason begin that the discriminator which is the next stage needs a constant amplitude FM voltage as its input for its satisfactory operation. The limiter stage may be deleted if a **ratio detector** is called to replace the discriminator. The discriminator or the detector removes the original modulating signal from the IF signal. Now the signal may be passed through de-emphasis to restore the original level of the signal. The A.F. amplifier now amplifies the signal, which is fed to the loud speaker, which converts this electrical signal into the original sound signal.

An FM receiver always has an RF amplifier. It reduces noise level as it has to handle large band width. The input impedance of the receiver should be properly matched with impedance of the antenna.

A *limiter* is type of 'clipper circuit', which checks that the input to the demodulator does not contain any 'spurious' signals which may be source of distortion. If need arises, two limiter circuits in cascade arrangement may be used.

The FM demodulator is a bit different from the AM demodulator. An FM demodulator is *frequency-to-amplitude* converter which converts the frequency deviation of the carrier into AF amplitude variations identical to the original signal.

In FM receivers, there is undesired amount of noise in the output in the absence of transmission on a given channel or between channels. To overcome this problem **muting circuits** are some times included in FM receivers.

Figure 9.16 shows superheterodyne FM radio receiver which is similar to AM receiver with the difference that an FM receiver has much more operating frequency. It has limiter, de-emphasize network, and different methods of obtaining gain control. A typical FM receiver operating at 87 MHz to 107 MHz may have an IF stage of 107 MHz and bandwidth of 200 kHz.

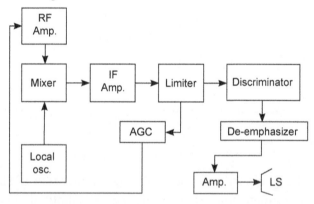

**Fig. 9.16**

## SUMMARY

1. FM reception is almost noise free.

2. The FM receivers also use Heterodyning principle. Their range is 400 MHz to 1000 MHz.

3. The FM detection takes place in two stages:

    (*i*)   The convertion of FM signal into corresponding AM signal by "discriminator"

    (*ii*)  Feeding this AM wave into a "diode detector".

4. The types of frequency discriminator/detector are: tuned discriminator, Foster seeley discriminator and ratio detector.

□□□

# Phase Modulation (PM)

The phase modulation is similar to frequency modulation, **as a change in phase of the carrier also causes a change in frequency** of the carrier at the same instant. Thus, changes in the frequency are proportional to the changes in phase producing an FM signal. The FM signal which is produced indirectly from a phase modulated signal is called "Equivalent FM".

## 10.1 PHASE MODULATION

In phase modulation, phase angle of the carrier voltage is changed in accordance with the instantaneous value of the modulating signal voltage, the amplitude and **frequency remaining the same**.

The FM and PM both belong to the general class of **Angle modulation or Exponential Modulation**. The FM and PM are not much different in the sense that variation in the phase of a carrier is accompanied by a corresponding change in the frequency. This is due to the relationship between phase angle $\phi$ and frequency $\omega$ of the carrier, $\omega = d\phi/dt$.

## 10.2 COMPARISON OF AM, FM AND PM

It will be interesting here to compare AM, FM and PM. The Fig. 10.1 shows a sinusodial carrier being modulated by a step voltage.

In case of AM [Fig. 10.1 (*a*)], the amplitude follows the step change, but frequency and phase remain constant. The change in the amplitude can be observed on a CRO.

In case of FM [Fig. 10.1 (*b*)], the amplitude and phase remain constant, where as, the frequency follows the step change. The change in the frequency can be observed on a **frequency counter**.

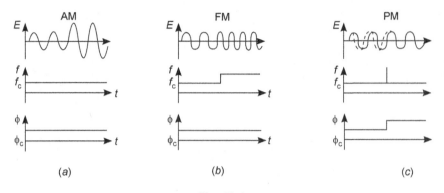

**Fig. 10.1**

In case of PM [Fig. 10.1 (c)], the phase angle follows the step change, while the amplitude remains constant. The change in the phase may be observed on a **phase meter** but not as direct as the measurement of amplitude or frequency. After the step change in the phase, the sinusoidal carrier appears as though it is a continuation of the dotted curve as shown. As the step changes in phase, the abrupt displacement of the waveform on the time axis makes it appear as though the frequency undergoes an abrupt change. This has been shown by the spike. The spike change in frequency can be directly measured by a frequency counter.

Here it is also to be informed that for the same value of modulating index, spectrum for the PM wave will be same as for a FM wave.

## 10.3 EXPRESSION FOR PM WAVE

Let the carrier be expressed as

$$e_c = E_c \sin (\omega_c t + \phi_0)$$

and the modulating signal as

$$e_m = E_m \sin \omega_m.t$$

The phase angle of the carrier before phase modulation is

$$\phi_c = \omega_c t + \phi_0$$

After phase modulation, the instantaneous phase angle of carrier is given by

$$\phi(t) = \omega_c t + \phi_0 + k_p. \, e_m$$

where $k_p$ is a constant called **phase deviation constant** and defined as the phase deviation per unit amplitude of the modulating signal. Its unit is rad/volt.

The phase angle $\phi_0$ is constant and plays no role in the process of phase modulation, thus may be omitted. Therefore, the equation of a PM wave may be expressed as:

$$e_{PM} = E_c \sin (\omega_c t + k_p.E_m \sin \omega_m t)$$

Putting $k_p . E_m = m_p$, the above equation can be written as:

$$e_{PM} = E_c \sin (\omega_c t + m_p \sin \omega_m t)$$

Where, $m_p$ is the modulation index for **phase modulation**, which may be **defined as maximum phase angle ($\phi_m$) produced by the modulating signal** and is proportional to $E_m$, the maximum amplitude of the modulating voltage (but independent of $f_m$).

**Notes :**

1. The maximum frequency deviation produced by PM is

   $$(\Delta_\omega)_{PM} = k_p \, \omega_m E_m$$

   and depends on modulating frequency $\omega_m$.

   The students can recall that frequency deviation in FM $= k_f . E_m$, where $k_f$ is a constant. Therefore for an equal bandwidth in FM and PM

   $$k_f = k_p . \omega_m$$

2. In FM, the modulation index $m_f$ is inversely proportional to the modulating frequency but in PM, the modulation index $m_p$ is directly proportional to the modulating voltage but independent of frequency.

3. If we integrate the modulating signal and then allow it to phase modulation, we obtain an FM wave.

4. The bandwidth in PM is given by Carson rule,

   $$(BW)_{PM} = 2(\Delta\omega)_{PM} = 2 \, k_p \, E_m . f_m$$

   where $k_p . E_m = m_p = \phi_m$, which is the phase modulation index.

5. It is possible to obtain FM from PM by the **Armstrong** system.

6. Strictly speaking, there are two types of analog modulation systems — amplitude modulation and angle modulation. The angle modulation may be subdivided into two distinct types–FM and PM, which are closely allied.

**Problem 10.1.** *A 25 MHz carrier is modulated by a 400 kHz signal. If carrier voltage is 4V and the maximum frequency deviation is 10 kHz. Write down the equation for the PM wave.*

**Solution.** The carrier frequency

$$\omega_c = 2\pi f_c = 2\pi \times 25 \times 10^6 = 1.57 \times 10^8 \text{ rad/sec.}$$

Modulating frequency

$$\omega_m = 2\pi \times 400 = 2513 \text{ rad/sec.}$$

Frequency deviation       $\Delta f = 10 \text{ kHz} = 10000 \text{ Hz}$

Carrier voltage       $E_c = 4V$

The modulating index for PM:

$$m_p = \frac{\Delta f}{f_m} = \frac{10,000}{400} = 25$$

The standard equation for PM wave

$$e_{PM} = E_c \sin(\omega_c t + m_p \sin \omega_m t)$$

or                        $e_{PM} = 4 \sin(1.57 \times 10^8 \, t + 25 \sin 2513 \, t)$ **Ans.**

**Problem 10.2.** *If a carrier $E_c \cos \omega_c t$ is phase modulated by a signal $5 \sin 2\pi$ $(15 \times 10^3)t$. Assume $k_p = 15$ kHz/Volt, calculate modulation index and bandwidth.*

**Solution.** Modulating signal $= 5 \sin 2\pi (15 \times 10^3)t$

Comparing this with          $e_m = E_m 2\pi f_m t$

We get                       $E_m = 5V$ and $f_m = 15 \times 10^3$ Hz

Given                        $k_p = 15$ kHz $= 15000$ Hz/V

    (*a*)  Modulation Index          $m_p = k_p . E_m$

       or,                          $m_p = 15,000 \times 5 = 75,000$ **Ans.**

    (*b*)  Bandwidth                   $= 2 k_p . E_m . f_m$

$$= 2 \times 15000 \times 5 \times (15 \times 10^3)$$

$$= 2250 \times 10^6 \text{ Hz}$$

$$= 2250 \text{ MHz  Ans.}$$

## 10.4  GENERATION, TRANSMISSION AND RECEPTION OF PM/FM WAVE

The same circuit can be employed for generation of PM as well as of FM wave. The Fig. 10.2 shows a basic PM/FM system.

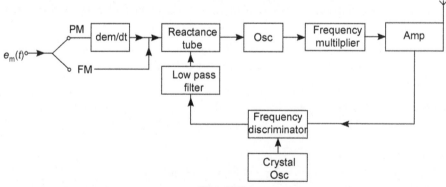

**Fig. 10.2**

    (*a*) **Transmission (Fig. 10.2):** Change in the oscillator frequency is accomplished by a reactance tube. The tube offers a reactance to an oscillator, the reactance being a function of input voltage as the input voltage changes, the reactance also changes and thus the oscillator frequency. If the voltage supplied to the reactance tube is modulating signal $e_m(t)$, the result is an FM wave, where as if the input voltage is a derivative of the modulating signal ($d_{em}/dt$) the result is a PM wave.

The output of the oscillator is now given to a frequency multiplier to obtain the desired frequency. The output of the multiplier is amplified and transmitted through an antenna.

At some point in the transmitter, the carrier frequency is sampled and compared with a frequency standard, preferably a crystal oscillator is used for this purpose. The output of the mixer is then given to a frequency discriminator.

(b) **Reception (Fig. 10.3):** The signal is received and amplified by RF amplifiers. The output of RF amplifier is given to frequency converter and intermediate frequency is obtained which is then amplified and fed

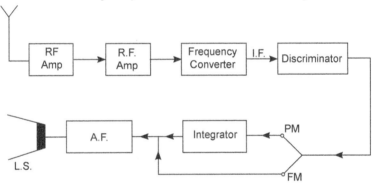

**Fig. 10.3**

to discriminator. The obtained PM/FM signal is amplified and fed to a loudspeaker which gives the original sound signal. Note that the PM signal before giving to audio amplifier is passed through an integrator net work.

## 10.5 COMPLETE PM SYSTEM

The Fig. 10.4 shows a block diagram of a complete PM system. Here $m(t)$ is the modulating input and $m_{eq}(t)$ is the equivalent receiver output. The Fig 10.4 is self explanatory. Note that:

(a) If modulating input is a triangular wave, the receiver output is a square wave.

(b) The square wave as the modulating input, gives an impulse output at the receiver.

(c)  A sinusoidal modulating input gives out a cosinusoidal output at the receiver.

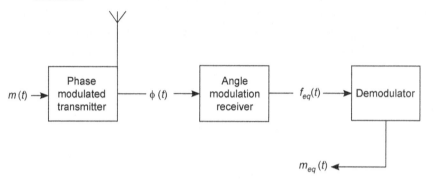

**Fig. 10.4**

## 10.6 GENERATION OF PM SIGNAL FROM FREQUENCY MODULATOR

It can be mathematically verified that, if the modulating signal is differentiated before it is applied to a frequency modulator, the output is a PM signal (Fig. 10.5).

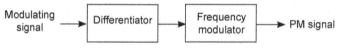

**Fig. 10.5**

In a frequency modulator, the instantaneous frequency of the modulating signal varies according to the signal applied at its input. In this case, the instantaneous frequency will vary as per differential of the modulating signal that is, the integral of instantaneous frequency will vary according to the modulating signal. In other words, the instantaneous phase is modulated in accordance to the modulating signal, the output, therefore, is a PM signal.

## 10.7 GENERATION OF FM SIGNAL FROM PHASE MODULATOR

It can be mathematically verified that if the modulating signal is integrated, before it is fed to a phase modulator the output is an FM signal.

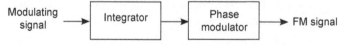

**Fig. 10.6**

In Fig. 10.6 the input to the phase modulator is the integral of the modulating signal. The instantaneous phase of the modulated signal varies in accordance with the modulating signal. In other words, derivative of the instantaneous phase varies as per the modulating signal. Since the rate of change of phase is frequency, we can say that the instantaneous frequency varies as per the modulating signal, which implies that output is an FM signal.

## 10.8 FM VS PM

**The FM and PM are compared below :**

1. If frequency modulating factor $m_f$ and phase modulating factor $m_p$ are equal, the FM and PM have identical spectra. However, instantaneous phase of the FM and PM waves for a given modulating signal will be different.

2. The frequency deviation in PM is directly proportional to the frequency of the modulating signal, whereas in FM it is independent of the modulating frequency. In both the cases however, the frequency deviation is proportional to the amplitude of the modulating signal.

3. For a given modulating signal, modulation index for PM is independent of the modulating signal frequency, while for FM it is inversely proportional to the frequency of the modulating signal.

4. In PM, bandwidth is extremely large *i.e.*, the side bands do not cover the increase in frequency, whereas in FM, side bands cover it rapidly.

5. In PM, the phase deviation is proportional to the amplitude of the modulating signal but independent of frequency.

   In FM also the frequency deviation is proportional to the amplitude of the modulating signal. Hence FM must be a form of PM.

6. The difference between FM and PM is not apparent for the same modulating frequency but it reveals when the modulating frequency is changed.

7. If a FM transmission is received on a PM receiver, the **Bass frequencies** would have considerably more deviation of phase than a PM transmission. Since the output of a PM receiver would be proportional to the phase deviation, the signal would appear **bass boosted.** Alternately, PM transmission received on an FM receiver would be **lacking in bass.** This is a practical difference between FM and PM, but it is clear that **one can be obtained from the other very easily.**

Keeping above in mind for transmission and reception of analog signals, FM is preferred over PM. Another reason for this decision is the difficulty of providing correct **reference phase** in the detection stage of the PM receiver. The PM thus is not suitable for commercial radio broadcasting and radio telephony.

**Table 10.1 :**    FM Versus PM

| S.No. | FM | PM |
|-------|-----|-----|
| 1. | Modulation index $m_f$ is proportional to the modulating voltage as well as to the modulating frequency. | The modulating index $m_p$ is proportional only to the modulating voltage. |
| 2. | The frequency deviation is proportional to the modulating voltage. | The phase deviation is also proportional to the modulating voltage. |
| 3. | Associated with the change in $f_c$, there is some phase change. | Associated with the change in phase, there is some change in $f_c$ (carrier frequency). |
| 4. | The FM can be received on a PM receiver. | It is also possible to receive PM on an FM receiver |
| 5. | The SN ratio is better than PM. | The SN ratio is poor than FM. |
| 6. | The noise immunity is better than AM and PM. | The noise immunity is better than AM but not than FM. |

## SUMMARY

1. In phase modulation, phase angle of the carrier is changed in accordance to the modulating signal, amplitude and frequency remains some.

2. The phase modulation can be obtained from frequency modulation and vice versa.

3. The equation for PM wave is

$$e_{PM} = E_c \sin (\omega_c.t + m_p.\sin \omega_m t)$$

4. The important components of a PM system are: PM Transmitter, Angle modulator, receiver and output.

5. The stages of obtaining PM signal from FM are: differentiation, frequency modulation and output.

6. The stages of obtaining FM signal from a PM signal are: Integration, phase modulation and output.

# More About Transmitters and Receivers

Transmitter and receiver are very important components of all communication systems.

This chapter furnishes advance topics on AM, SSB, FM and PM transmitters and receivers.

## 11.1 BASIC REQUIREMENT OF AM TRANSMITTER: FLYWHEEL EFFECT

To generate AM waves, we supply current pulses to a tuned (tank) circuit. Each pulse is made proportional in amplitude to the size of the modulating sine wave. This will be followed by the next sine wave proportional to the next applied pulse and so on. **At least 10 times pulses per cycle should be fed to the circuit; and if the current pulses are made proportional to the modulating voltage, we get a good AM wave.**

This is called "Flywheel Effect" of the tuned circuits and holds good for a tuned circuit whose $Q$ factor is of moderate value.

## 11.2 NEGATIVE FEEDBACK IN AM TRANSMITTERS

Generally, negative feedback is provided in AM transmitters in relation to the modulating signal. The radiated signal is picked up from the antenna, demodulated, combined with the modulated signal and used as negative feedback at the appropriate point. The negative feedback can be used with low as well as high level modulators.

Advantages of using negative feedback with transmitters are same as with amplifiers. The negative feedback reduces amplitude as well as frequency distortion thus increases faithfulness of the transmitter. The negative feedback also reduces the undesirable hum and noise in the transmitter.

## 11.3 AM VS FM BROADCASTING

This has been compared below:

| S.No. | AM Broadcasting | FM Broadcasting |
|-------|-----------------|-----------------|
| 1. | It requires smaller transmission bandwidth. | It requires larger bandwidth. |
| 2. | It can be operated in low, medium and high frequency bands. | It needs to be operated in high and very frequency bands. |
| 3. | It has wider coverage. | Its range is restricted to 50 km. |
| 4. | The demodulation is simple. | The process of demodulation is complex. |
| 5. | The stereophonic transmission is not possible. | In this, stereophonic transmission is possible |
| 6. | The system has poor noise performance. | It has an improved noise performance. |
| 7. | The AM signal reception does not have any threshold in the useful range of signal noise ratio (SNR). | The FM signal reception exhibits useful range of signal noise ratio. |

## 11.4 FREQUENCY DRIFT

It is necessary that the carrier frequency of a "frequency Transmitter/ modulator" is maintained almost constant, even though the instantaneous frequency of the modulator may vary with the modulating signal. Any variation in the frequency is called **Frequency drift**.

The drift is usually specified in **PPM** (Part per million).

In an oscillator, the frequency of oscillation is close to the "resonant frequency" of the Tank circuit (See Fig. 11.1)

The frequency of oscillations is given by:

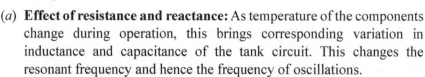

$$f = \frac{1}{2\pi}\sqrt{\frac{1}{LC} - \frac{R^2}{L^2}}$$

Neglecting $R$,     $$f = \frac{1}{2\pi\sqrt{LC}} \text{ Hz.}$$

But the exact value of the frequency is influenced by the following factors:

**Fig. 10.1**

(a) **Effect of resistance and reactance:** As temperature of the components change during operation, this brings corresponding variation in inductance and capacitance of the tank circuit. This changes the resonant frequency and hence the frequency of oscillations.

The variation in inductance and capacitance of the tank circuit may be reduced by enclosing the tank circuit in a **chamber**, which will keep the temperature constant.

(b) **Effect of $Q$ of the tuned circuit:** The frequency variations vary inversely of the $Q$ factor of the tuned circuit, hence the $Q$ factor needs to be kept as large as possible.

Note that Q factor of a resonant circuit is

$$= \frac{\text{Inductive rectance}}{\text{Resistance}} = \frac{X_L}{R}$$

(c) **Effect of collector voltages:** The variations in the collector voltage results in variations in a.c. currents flowing between tank circuit and transistor of the oscillator and thus affects the frequency. Such a variation can be reduced by using high $Q$ tank circuits. For this, **Hartley oscillator** is preferred

Such variations can however, be eliminated completely by inserting a **stabilising capacitor** $(C_s)$ in collector as well as in the base circuits of the Hartley oscillator as shown in Fig. 11.2.

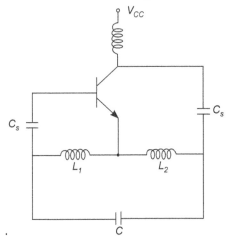

**Fig. 11.2**

The value of $C_s$ may be calculated assuming the $Q$ factor of the tank circuit $(L_1, L_2, C)$ as infinitely large.

(d) **Effect of harmonics:** The harmonics of the frequency of oscillations destroy the freqeuncy stability. Such a variation can be minimised by (i) using a tank circuit with high $Q$ and (ii) by adjusting the amplitude of harmonic voltage and currents at minimum.

All this necessitates to set up a **frequency stabilisation** system in the transmitter to maintain the drift within 2 kHz (as specified).

## 11.5 FREQUENCY STABILISATION

The most important requirement of a transmitter is stabilisation of the carrier frequency. It is necessary that the average carrier frequency of a **frequency modulator** be maintained nearly constant even if the instantaneous frequency of the modulator varies with the modulating frequency. Slight **drifts** in the operation of the device is accompanied by an appreciable change in the average frequency.

The Fig. 11.3 shows a typical frequency stabilisation system. The carrier frequency $(f_c)$ of the FM signal is compared with the frequency output $(f_0)$ of the crystal oscillator.

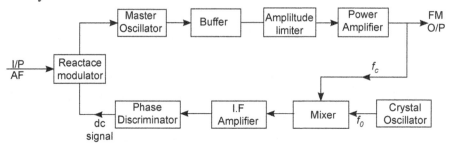

**Fig. 11.3**

The output of the reactance modulator is isolated by a buffer stage. The output of buffer is given to an amplitude limiter and then to a (class $C$) power amplifier. A small part of the output $(f_c)$ is given to a mixer, where it is mixed with the output signal of crystal oscillator $(f_0)$. The intermediate frequency (IF) signal obtained from the mixer is amplified and given to a "phase discriminator". The output of the discriminator is a dc signal, which is applied to the reactance modulator as a correcting voltage to compensate for any **drift** in the average frequency of the modulator.

The discriminator is such that it reacts to the changes in the incoming frequency but not to the frequency variation due to "frequency modulation" process. The discriminator gives a positive output dc voltage, if the input frequency exceeds the discriminator's tuned frequency and it gives a negative output dc voltage, if the input frequency is lower than discriminator's tuned frequency.

## 11.6 RADIO TELEPHONE TRANSMITTERS

The radio telephone transmitters are intended for **point to point communication.**

The maximum modulating frequency allowed is 3 kHz. The short waves (SW) are used for long distance radio telephony. The SW radio telephone transmitters are similar to radio broadcast transmitters. For short distance, U.H.F. (ultra high frequency) communication is used. The output power is about 1 to 10 kW.

The radio telephone transmitters have the following important features.

(*a*) **Volume Compressor:** Volume compressors are used almost in all S.W. radio telephone transmitters. The volume compressor compresses the volume range, as to have a ratio of maximum to minimum volume as 20 dB. This has advantages as : raised modulation level, elimination of overmodulation and improvement of signal-noise ratio.

The Fig. 11.4 shows block diagram of the volume compressor.

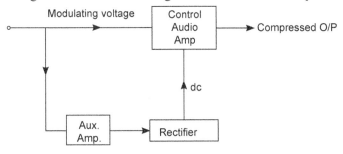

**Fig. 11.4**

(*b*) **Diode clipper:** This has almost similar function as of the volume compressor. It also prevents overmodulation. The only difference is that, whereas the volume compressor starts its job from the beginning, The clipper starts functioning, when the modulating voltage reaches a predetermined value.

The Fig. 11.5 (*a*) shows a typical diode clipper and Fig. 11.5 (*b*) shows the clipped output.

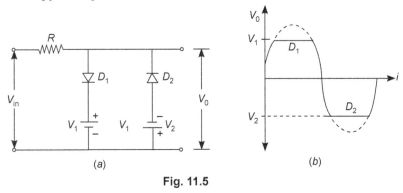

**Fig. 11.5**

No doubt, that the clipping distorts the output signal but the distortion is not severe.

(*c*) **Pre-emphasis:** Pre-emphasis is the process in which the high frequency terms are **boosted** before modulation; because high frequency terms have small amplitude and therefore have lesser power as compared to low frequency terms. In other words, low frequency terms which carry more power but less intelligence are attenuated. This is done when intelligence is more important, than the natural reproduction. The Fig. 11.6 (*a*) shows pre-emphasis circuit and Fig. 11.6 (*b*) shows output.

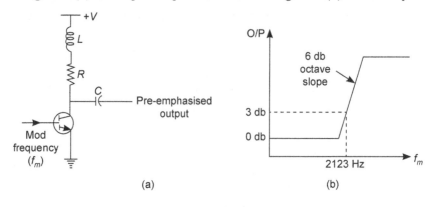

**Fig. 11.6**

Note that in the receiver, reverse process is carried out, which is called "de-emphasis".

(*d*) **Voice operated device Anti singing (VODAS):** Till now "Echo suppressor" was used in line telephony to suppress the echo; but the circuit can generate self oscillations when either subscriber speaks. These oscillations are called **singing** which can cause the relays to close. So a **singing suppressor (VODAS) should** be used instead of "Echo suppressor".

The "singing suppressor" supresses echo as well as singing.

Since such a singing suppressor operates by voice from one of the subscribers, the system is called **voice operated device antisinging** (VODAS).

## 11.7 AM RADIO TRANSMITTERS

The Fig. 11.7 shows block diagram of an A.M. Radio transmitter. The function of each block is given below:

1. **Master oscillator:** It generates oscillations of desired frequency. The generated frequency is required to remain constant within limits, inspite of variations in the supply voltage, ambient temperature etc. The frequency variation with time and age are also to be avoided.

2. **Buffer or Isolating amplifier:** A buffer or isolating amplifier is placed between master oscillator and harmonic generator. The buffer amplifier does not draw any input current hence does not cause any loading on

the master oscillator; therefore changes in carrier frequency due to variation in loading are avoided.

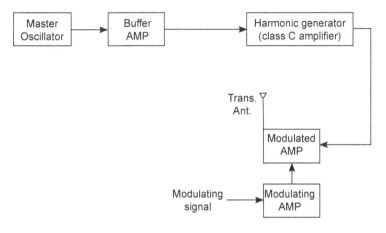

**Fig. 11.7**

3. **Harmonic generator/Class C amplifier:** The master oscillator generates voltage at a frequency, which is a submultiple of the carrier frequency. The harmonic generators basicially are class C amplifiers, in which the tuned circuit selects the desired harmonic frequency signal and amplifies the same.

   There is a chain of class C amplifiers which raise the power to the required level.

4. **Modulated amplifier:** This is a class B push pull power amplifier and is modulated by a modulating audio amplifier as shown. In small **transistorized** radio transmitters, collector modulation or base modulation or both may be used. In high power radio broadcast or radio telephone transmitters "Plate modulation" may be used.

5. **Modulating amplifier:** This modulates the modulated amplifier. This is a class B push pull amplifier. In low power transmitters, class A amplifier may also be used.

   The low power transmitters with output power upto 1 kW or so may be transistorized, but the higher power transmitters use vacuum tubes.

## 11.8 FREQUENCY SCINTILLATION

The **Frequency scintillation** is caused by **abrupt changes** in load on the master oscillator. Any abrupt change in the load causes a change in the resistance and reactance coupled into "tank circuit" of the master oscillator and causes change in the frequency of oscillations.

In order to avoid frequency scintillation, the "master oscillator" is made to drive "buffer amplifier", such that the latter draws little power from the former, so that it produces little loading on the former. In this way, though the load on the transmitter changes, the loading of the master oscillator does not change. As a result, no abrupt change in frequency of the master oscillator occurs and the frequency scintillation is avoided.

## 11.9  PRIVACY DEVICES IN RADIO TELEPHONY

(*a*) Some communication systems such as **telephony** utilizes radio waves that are radiated in space. In the route towards the receiver these waves are intercepted causing encroachments in private conversation, so to avoid this encroachment, some sort of privacy is introduced. That is, signals are delivered **unintelligible** (Not understandable *i.e.*, not clear) at receivers of the unauthorised persons.

These are used in wireless telephones of police and of other investigating agencies.

(*b*) **Requirements of privacy devices:** The privacy devices should fulfil the following requirements:

1. Conversation should be delivered unintelligible to unauthorised persons, even if they employ special receiving equipment.

2. Performance of the circuit should not be affected by the use of a privacy device.

3. Bandwidth of the channel must be maintained same.

4. Time delay of privacy device is maintained as low as possible.

(*c*) **Types of privacy devices:** The different types of privacy devices are:

1. **Speech inversion privacy device:** In this method, the speech band is **inverted** before modulation. This inverted speech is unintelligible and can be used to modulate carrier, then fed to antenna and radiated. In receiver, the signal is demodulated and the original signal is collected by adopting the same procedure as was used for inverting the speech.

   **Example:** If speech band contains frequencies of 200 Hz – 2800 Hz and is used to modulate a carrier of frequency 3000 Hz. This signal contains lower sidebands of frequency 2800 – 200 Hz and upper sidebands from 3200 – 5800 Hz. The lower sideband frequency is an inversion of the original speech. That is 200 Hz is represented by 2800 Hz and 2800 Hz is represented by 200 Hz. Now this signal is used to modulate the carrier of 3000 Hz and radiated through antenna, but the upper sidebands are filtered. In receiver, it is demodulated and corrected.

2. **Split band privacy device (or scrambler system) (See Fig. 11.8):** This system provides an ideal system of privacy which is sufficient for most of the applications. Here the whole speech (0–2800 Hz) is divided into four equal bands and the bandwidth of each band is 2800/4 = 700 Hz.

The first band 1 extends from 0–700 Hz.

The second band 2 extends from 700–1400 Hz.

The third band 3 extends from 1400–2100 Hz.

The fourth band 4 extends from 2100–2800 Hz.

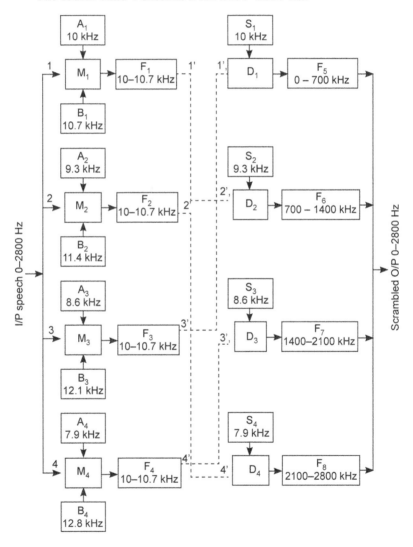

**Fig. 11.8**

With the help of modulators and demodulators, each of these bands is displaced from its original position to a predetermined new position. In Fig. 11.8, the bands 1 and 2 are **transposed**, similarly the bands 3 and 4 are transposed. In this process of displacement, the bands may be inverted, if necessary.

Consider the block diagram shown in Fig. 11.8 in which the speech band is applied parallel to four modulators $M_1$, $M_2$ and $M_3$ and $M_4$ each of which combines with this signal with one of the two associated oscillators $A$ ($A_1$, $A_2$ .........) or $B$ ($B_1$, $B_2$ .........). The output of each modulator is applied to filter' (F1, F2...) which has a pass band of 10 kHz-10.7 kHz in every case.

If oscillator $A_1$ is used, the sidebands of 10 kHz to 12.8 kHz a 10 to 7.2 kHz will be produced at the output of modulator $M_1$ and filter $F_1$ except of 10–10.7 kHz, which is the OH of filter. That is, the speech band 0–700 Hz is converted into 10–10.7 kHz. Similarly $M_2$, $M_3$ and $M_4$ convert 700–1400 Hz, 1400–2100 kHz and 2100–2800 Hz into 10–10.7 kHz respectively. By using $A_1$, $A_2$, $A_3$ and $A_4$, modulating band is contained in upper sidebands. If oscillator $B_1$ is used the output produces sidebands of 10.7–13.5 kHz and 7.9–0.7 kHz. In this case also filter $F_1$ passes a band of 10–10.7 kHz, but the band is inverted.

The second stage of this system consists of four detectors $D_1$, $D_2$, $D_3$ and $D_4$ and four corresponding oscillators $S_1$, $S_2$, $S_3$ and $S_4$. The frequencies of these oscillators are so chosen that output of the four detectors occupy frequency bands 1, 2, 3 and 4 respectively. Filters $F_5$ to $F_8$ select these bands and output of all the four filters are added up to produce the complete speech band of 0–2800 Hz. The connection between points 1', 2', 3', 4' and 1', 2', 3', 4' are made randomly.

This system gives excellent privacy, does not extend the frequency band and involves no time lag. But the system produces poor quality of speech. Also, an elaborate equipment is required and a knowledge of **code jumping** is needed.

3. **Wobbling speech privacy system:** The interception of speech can be prevented by wobbling (moving to and fro) the carrier of the radio transmitter at about ±500 Hz at a very low rate, usually 2 or 3 times per second. Wobbling can be done by placing a rotating condenser at a low rate across tank circuit of the master oscillator. A simple detector cannot receive such wobbled signals. The signal is received by proper method, the carrier is eliminated and the privacy equipment is used to reconstruct the speech.

4. **Displaced speech privacy system:** In this system, the speech band is displaced in frequency by an amount equal to the highest audio frequency in the speech band and then to modulate the carrier frequency. If speech band is from 200–2800 Hz, this band is displaced by 3 kHz, which results the displaced band from 3.2 to 5.8 kHz.

This displacement is achieved by modulating 3 kHz sub-carrier with the speech band of 200-2800 Hz, selecting only the upper sideband (USB) and rejecting the lower sideband (LSB) with a suitable filter. The system provides good degree of privacy, but band width requirement is doubled.

## 11.10 IMAGE FREQUENCY REJECTION

(*a*) In radio receivers, frequency of the local oscillator is kept equal to the signal frequency **plus** the intermediate frequency (IF) at all times. The value of IF is taken as 455 kHz.

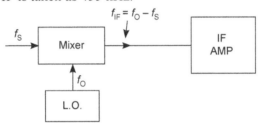

**Fig. 11.9**

If $f_s$ is the signal frequency, $f_{IF}$ is the intermediate frequency, the frequency of the local oscillator (L.O.) should be

$$f_O = f_s + f_{IF}$$

In the mixer, $f_o$ and $f_s$ are mixed resulting in various frequencies in which $f_{IF}$ is one. Only $f_{IF}$ is passed through the I.F. amplifier and all others are rejected.

If somehow a spurious intermediate frequency $f_{SIF} = f_O + f_{IF}$ manages to enter the mixer, then this frequency will also mix with $f_O$ to produce $f_{IF}$ and will also be amplified by the I.F. amplifier, resulting in an "Interference"; *i.e.*, two neighbouring stations will be received simultaneously. The $f_{SIF}$ is called IMAGE FREQUENCY (Spurious) and is equal to the signal frequency plus twice the intermediate frequency *i.e.*,

$$f_{SIF} = f_S + 2f_{IF}$$

(*b*) **Rejection of image frequency.** The image frequency rejection is one of the most important characteristics of RF amplifiers.

The rejection of an image frequency (by a single tuned RF circuit) is defined as the ratio of the gain at the signal frequency to the gain at the image frequency. This ratio is represented by $\alpha$ and given as

$$\alpha = \sqrt{1 + Q^2 \rho^2}$$

where,

$$\rho = \frac{f_{SIF}}{f_S} - \frac{f_S}{f_{SIF}}$$

$f_s$ = signal frequency

$f_{SIF}$ = Spurious Image Frequency $= f_s + 2f_{IF}$

$Q$ = loaded $Q$ factor of the tuned circuit/ antenna

If there are two tuned circuits, in that case the rejection of each circuit is calculated and the total rejection is found by multiplying the both.

The image rejection is governed by the "front end selectivity" of the receiver and the image frequency must be rejected before the IF stage. If the spurious (image) frequency manages to enter the first stage IF amplifier, it becomes impossible to remove it afterwards. If $f_{SIF}/f_S$ is kept large (as in broadband), the RF section can be eliminated, however, the RF section is indispensable in short wave range and above.

The image frequency rejection is not a problem in receivers without RF section, but care must be taken at L.F.

## 11.11   TRACKING AND ALIGNMENT OF RECEIVERS

(*a*) **Tracking.** This is defined with reference to the tuned circuits. Two tuned circuits are said to be "tracking" each other when their resonant frequencies can be made to vary in the same proportion by rotating a common shaft. In a communication receiver, the tuned circuits of RF section and of the local oscillator "track" together.

(*b*) **Alignment.** This is also defined with reference to tuned circuits. Different tuned circuits are said to be aligned together, when they are resonant at the same frequency. Different tuned circuits in the IF section of a communication receiver are aligned.

We should discuss their procedure:

## 11.12 PROCEDURE FOR TRACKING

A superheterodyne receiver has a number of tuned circuits, which have to be properly tuned so as to have a perfect reception. This results in a mechanically coupled system which requires only one knob for tuning. Irrespective of the received frequency, the tuned circuits of RF section and of the mixer / local oscillator must be tuned to the incoming frequency. The local oscillator has to be tuned to a frequency equal to the sum of received frequency and intermediate frequency. If there is any error in frequency difference, a wrong I.F. will be generated. This will give "**Tracking error**."

In fact, it is impossible to be exactly same the tuning of RF oscillator and of the local oscillator. In practice, they are designed to be in step at either end at the centre of the broadcast band. Such alignment is accomplished with **Padder** capacitor associated with local oscillator or **trimmer** capacitor mounted on the gang capacitor or a combination of both may be used.

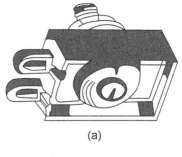

(a)

The Fig. 11.10 shows external appearance of Padder and Trimmer respectively.

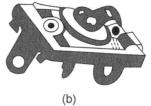

(b)

The Fig. 11.11 (*a*) shows Padder tracking (*b*) shows Trimmer tracking and (*c*) shows the combination method. The associated tracking error is also shown with each method. Here $f_s$ is RF amplifier frequency and $f_o$ is local oscillator frequency. The $P$ stands for Padder and $T$ stands for Trimmer.

**Fig. 11.10**

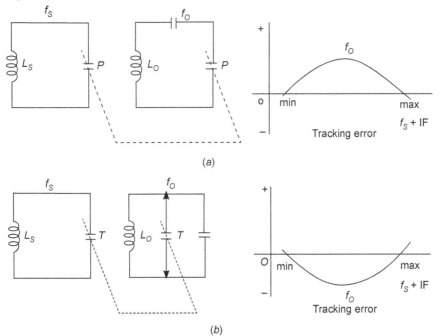

(a)

(b)

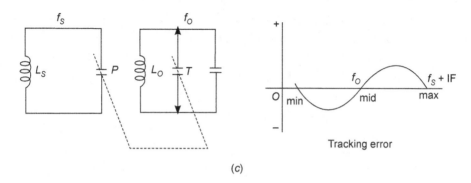

(c)

**Fig. 11.11**

The combination method can be adjusted to give error at three points across the band at each end at the middle. It is also called "Three print Tracking."

The Fig. 11.12 shows curves for correct, misaligned and badly misaligned situations with correct tracking. A tracking error of as low as 2 kHz can be obtained which is acceptable. To obtain correct tracking, coil of the local oscillator should be correctly adjusted, as the capacitor (connected in its series) has a fixed value and cannot be varied.

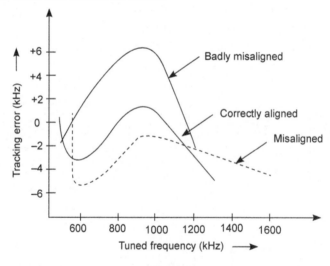

**Fig. 11.12**

## 11.13 PROCEDURE FOR ALIGNMENT

For expeditious and simple tuning, all the tuned circuits must be tuned by movement of a single dial. Such a "single dial tuning" requires that rotor of all these tuning capacitors should be mounted on the same shaft, *i.e.*, the tuned circuits of RF amplifiers and oscillators must be aligned.

1. **Alignment of Tuned circuits of RF amplifier:** Proceed in the following steps:

    (*i*)   Feeding at the receiver input, a signal of selected frequency at the higher frequency end of the tuning dial and tuning the receiver to this frequency and adjusting the RF circuit trimmer to get maximum power at the receiver's output.

    (*ii*)  Repeating the above procedure by feeding a signal of another selected frequency at the lower frequency end of the tuning dial.

    (*iii*) Repeating the above two steps alternately to get the maximum output power at the receiver simultaneously at both these frequencies.

    For M.W. band (extending from 550 kHz – 1650 kHz) we may choose 600 kHz and 1600 kHz as the two signal frequencies for RF tuned circuits alignment.

2. **Procedure for tuned circuits of oscillator:** Proceed in the following steps:

    (*i*)   Feeding at the receiver input, a signal of a selected frequency at the upper end of the tuning band and adjusting the oscillator circuit trimmer to get the maximum power at the receiver's output.

    (*ii*)  Feeding at the receiver input, a signal of another selected frequency at the lower end of the tuning band and adjusting the oscillator circuit padder to get maximum power at the receiver output.

    (*iii*) Repeating the above two steps alternately to get maximum receiver power at both the frequencies.

## 11.14 FREQUENCY CONVERSION/MIXING

(*a*) By frequency conversion, we mean changing of frequency of a carrier by modulation from one frequency to other (Fig. 11.13). This takes place when one signal is "mixed" to the second signal such that output contains product of the two signals. One of the products contains sum and difference frequencies of the two input signals. Other products are also present but we are interested in the first product, as the latter are removed by "Filtering". The output frequency is called intermediate frequency (IF).

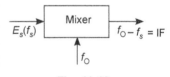

**Fig. 11.13**

(*b*) **Conversion Trans conductance ($g_c$).** The input to a mixer is the signal voltage which has a magnitude $E_s$ at frequency $f_s$. The output is usually a current component ($I_{IF}$) at IF frequency ($f_0 - f_s$). This current is proportional to $E_S$.

i.e., $$I_{IF} = g_c . E_S$$

or $$g_c = \frac{I_{IF}}{E_S}$$

$g_c$ is called "conversion" trans-conductance and is defined as a ratio of output current component and the input signal voltage.

## 11.15  TYPES OF MIXING

The types of mixing are:

 (*i*)  Additive mixing

 (*ii*)  Multiplicative mixing.

We shall study these mixing and mixers.

## 11.16  ADDITIVE MIXING

The input signal frequency with sidebands is simply added to the output of a local oscillator (L.O.) and then $E_s$, $f_s$ passed through a non linear (square law) device such as diode, this gives an almost proportional relationship between input and output signals. The output contains many products such as: same frequency, difference, frequency harmonics and as such it is passed to an IF amplifier, which also, acts as a bandpass filter (Fig. 11.14).

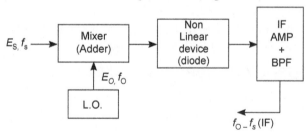

**Fig. 11.14**

The magnitude of $E_o$ is kept very much higher than $E_s$ to obtain a large transconductance and the output is proportional to the input in amplitude. The sidebands are also reproduced around the new carrier frequency without any (amplitude or phase) distortion.

The output of the diode contains a component, which is proportional to the square of the input. The diode may be a separate device from the mixer (adder) or it may be a base emitter junction diode of a transistor.

## 11.17 ADDITIVE MIXERS

Various additive mixers are described below:

(*a*) **Separately excited mixers**

(*i*) **Transistor mixer.** The Fig. 11.15 shows bipolar transistor used to mix (add) input signal with the signal from a *separate local oscillator. The circuit is an example of additive mixing, as the total BE signal voltage is just* sum (or addition) of signal voltage and the local oscillator's voltage.

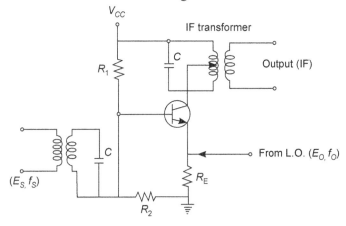

**Fig. 11.15**

The local oscillator voltage $(E_0)$ is kept much more than the signal voltage $(E_s)$ but such that it does not cause clipping. The output current component of the IF output drives primary of a double tuned IF transformer of the IF amplifier. The transistor bias is adjusted by resistors $R_1$ and $R_2$, such that the transistor conducts near **cut off point**.

(*ii*) **Diode mixer.** The Fig. 11.16 shows a diode mixer. The diode is used as a switch and therefore acts a non linear device. This non linearity is used to operate them as mixer.

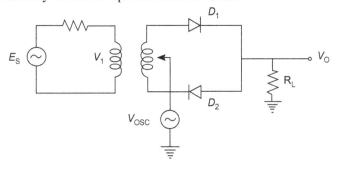

**Fig. 11.16**

The oscillator voltage ($V_{OSC}$) and frequency is much larger than of $V_1$. The oscillator voltage is applied at the centre tap of secondary winding of the input transformer. The diode $D_1$ is forward bias for positive half cycle of the $V_{OSC}$ and $D_2$ is forward bias for negative half cycle of $V_{OSC}$.

For positive $V_{OSC}$, $D_1$ is conducting and $D_2$ is off, the output voltage is given by:

$$V_0 = V_1 + V_{OSC}$$

For negative $V_{OSC}$, $D_2$ is conducting and $D_1$, is off, the output voltage is given by:

$$V_0 = V_1 - V_{OSC}$$

(*iii*)  **Self Excited Additive mixer (FET MIXER).** The Fig. 11.17 shows a self excited mixer. In this configuration, the transistor is used to build an **Hartley oscillator**, whereas the FET is used as a mixer due to square law characteristic of its drain current. The oscillator and mixer (adder) are tuned in "unison" through a gang capacitor. The output of the mixer (adder) is taken through a double tuned intermediate frequency transformer (IFT) and fed to the I.F. amplifier (IFA).

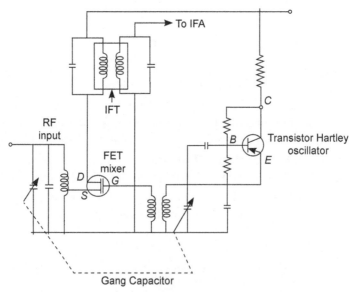

**Fig. 11.17**

For this mixer also, we can use a bipolar transistor instead of FET.

The conversion transconductance of a self excited transistor mixer is of the order of 6 mho.

## 11.18 MULTIPLICATIVE MIXING

In additive mixing, we simply "add" the input signal and local oscillator signal, whereas in multiplicative mixing, we "multiply" the two signals (Fig. 11.18).

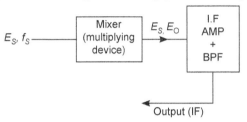

**Fig. 11.18**

The multiplicative mixing occurs when the trans-conductance of the mixer circuit is made to vary with the local oscillator's voltage so that output current is a "function" of the product of $E_S$ and $E_O$.

## 11.19 MULTIPLICATIVE MIXER

The Fig. 11.19 shows a multiplicative mixer which uses a dual gate metal oxide silicon field effect transistor (MOSFET) as the multiplying device.

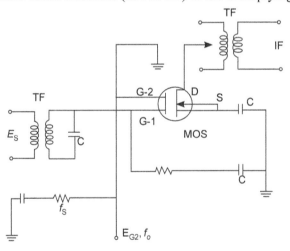

**Fig. 11.19**

As shown, the MOSFET has 2 gates G-1 and G-2 and terminals: source ($S$) and drain ($D$). The input signal ($E_s$) is fed to Gate-1 (G-1), which is biased into active region just as cut off. The oscillator's output ($E_{G2}, f_o$) is fed to the gate-2 (G-2). The bias is also applied on gate 2. In the Fig. TF stands for Transformer. The Fig. 11.20 shows variation of signal gate conductance ($g_m$) as a function of the oscillator gate voltage $E_{G_2}$. Note that this is approximately a straight line.

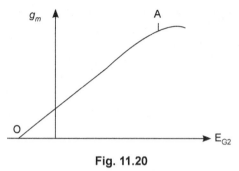

**Fig. 11.20**

## 11.20 RF AMPLIFIERS (RFA)

(*a*) The RF (Radio frequency) amplifiers are used in radio receivers for the following purposes:

    (*i*)   To provide front end selectivity.

    (*ii*)  Amplification.

    (*iii*) To separate incoming signals from the antenna.

    (*iv*) To provide precise band pass filtering required in I.F. amplifiers.

    (*v*)  To provide harmonic filtering.

(*b*) At higher frequencies, frequency stability is a problem, so amplifier with low internal feedback (such as common base amplifiers) are preferred. Also we need to provide neutralisation (compensation) to maintain stability under operating conditions. The R.F. amplifiers operate at 50 kHz to 1000 MHz.

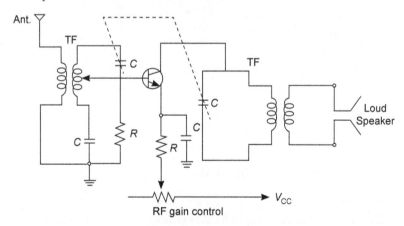

**Fig. 11.21**

A parallel (or series) resonant circuit is the "heart" of an RF amplifier. The gain of the amplifier is related to the output load (such as loudspeaker) Fig. 11.21.

Most commonly used RF amplifies are single tuned and transformer coupled type; as shown in Fig. 11.21.

(c) **Advantages of RF amplifiers:**

   (i)   High gain.

   (ii)  Better sensitivity and selectivity.

   (iii)  Image frequency rejection.

   (iv)  Better SN ratio.

   (v)   Coupling of receiver to the antenna.

The tuned circuit of RFA selects only the tuned frequency and rejects all others including image frequency. So an RF amplifier gives better "Image frequency rejection" (IFR). The gain provided by the RFA results in improved SN ratio, because the incoming weak signal is raised to a higher level before it is fed to mixer, which contributes most of the noise generated by the receiver.

(d) **Neutralisation by RFA**

The process of neutralisation consists of taking some of the output of RFA and feeding it back to the input in such a way as to cancel the positive feedback, that causes oscillations.

Note that neutralisation is not required always. It depends upon many factors such as actual radio frequency, circuit design, shielding methods etc.

Here we will study about methods types of neutralisation.

## 11.21 METHODS/TYPES OF NEUTRALISATION

1. **Hazeltine neutralisation.** In this method, $V_{cc}$ is applied at a tapped point in the resonant/tank circuit. This places the tapped point at RF ground. This makes opposite ends of the tank circuit at 180° out of phase with each other (Fig. 11.22).

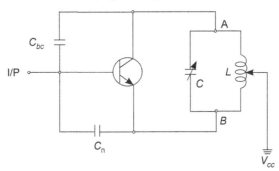

**Fig. 11.22**

The signal at the point $A$ is the collector output and is leaking back through capacitor $C_{bc}$, and causing oscillations. By this arrangement, the leakage signal is cancelled. The neutralising capacitor $C_n$ can be adjusted to control the amount of feedback.

2. **Transformer coupled neutralisation.** The secondary of the transformer provides a signal 180° out of phase of the collector output. This secondary voltage is fedback as the neutralising voltage (Fig. 11.23).

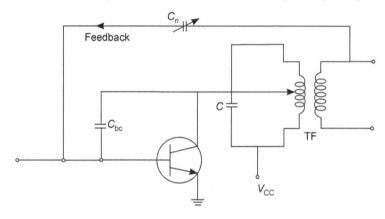

**Fig. 11.23**

3. **Grid (Rice) neutralisation.** In accomplishes neutralization by placing a ground tap on the grid tank circuit. This makes the ends of the tank circuit out of phase of each other (Fig. 11.24).

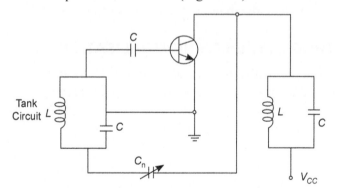

**Fig. 11.24**

4. **Push pull/cross neutralisation.** It is accomplished, as signals at collectors of the two transistors are out of phase with each other. The fed back signal is out of phase with the leaked signal.

As seen, there are two transistors connected in series and two neutralising capacitors, one with each transistor (Fig. 11.25).

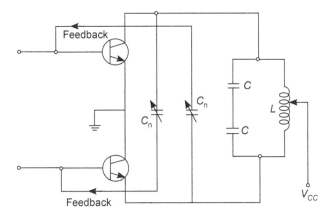

**Fig. 11.25**

## 11.22 INTERMEDIATE FREQUENCY (IF)

Following factors decide the choice of I.F.

(*i*) The IF should not be very high as it creates tracking problems. A high IF is also responsible for poor selectivity and poor adjacent channel rejection.

(*ii*) If IF is low, image frequency rejection (IFR) becomes poor and goes poorer as the ratio of image frequency to signal frequency goes further low.

(*iii*) At a very low value of IF, selectivity becomes too sharp; cutting off the sidebands.

(*iv*) The IF should not fall within the tuning range of the receiver, otherwise instability will occur; and tuning to a frequency adjacent to IF will be impossible.

## 11.23 IF AMPLIFIER (IFA)

An intermediate frequency amplifier amplifies intermediate frequencies. An IFA is a fixed frequency amplifier and rejects the unwanted adjacent frequencies. It has a frequency response with steep cut offs. They may have single tuned or double tuned circuits. When flat top response is desired, double tuned or staggered tuned are preferred.

For fabricating them, vacuum tubes, bipolar transistors, FETs or I.C.s. can be employed. The Fig. 11.26 shows a single tuned two stage IFA for a domestic receiver.

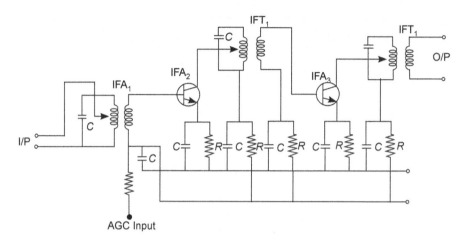

**Fig. 11.26**

The single tuned circuit gives more gain and better sensitivity whereas double tuned circuit gives better "adjacent frequency rejection", but still single tuned circuits are fabricated by bipolar transistors. In the Fig, ACG = Automatic gain control, IFT = IF transformer.

## 11.24 QUADRATURE AMPLITUDE MODULATION

(*a*) The Quadrature Amplitude Modulation (QAM) is also called "Quadrature Carrier Multiplexing" (QCM). and also as "*Band width conservation Scheme*" (BCS).

Further, this scheme enables two DSBSC (Double sideband with suppressed carrier) modulated signals to occupy the same B.W. and therefore, it allows the two signals separation at the receiver; therefore,

(*i*) **QAM Modulator/Transmitter.** [Fig 11.27(a)]. It consists of two separate balanced modulators, which are supplied with two carrier waves of the same frequency and differing by quadrature (*i.e.*, by 90°).

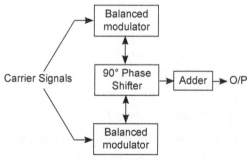

**Fig. 11.27** (a)

The output of the balanced modulators is given to an "Adder" circuit, and we get the output (multiplexed) signal, which is transmitted. This signal consists of "in phase" and the "quadrature phase" components.

(*ii*) **QAM detector/Receiver.** [(Fig 11.27(b)]. The transmitted (multiplexed) signal is applied simultaneously to two separate coherent detectors, which are supplied with two local carriers of same frequency but differing in phase by 90°. The outputs of the detectors are given to low pass filters (LPF) which gives the two original signals.

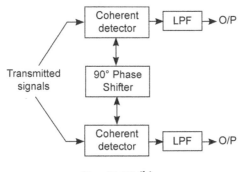

**Fig. 11.27** (b)

For satisfactory operation of the receiver, the phase and frequency of the local oscillators at transmission and reception should **be coherent (or aligned)** to each other.

(*b*) **Application.** The QAM system is used exclusively for digital modulation of analog carriers in data modems (modulator–demodulator) to convey data through public telephone network (PTN). The QAM is also used for digital satellite communication systems and in colour televisions.

## 11.25 FM CHANNEL ALLOCATION

The FM broad band comprises a RF spectrum from 88 MHz to 108 MHz (108 – 88 = 20 MHz = 20000 kHz) dividing into 100 channels of 200 kHz each (20000/100 = 200 kHz).

The available channels are allotted a number from 201 (88 MHz) to 300 (108 MHz). The channel number $n$ is related to carrier frequency $f_c$ by the following relation:

$$n = 5(f_c - 47.9) \text{ MHz}$$

The channels are designed as class *A*, class *B* and class *C*. The class *A* allocation is designated to render service to small city, town and surrounding area. These stations are not authorised to operate with radiated power more than 3 W and antenna height more than 300 feet.

The table 11.1 gives class A channels.

**Table 11.1**   Class A Channel Allocation

| Frequency (MHz) | Channel Number | Frequency (MHz) | Channel number |
|:---:|:---:|:---:|:---:|
| 92.1 | 221 | 100.1 | 261 |
| 92.7 | 224 | 100.9 | 265 |
| 93.5 | 228 | 101.7 | 269 |
| 94.3 | 232 | 102.3 | 272 |
| 95.3 | 237 | 103.1 | 276 |
| 95.9 | 240 | 104.9 | 285 |
| 96.7 | 244 | 105.5 | 288 |
| 97.7 | 249 | 106.3 | 292 |
| 98.3 | 252 | 107.1 | 296 |

Except class *A* channels listed above, all channels (such as 222, 223, 225, 226, 227, 229, 230, 231 etc.) are designated as class *B* and class *C* channels. They render service to bigger cities, towns and the surrounding area. The class *B* and *C* channels are not authorised to operate power more than 50 kW and 100 kW respectively.

## 11.26  STEREO FM TRANSMITTER AND RECEIVER

(*a*) **Stereo FM transmitter:** Stereo FM transmission is a modulation system, in which sufficient information is sent to the receiver to enable it to reproduce original stereo material.

The audio is picked up by two microphones located to emphasize different sections (Left and right). The microphone output are combined in two distinctly different manners. The monaural (Mono phonic) mixer is a straight forward mixer of the two channels. This results is sum of the channels $(L + R)$ in the frequency range 0 to 15 kHz.

The second channel receives a more sophisticated treatment. The output of the stereo mixer is the difference of the two microphones $(L - R)$ in the frequency range 0 to 15 kHz . This is fed to the balanced modulator. Also fed to the balanced modulator is a 38 kHz carrier provided by a sub carrier generator. The balance modulator mixes the audio and 38 kHz carrier and creates ± sidebands mixing 0 to 15 kHz with 38 kHz resulting in a lower sideband of 23 to 38 kHz and upper sideband of 38 to 53 kHz. At the same time, the original 38 kHz carrier is eliminated (Fig. 11.28).

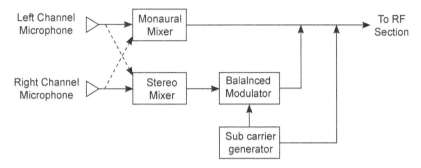

**Fig. 11.28**

The final frequency groups modulating the FM stereo transmitter 0 to 15 kHz representing the sum of the two channels; 23 to 53 kHz, representing the difference of the two channels; and the 19 kHz signal derived from the 38 kHz sub carrier and used as a synchronizing or local oscillator carrier at the receiver. The above stereo system permits compatible monophonic reception by FM receivers that do not have stereo capability.

(b) **Stereo FM receiver:** The stereo receiver reverses the process of transmitter. The $(L + R)$ signals confined in $0 - 15$ kHz frequency range is directed to a low pass filter (Fig. 11.29).

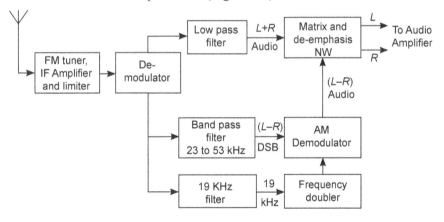

**Fig. 11.29**

The $L - R$ signal contained in the 23 to 53 kHz frequency range is extracted by an appropriate band pass filter. The 19 kHz pilot carrier is extracted by appropriate filter and converted to 38 kHz in a frequency doubler. The 38 kHz carrier is mixed with $L - R$ signal in an AM demodulator. Here it functions as the carrier for the $L - R$ double sideband (DSB). The output of the AM demodulator is $L - R$ audio signal. The output combined with the $L + R$ audio signals is mixed in a de-emphasis network. The output of this stage are the original $L + R$ channel information.          .

## 11.27  SSB RECEIVERS

The SSB receivers demodulate the SSB signals and process them. The SSB receivers are not used as broadcast receivers. They are used to receive signal in crowded frequency bands such as short wave bands. So, these are usually made double (or multi) conversion type. Their special qualities are:

  (*i*)  high reliability.

 (*ii*)  simple maintenance.

(*iii*)  ability to demodulate SSB signals.

(*iv*)  suppression of adjacent channel signals.

  (*v*)  high SN ratio.

(*vi*)  In case of independent sideband (ISD) receivers, it should be capable to separate independent sidebands.

## 11.28  GENERAL SSB RECEIVER

The Fig. 11.30 shows block diagram of SSB receiver. The input sideband is received from receiving antenna and fed to RF amplifier. The RF amplifier amplifies the SSB signal with frequency band of 15.100–15.105 MHz (USB).

The output is fed to first mixer where the signal is hetrodyned with the input voltage of the first local oscillator (L.O.) of frequency 12 MHz and produces the output with a frequency of 3.100–3.105 MHz.

The output of the first mixer is amplified in the **first I.F. amplifier and** fed to second mixer.

The second mixer is also fed with output of **second local oscillator** (L.O.) of frequency 3 MHz, so that it produces an IF signal of 100–105 kHz.

This 100–105 kHz IF signal is amplified by second IF amplifier and the signal is fed to final detector and crystal filter.

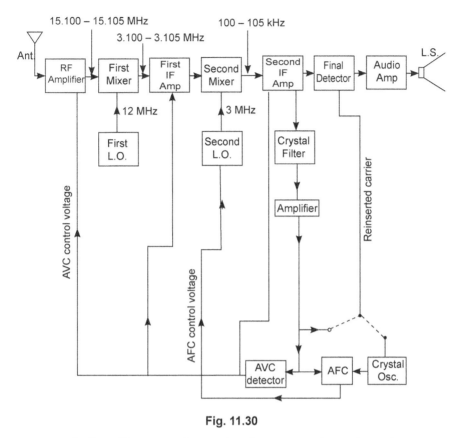

**Fig. 11.30**

The crystal filter filters out the carrier frequency of 100 kHz from upper sideband (USB) and is amplified by amplifier and fed to automatic frequency control (AFC) system.

The AFC system produces AFC control voltage with frequency difference between the **reinserted carrier** and local oscillator signal. The AFC control voltage changes the second local oscillator frequency to maintain the **synchronism** between the crystal oscillator frequency and received carrier frequency. When this condition is achieved, the received signal is fed to final detector.

The final detector reconstructs the original signal and fed to audio amplifier.

The audio amplifier raises the strength of the detected signal to the required level which is fed to the loudspeaker.

## 11.29  SSB RECEIVER WITH SQUELCH AND BFO (DOUBLE CONVERSION SYSTEM)

(*a*)  A double conversion system has two mixers. When a communication receiver has two I.F.s, it is said to have a double conversion system.

The communication receivers which require a very high quality performance, use double conversion system. The first I.F. is quite high (in the range of several MHz) and the other is quite low (in the range of few kHz). The output of RF amplifier is mixed into HF mixer with the frequency of the local oscillator I. The output of HF mixer is higher than 455 kHz. This frequency is amplified by a high frequency amplifier (HFA) and its output is given to LF mixer, where it is mixed with the frequency of local oscillator II. The output of the LF mixer is a low IF signal, which is detected as usual. The Fig. 11.31 shows a double conversion type receiver.

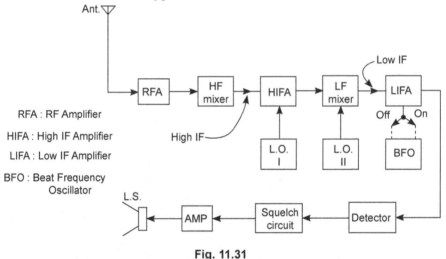

**Fig. 11.31**

The high IF gives image frequency rejection and allows its better attenuation. The low IF gives a good selectivity and adjacent channel rejection. Note that high IF should come first, otherwise image frequency rejection will not be sufficient and will be mixed with the signal. Any number of IF stages after this, will not be able to reject the image frequency.

The method is not used for domestic receivers or receivers working on medium frequency band. The short wave band receivers however can use this system.

(b)  The additional systems used in double conversion receivers are:

   (i)  **Squelch (Mute or Quiet) System:** When a signal is absent *i.e.*, no transmission, a sensitive receiver will produce high noise. The reason is that in the absence of a signal, AGC (automatic gain control) disappears, causing the receiver to operate at its maximum sensitivity. The low level of the noise at the input of the receiver gets amplified and which is very much uncomfortable specially in case of police or ambulance receivers. To solve this problem, a squelch or mute or quiet circuit is used in the receiver; which enables the receiver's output to remain cut off unless the signal is present.

(*ii*) **BFO (Beat Frequency Oscillator):** The communication receivers should be able to receive transmission in morse code. In order to make **dots**, **dash** and **spaces** in morse code audiable, we use a BFO. This is a simple hartley type oscillator which operates at a frequency of 1 kHz above or below the last intermediate frequency. When the IF is present, a whistle is heard in the loud speaker and dots or dashes can be heard. To avoid interference, the BFO is switched off otherwise. The Fig. 11.32 shows block diagram of B.F.O. It gives a very large frequency range with a single dial rotation.

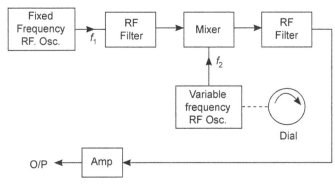

**Fig. 11.32**

At the output of the mixer, we get sum and difference of $f_1$ and $f_2$. The BFO is usually affected by spurious beat notes called "whistles". These whistles appear, when IF is obtained.

## 11.30  TYPES OF SSB RECEIVERS

The SSB receivers are of the following types:

1. Pilot carrier receiver

2. Suppressed carrier receiver.

These are discussed below:

## 11.31  PILOT CARRIER SSB DEMODULATOR/RECEIVER

(*a*) **Principle.** In this modulator (Fig. 11.33), 'a pilot carrier' is used at the receiver to synchronise the local oscillator used for demodulation to the original carrier, this improves operation of the demodulator.

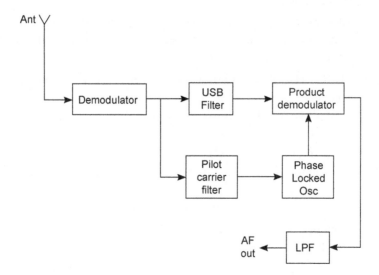

**Fig. 11.33**

At the receiver 3 MHz signal is picked up, amplified and converted to I.F. at $100 \pm 5$ kHz, producing pilot carrier as 100 kHz and upper sideband (USB) in the range of $100 - 104$ kHz. The output of receiver is passed through USB filter and then to the product demodulator. A "phase lock oscillator" produces 100 kHz carrier for the demodulator. A final low pass filter (LPF) removes the sum component of the demodulator, leaving 0 to 4 kHz AF signal.

(b) **Block diagram.** This receiver is a double conversion type with a squelch (mute) circuit (Fig. 11.34). It also has an AFC system which provides a good frequency stability. The local oscillator is used with a multiplier.

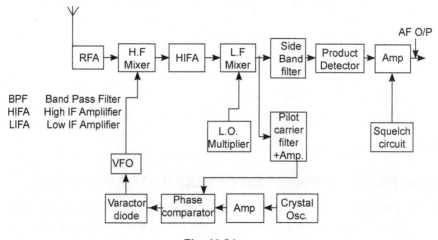

**Fig. 11.34**

The output of L.F. mixer contains the wanted sideband and the pilot carrier. The sideband goes to filter then to product detector and then to the amplifier. The pilot carrier is filtered and amplified separately. A phase comparator is also provided which has two inputs: one from pilot carrier amplifier and other from the crystal oscillator. The comparator compares these two input frequencies, and gives an output, which will be zero, if the frequencies are equal. This ideal condition gives very good frequency stability. The output of the comparator is given to a varactor diode and then to VFO (Variable Frequency Oscillator) and then to the H.F. mixer.

## 11.32 SUPPRESSED CARRIER/ISB RECEIVER

The Fig. 11.35 shows block diagram of a suppressed carrier SSB receiver which is used for ISB (Independent sideband) reception. Its RF amplifier is a wideband amplifier covering a range of 15 kHz to 30 MHz. The first intermediate frequency is very high say 35 MHz. A high IF provides much higher image frequency rejection, which is very important in SSB receivers. Frequency synthesiser provides a high frequency stability. Upto low frequency IF stage, this receiver is similar to an ordinary double conversion type receiver, after this, difference arises due to the presence of two independent sidebands (ISBs), which are separated after the LF mixer. The upper ISB is filtered by a bandpass filter and IF is amplified and after the product detector stage, audio frequency (AF) output (say channel A) is obtained.

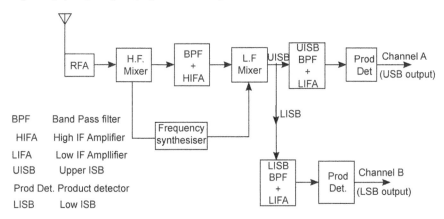

**Fig. 11.35**

The lower ISB is filtered through a separate band pass filter, amplified and after the product detector stage, AF output (say channel B) is obtained. In this way, both independent sidebands are obtained, note that carrier is absent in both the outputs.

## 11.33  TRANS-RECEIVER FOR SSB SIGNALS

This circuit can be used as a modulator at transmitter side and as a demodulator at the receiver side, and can be switched on to either side as per requirement, so it is preferred to a product demodulator. This is used in trans-receivers. (Transmitter plus receiver)

See Fig. 11.36. When used as a demodulator, SSB signal is fed at terminals (1, 2). The circuit behaves as a 'non linear resistance' giving sum and difference frequencies. The transformer blocks the RF frequencies and allows only AF frequencies at terminals (3, 4).

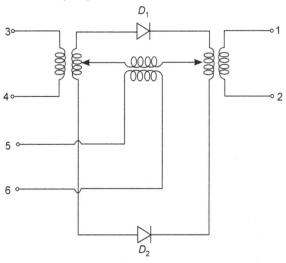

**Fig. 11.36**

When used as a modulator input is given at terminals (5, 6) from crystal (local) oscillator, which has a frequency of about 200 kHz. The modulated output is obtained across terminals (1, 2).

|            |               |
|------------|---------------|
| demodulator | Input at 1, 2 |
|            | Output at 3, 4 |
| modulator  | Input at 5, 6 |
|            | Output at 1, 2 |

## 11.34  COHERENT AND NON COHERENT SSB DETECTION

In Coherent detection phase of the carrier generated by the local oscillator of the receiver should be identical (coherent or synchronised) with the carrier generated by the local oscillator of the transmitter. Otherwise, the detected signal will be distorted; therefore, this method is called "Coherent or synchronous" method of detection.

In non coherent (Non synchronous) detection, phase of the local oscillators on the two sides is not identical and the detected signal is distorted.

### (a) Coherent Detection of Single Tone DSBSC Wave

The Fig. 11.37 shows a Coherent SSB detector. The modulated SSB signal is first multiplied with the locally generated carrier and then passed through a low pass filter (LPF). The output is the original modulating signal.

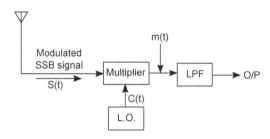

**Fig. 11.37**

The multiplier output is given by

$$m(t) = S(t) \times c(t) \qquad \qquad ...(1)$$

Where $\qquad S(t) = $ SSB wave

$$= \frac{1}{2} V_m V_c [\cos 2\pi (f_c + f_m)t + \cos 2\pi (f_c - f_m)t]$$

and $\qquad c(t) = $ local carrier output

$$= \cos (2\pi f_c t)$$

Substituting the expressions for $S(t)$ and $c(t)$ in eq. (1) we obtain

$$m(t) = \cos (2\pi f_c t) [\frac{1}{2} V_m V_c \cos \{2\pi(f_c + f_m)t\} + \cos 2\pi (f_c - f_m)t]$$

Solving above we get,

$$m(t) = \frac{1}{2} V_m V_c [\cos \{2\pi(2f_c + f_m)t\} + \cos (2\pi f_m t)$$
$$+ \frac{1}{2} V_m V_c [\cos 2\pi (2f_c - f_m)t] + \cos (2\pi f_m t)]$$

The output

$$V_0(t) = \frac{1}{2} V_m V_c \cos(2\pi f_m)t$$

Note that frequencies $(2f_c + f_m)$ and $(2f_c - f_m)$ are removed by low pass filter.

(*b*)  Non **Coherent detection:**

**The disadvantage** of coherent method is that, it requires an additional system at the receiver to ensure the synchronisation. This makes the receiver complex and costly. So this method is not used.

## SUMMARY

1.  The AM transmitters are provided with negative feedback.

2.  The frequency of modulator may vary, this is called "Frequency drift".

3.  In radio telephony, privacy devices are provided so that the conversation is not reached to unauthorised persons.

4.  The important types of privacy devices are:

    (*i*)    Speech inversion privacy devices

    (*ii*)   Split band privacy devices.

5.  Image frequency rejection is one of the most important characteristics of RF amplifiers.

6.  The mixers may be

    (*i*)    Additive mixers

    (*ii*)   Multiplicative mixers

7.  The SSB receivers may be

    (*i*)    Coherent receivers

    (*ii*)   Non coherent receivers

❑❑❑

# Analog Pulse Modulation

The pulse communication is one of the fastest area of electronic communication. Its concept goes back as far as to 1812 A.D., but its growth has been only in the last decade after availability of inexpensive I.C.s.

## 12.1 PULSE

A pulse is an abruptly changing voltage or current wave which may or may not repeat itself. The simplest non repetitive pulse is a stepped up voltage or current shown in Fig. 12.1 (*a*) which can be obtained by connecting a voltmeter across a battery through a switch and then suddenly closing the switch. The voltmeter will read Zero upto a time when the switch is closed, where upon the voltage will suddenly rise to its maximum value and will stay there.

The Fig. 12.1 (*b*) shows a repetitive pulse train and Fig. 12.1 (*c*) shows a pulse with its trailing leading edge.

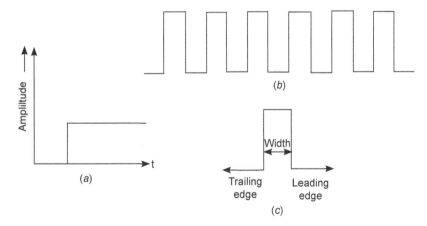

**Fig. 12.1**

## 12.2  PULSE MODULATION

It may be defined as a modulation system in which some parameter of a pulse train is varied in accordance with the instantaneous value of the modulating signal i.e. a pulse train is used as the carrier.

In this system, waveforms are sampled at regular intervals and the information is transmitted at the sampling rate.

The parameters of the pulses which may be varied are : amplitude, width (or duration) position and time.

## 12.3  QUANTIZING/QUANTIZATION

In pulse modulation, pulses result from sampling the modulating signal wave.

In other words, the modulating wave is sliced into small units, the process is called quantizing or quantization. These quantum points are then converted into digital binary codes, which represents amplitude of the wave at that point (Fig. 12.2).

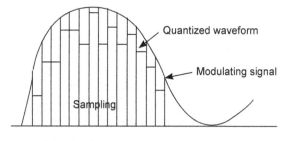

**Fig. 12.2**

## 12.4  PULSE COMMUNICATION SYSTEM

The AM, FM and PM systems are called CW (continuous wave) modulation systems, in which some parameter (amplitude, frequency, phase) of a continuous high frequency carrier wave is varied in accordance to the modulating signal. But in pulse modulation systems, instead of a continuous wave, a train of pulse is employed and some parameter of the pulse is varied in accordance with the instantaneous value of the modulating signal.

The pulses are quite short as compared to the time in between, so a pulse modulated wave **remains off** most of the time. The time interval between the pulses may be filled with sample of waves from other messages, so we can send many messages at a time on a pulse communication system.

So long, we have used sinusoidal carrier. In pulse communication, we use *rectangular pulses* as carrier and one of the parameters of the pulse (amplitude, width, position) is varied according to the signal. The characteristic of pulse communication is that the information remains at base band and is not translated to higher frequency carrier.

One of the chief advantages of pulse modulation is that if we combine pulse modulation with continuous wave modulation (AM, FM, PM), we can obtain "multi channel" communication system, a desirable feature for "data transmission". Recall that transmission of more than one channel at one carrier is called "multiplexing"; we are aware of "Time" and "frequency" division multiplexing (TDM and FDM).

## 12.5 CONCEPT OF SAMPLING

In electronic communications, **sampling** is the process of taking periodic samples of the wave form to be transmitted and then transmitting these samples. If enough samples are sent, the complete wave form can be received at the receiver.

The concept of sampling can be explained by the following example (Fig. 12.3 ).

Suppose a factory has four processing vats, each having a thermometer which is to be carefully monitored. This can be done in two ways.

   (*a*)  To receive **continuous** data of thermometers, four workers are required. Since any change in the temperatures will be gradual, this will not be economical. [Fig 12.3(*a*)]

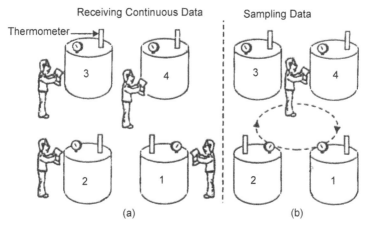

Fig. 12.3

   (*b*)  The **sampling** data will be more economical, as only one worker can monitor the data of all the four thermometers. If this single worker can take the data samples faster, that the thermometers reading can change, the same effect of continuous sampling can be achieved at a reduced cost (In this case one fourth). [Fig 12.3(*b*)]

The situation is similar to transmitting and receiving in electronic pulse communications. Clearly, sampling data, rather than continuous monitoring produces more efficiency and makes it possible to send more than one information on one single carrier. Note that only one transmitter and one receiver shall be needed.

## 12.6  SAMPLING ELECTRONIC SIGNALS (OR TDM)

Let us think that three different signals are to be sent over a single wire. By sampling technique, this can be done as shown in Fig. 12.4. The switch of **transmitter** as well as of the **receiver** are changing in such a manner that both attain same position at a time *i.e.*, when transmitter is at position A, the receiver should also be at *A* and so on. The **recorders** should be of slow response time *i.e.*, much less than the sampling rate.

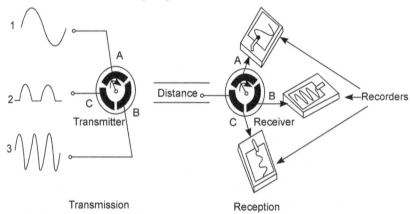

**Fig. 12.4**

This sampling is known as **Time Division Multiplexing (TDM)**, because multiple signals are sent by sampling them at different times.

## 12.7  SAMPLING THEOREM FOR LOW PASS SIGNALS: NYQUIST THEOREM

(*a*) If more no. of samples are taken, the information can be reproduced correctly. The other side is also correct, *i.e.*, if fewer samples of one information are taken, in between other information can also be sent. This is similar as in our example in which one person is reading temperature of several thermometers, lesser the time the person spends reading one thermometer, more time he has left to read other thermometers, or we can say to get other information.

Here, **Nyquist theorem** is to help, it says: *In order to convey an information completely, The minimum sampling frequency of a pulse modulated system, should be equal to (or more than twice) the highest signal frequency. Mathematically.*

$$f_s \geq 2f_m$$

Where  $f_s$  =  Minimum sampling frequency to convey complete information.

$f_m$  =  Maximum frequency component present in the information signal.

*e.g.* The minimum sampling frequency to transmit a pure sine wave of 2 kHz should be

$$= f_s = 2f_m = 2 \times 2 \text{ kHz} = 4 \text{ kHz}$$

(*b*) **Proof:** The Fig. 12.5 (*a*) shows continuous time signal $x(t)$. The Fig. 12.5 (*b*) shows its frequency spectrum. The Fig. 12.5 (*c*) shows impulse train. The sampling of $x(t)$ at the rate of $f_s$ Hz (samples) per second can be done by multiplying it with the impulse train. The impulse train consists of impulses repeating periodically every $T_s$ (sampling time) seconds, where $T_s = 1/f_s$. If. The Fig. 12.5 (*d*) shows the resulted sampled signal. The Fig. 12.5 (*e*) shows fourier form of the sampled signal.

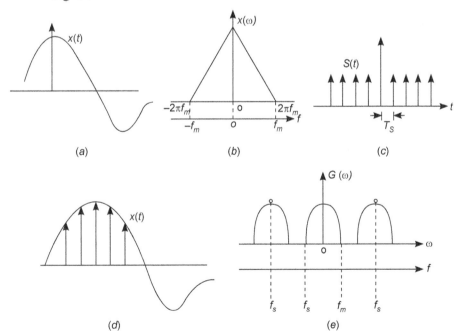

**Fig. 12.5**

The Fourier expansion of impulse train can be expressed as (Fig. 12.5)

$$S(t) = \frac{1}{T_s} [1 + 2 \cos \omega_s t + 2 \cos 2 \omega_s t + 2 \cos 3 \omega_s t + ...] \quad ...(1)$$

The sampled signal can be expressed as :

$$g(t) = x(t). S(t) \qquad ...(2)$$

Putting value of pulse train from eq. (1) in equation (2), urgent

$$g(t) = \frac{1}{T_s} [x(t) + 2x(t) \cos \omega_s t + 2x(t) \cos 2 \omega_s t + ...] \quad ...(3)$$

By taking Fourier transformation, the equation (3) becomes as:

$$G(\omega) = \frac{1}{T_s} [x(\omega) + x(\omega - \omega_s) + x(\omega + \omega_s) + x(\omega - 2\omega_s) + x(\omega + 2\omega_s t + ...)] ... (4)$$

$$= \frac{1}{T_s} \sum_{n=-\infty}^{\infty} x(\omega - n\omega_s) \qquad ...(5)$$

From Eqs. (4) and (5), it is clear that the spectrum $G(\omega)$ consists of $x(\omega)$ repeating periodically with period $\omega_s = 2\pi/T_s$ rad/sec. or $f_s = 2f_m$ Hz.

It shows that a signal where spectrum is band limited to $f_m$ Hz (*i.e.* the signal has no frequency components beyond $f_m$ or it has maximum frequency $f_m$) can be reconstructed from its samples taken at the rate $f_s > 2f_m$ Hz.

This proves the sampling theorem.

## 12.8 EFFECTS OF SAMPLING RATE ON A FREQUENCY SPECTRUM

The Fig. 12.6 shows frequency spectrum of the modulating wave. Note that higher order harmonics are smaller in amplitude.

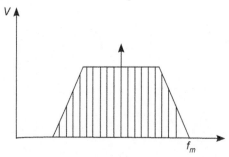

**Fig. 12.6**

(*a*) **When $f_s = 2f_m$:** The Fig. 12.7 (*a*) shows what happens when the sampling frequency $f_s$ is twice the maximum frequency component ($f_m$) present in the modulating wave. Theoretically the harmonics of the sampling extend to infinity, but practically the resulting spectrum needs only be passed through a **low pass** filter.

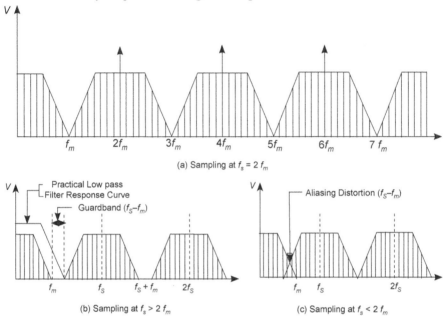

(a) Sampling at $f_s = 2\,f_m$

(b) Sampling at $f_s > 2\,f_m$

(c) Sampling at $f_s < 2\,f_m$

**Fig. 12.7**

(*b*) **When $f_s > 2f_m$:** If the sampling rate is made slightly larger than $2f_m$, a practical low pass filter can be used to pass only the maximum frequency component ($2f_m$) and not any other frequency component. The high sampling rate creates a **guard band** between $f_m$ and the lowest frequency components ($f_s - f_m$) of the sampling harmonics. See Fig. 12.8 (*b*).

(*c*) **When $f_s < 2f_m$ :** If the sampling rate is lower than $2f_m$, a distorted sampling spectrum is obtained. The distortion is known as **Aliasing distortion**. See Fig. 12.8 (*c*).

## 12.9 SAMPLING TECHNIQUES

There are three sampling techniques:

1. Ideal, Instantaneous or Impulse sampling.

2. Natural sampling.

3. Flat top sampling.

A comparison of the three is given below :

| S. No. | Parameters | Ideal sampling | Natural sampling | Flat top sampling |
|---|---|---|---|---|
| 1. | Principle | It uses multiplication | It uses multiplication or chopping principle. | It uses sample hold principle. |
| 2. | Generation circuit | | | |
| 3. | Waveform | | | |
| 4. | Feasibility | This method is not practically feasible. | This is used practically. | This is most popular method. |
| 5. | Sampling rate | Sampling rate may be infinity. | The sampling satisfies the Nyquist theorem. | The sampling satisfies the Nyguist theorem |
| 6. | Noise | Maximum. | Minimum. | Minimum. |

## 12.10  CLASSIFICATION OF ANALOG PULSE MODULATION SYSTEMS

Analog pulse modulation may be of two types: quantized or unquantized.

Considering either category, there are three parameters of a pulse, that can be varied.

(*i*)  Amplitude

(*ii*)  Width

(*iii*)  Position.

Accordingly, the Pulse modulation may be subdivided as: (Fig. 12.8)

1. **Pulse amplitude modulation (PAM):** In this, amplitude of a pulse train (carrier) is varied according to the amplitude of the modulating signal.

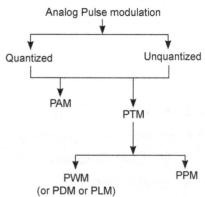

Fig. 12.8

2. **Pulse time modulation (PTM):** In this, timing of the pulse train is varied. There are two type of PTMs.

    (*a*) **Pulse width modulation (PWM):** This is also called as Pulse duration modulation (PDM) or pulse length modulation (PLM).

        In this, width (depth/length) of the carrier pulse train is varied according to the modulating signal.

    (*b*) **Pulse position modulation (PPM):** In this, position of pulses of the carrier pulse train is varied according to the modulating signal.

        These are discussed below in brief:

## 12.11 PAM (PULSE AMPLITUDE MODULATION)

(*a*) The PAM (Pulse amplitude modulation) is a modulation mode, in which amplitude of the carrier pulse train is varied according to the amplitude of the modulating signal.

(*b*) **Types of PAM:** There are two types of PAM :

    1. **Natural PAM:** Natural PAM sampling occurs when finite width pulses are used in the modulation and tops of pulses follow the modulating signal.

    Since a pulse is composed of infinite number of components, the circuit which processes a pulse must have infinite bandwidth. For practical purposes, the bandwidth should be such that it includes all the frequency components which have some bearing on the intelligence to be carried. [Fig 12.9(*a*)]

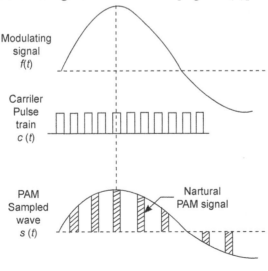

**Fig. 12.9(a)**

2.  **Flat top PAM:** The circuit for natural PAM is complicated. As the pulse top shape is to be maintained; so flat top PAM is preferred. In this, tops of the pulses are flat. The pulses have a constant amplitude within the pulse interval. [Fig. 12.9 (*b*)]

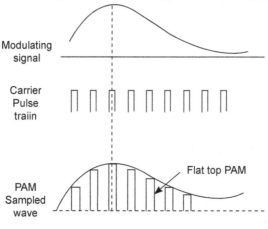

**Fig. 12.9(b)**

When the modulated waveform is reduced from the flat top waveform, through low pass filtering, it shows distortion. This distortion is negligible when the width is much less than the sampling period.

The flat top PAM is most popular and widely used, because during transmission the noise interferes with the top of the pulses. This noise can be removed easily if pulses have a flat top.

## Pam Transmitter/Generator

(*a*) **Natural Transmitter:** The Fig. 12.10. (*a*) shows schematic arrangement of natural PAM transmitter. A continuous modulating signal $f(t)$ can be sampled by a train of pulses represented by $C(t)$ in a sampler (or multiplier). The output is a PAM signal represented by $S(t) = f(t). C(t)$.

The sampled signal can be directly transmitted through a pair of wires and no further necessary processing. But if it is to be transmitted through the space using an antenna, it should be first amplitude (or frequency or phase) modulated

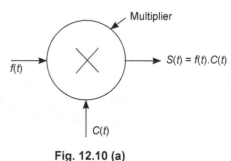

**Fig. 12.10 (a)**

by a high frequency carrier and then can be transmitted. The overall system is then termed as PAM-AM ; PAM-FM or PAM-PM respectively. At the receiving end AM, FM or PM detection is done to get PAM signal and the message is recovered.

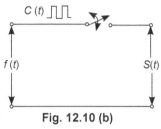

The Fig. 12.10 (*b*) shows a functional diagram of a natural transmitter. A sampled signal $S(t)$ is obtained by multiplication of input signal $f(t)$ and the carrier $C(t)$.

**Fig. 12.10 (b)**

(*b*) **Sample Hold (SH) circuit for generation of (flat top) PAM:** The Fig. 12.11 shows a SH circuit for generation of (Flat top) PAM signal.

The circuit consists of two FET switches and a capacitor, one FET (field effect transistor) acts as a sampling switch ($S_1$) and the other FET acts as a discharge switch ($S_2$). When $S_1$ is closed for short duration by a sampled pulse of the modulating signal applied at its gate (G), the capacitor $C$ is quickly charged upto a voltage equal to the instantaneous value of the incoming modulating signal. Now $S_1$ is opened and the capacitor holds the charge.

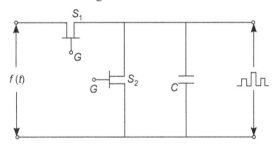

**Fig. 12.11**

Now the switch $S_2$ is closed by next pulse applied at its gate. The capacitor is discharged completely. The $S_2$ is now opened and we get the (flat top) PAM pulse.

## PAM Modulator and Demodulator

(*a*) **PAM modulator (Fig 12.12).** The PAM modulator is a simple "Emitter Follower" circuit. The modulating signal is applied at the input. At the base a, CLOCK signal is applied. The frequency of the clock signal is made equal to the frequency of carrier pulse train.

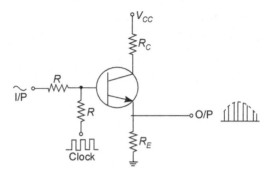

**Fig. 12.12**

When the CLOCK signal is "high", the circuit behaves as "Emitter follower" and the output follows the input (modulating) signal, when the CLOCK is "low", the transistor is "cut off" and the output is zero. In this way, at the output we get PAM signal.

(b) **PAM demodulator:** As we know, demodulation is reverse of modulation and is the process to recover the original modulating signal.

For this, the PAM signal is given to a sample and holding circuit (SH) as shown in Fig. 12.13.

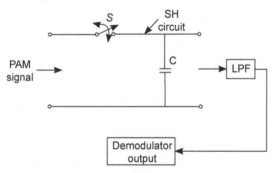

**Fig. 12.13**

The switch $S$ of the sample holding circuit is closed after arrival of a pulse of the PAM signal and is opened at the end of the pulse. In this way, the capacitor $C$ is charged and it holds this charge till the next pulse arrives. The output of the holding circuit is smoothened by passing it through a low pass filter (LPF) to get the final demodulated output.

(c) **Disadvantages of PAM:** The advantages have already been discussed. Following are the disadvantages of pulse amplitude modulation.

(i) The BW required for transmission of a PAM signal is very large as compared to the maximum frequency present in the modulating signal.

(*ii*) As the amplitude of the carrier pulses is varied according to the amplitude of the modulating signal, noise is maximum in the PAM wave.

(*iii*) As the amplitude of the PAM signal is varied, this also varies peak power required by the transmitter.

## 12.12 PULSE TIME MODULATION (PTM)

(*a*) In PTM, amplitude of the pulse is kept constant, where as width or position of the pulse is made proportional to the amplitude of the signal at the sampling instant. As amplitude is kept constant and does not carry any information, amplitude limiters can be used. The limiter similar to that used in FM will clip off the portion of the signal corrupted by noise and provides a good degree of noise immunity.

(*b*) **The PTM has two types:**

1. Pulse width modulation (PWM)
2. Pulse position modulation (PPM)

In both PWM and PPM, the amplitude is kept constant.

In PWM, long pulses expend considerable power while bearing no additional information. If this unused power is substrated from PWM, so that only time transitions are preserved we obtain PPM. In this way, PPM is more efficient than PWM. We will discuss both one by one:

## 12.13 PWM (PDM OR PLM)

In pulse width modulation (PWM), the width of the carrier pulse is varied according to the instantaneous value of the modulating signal, while the amplitude of the carrier remains constant. This system is also called "Pulse duration modulation" (PDM) or "Pulse length modulation" (PLM).

In this system, the amplitude of the modulating signal (*i.e.* intelligence) restricted, so that one pulse may not run into the next pulse, causing an overlap.

**We can obtain PWM** from **PAM** because when compared to PAM, we see that the information has been converted to "Time amplitude".

The PWM may be classified as:

1. **Symmetrical PWM Fig. 12.14 (*i*)** : In this type, the trailing as well as leading edges of the pulses are varied in accordance with the amplitude of the modulating signal. When the modulating signal is zero, the pulse has a **reference width**. So long the signal is positive, the pulse width increases. When the signal is negative the pulse width decreases, however the spacing between the pulses remains same.

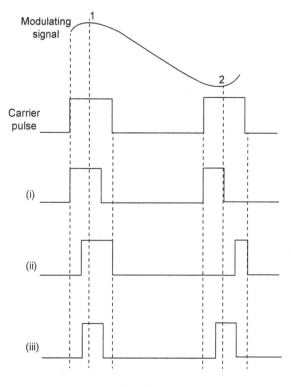

**Fig. 12.14**

2. **Trailing Edge PWM Fig. 12.14 (*ii*)** : In this type, the trailing edge of the pulses is varied according to the amplitude of the modulating signal. The leading edge of the pulses, however remains at a fixed rate w.r.t. each other, the timing between each leading pulse edge, therefore is constant.

3. **Leading Edge PWM Fig. 12.14 (*iii*):** In this type of PWM, the leading edge of each pulse changes according to the amplitude of the modulating signal, the trailing edge of each pulse however, is fixed and timing in between the trailing edges is constant.

## Modulation and Demodulation of PWM

(*a*) The Fig. 12.15 shows a method of generating PWM. A saw tooth waveform whose period is equal to the largest pulse width to be produced is added to the information signal. A level detector and a.c. pulse shaper are used to produce PWM wave.

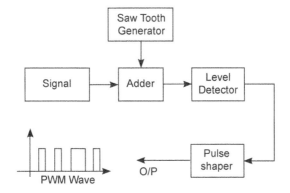

**Fig. 12.15**

(*b*) **Demodulation of PWM: (Fig 12.16)** The PWM wave is first given to a differentiator. The output of the differentiator is fed to positive and negative clippers. The waves obtained as output from the clippers are given to a ramp generator which when ON, produces a saw tooth output. (When it is OFF the output returns to zero).

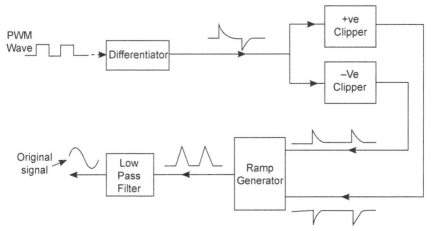

**Fig. 12.16**

This saw tooth output is given to an Integrator (or low pass filter) and the original modulating signal is obtained.

(*c*) **Advantages of PWM:**

    (*i*)  Noise is less in PWM as the amplitude is kept constant. Note that PAM is noisy.

    (*ii*)  The signal and noise separation is easy.

    (*iii*)  The PWM does not require synchronization between transmitter and receiver.

(*d*) **Disadvantages:**

 (*i*)  Large BW is required for PWM communication as compared to PAM.

 (*ii*)  The transmitter should be able to handle more power (equal to the power of the maximum width of pulse).

## 12.14 PPM (PULSE POSITION MODULATION)

In this system, amplitude and width of the carrier pulses are kept constant while position of each pulse with respect to the position of a reference pulse is varied in accordance to the instantaneous sampled value of the modulated signal. For this, the transmitter is to send "synchronizing pulses" to keep the transmitter and receiver in synchronism (Fig. 12.17).

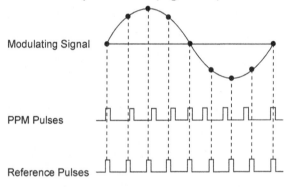

**Fig. 12.17**

## Modulation and Demodulation of PPM

 (*a*) **Modulation of PPM signal:** For generating a PPM signal (Fig. 12.18) the modulating signal is first converted into the pulse width modulation (PWM) signal by sending it through a PWM converter. The PWM signal is then fed to a differentiator. At the output, positive and negative spikes are obtained. These spikes are passed through a positive clipper which removes the positive spikes and gives negative spikes at the output. The negative spikes are now made to trigger a monostable multivibrator and finally the PPM signal is obtained.

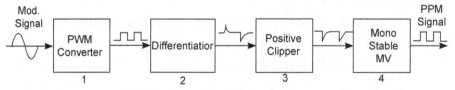

**Fig. 12.18**

1.  Note that the PPM modulation is a trailing edge modulation.

2.  This makes the leading edge fixed, which in turn makes the positive part of the differentiated wave unchanged by the modulating signal, and the positive part is therefore clipped.

3.  The negative part of the differentiated wave contains original information in terms of relative position.

4.  The monostable multivibrator causes the pulse of equal duration to appear at the output. The position of the output pulse carries the original information.

(*b*) **Demodulation of PPM signal:** Demodulation of PPM signal is carried out in the following stages (Fig. 12.19)

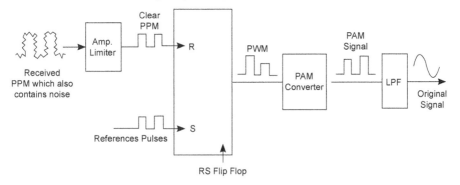

**Fig. 12.19**

(*i*)   The PPM signal is passed to an amplitude limiter, which removes amplitude variations (if any) due to noise.

(*ii*)  Now the clear PPM signal obtained from the limiter along with reference pulses is given to a RS Flip, which converts the PPM signal into a PWM signals.

(*iii*) The PWM signal is now converted into PAM signals by a PAM converter.

(*iv*)  The PAM output is now passed to a Low Pass Filter (LPF) to get the original signal.

The Figure 12.20 shows the above process of demodulation by showing the different outputs.

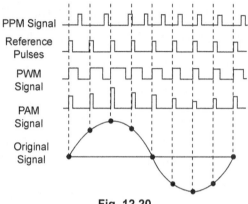

**Fig. 12.20**

(*c*)  **Advantages of PPM:**

(*i*)   As the amplitude and width is kept constant, the transmitter handles constant power.

(*ii*)  As amplitude is constant. It is less noisy.

(*iii*)  The signal and noise separation is easy.

(*iv*)  Due to constant pulse width and amplitude, the transmitted power for each pulse is same.

(*d*)  **Disadvantages:**

(*i*)   It needs synchronization between transmitter and receiver.

(*ii*)  Large B.W. is required as compared to PAM.

**Table 12.1**   A comparison of PAM, PWM and PPM

| S. No. | PAM | PWM (or PDM, PLM) | PPM |
|---|---|---|---|
| 1. | Amplitude of the carrier pulse is proportional to the amplitude of the modulating signal. | The width (or duration or length) of the carrier pulse is proportional to the amplitude of the modulating signal. | The relative position of the carrier pulse is proportional to the position of the modulating signal. |
| 2. | The B.W. of the transmitting channel depends upon the width of the pulse. | The B.W. of the channel depends upon rise time of the pulse. B.W. $= \dfrac{1}{2t}$, where $t$ is the rise time. | The B.W. of the channel depends upon rise time of the pulse. B.W. $= \dfrac{1}{2t}$, |
| 3. | The instantaneous power of the transmitter varies. | The instantaneous power of the transmitter varies. | The instantaneous power of the transmitter remains constant. |

| 4. | Noise interference is high. | Noise interference is low. | Noise interference is low. |
| 5. | Similar to amplitude modulation (AM). | Similar to frequency modulation (FM). | Similar to phase modulation (PM). |

## SUMMARY

1. A pulse is an abruptly changing voltage or current wave, which may or may not repeat itself.

2. In pulse modulation, some parameter of a pulse train is varied according to the instantaneous value of the modulating signal.

3. The pulse modulation may be analog or digital modulation.

4. The types of analog pulse modulation are: Pulse amplitude, pulse width and pulse position modulations.

5. In pulse modulation systems, periodic samples of the modulating signal wave are transmitted and the complete wave is received at the receiver.

6. The sampling theorem says: "In order to convey an information completely, the minimum sampling frequency of a pulse modulation system should be equal to (or more than twice) the highest signal frequency.

7. In pulse amplitude modulation, the amplitude of a pulse train is varied according to the amplitude of the modulating signal.

8. In pulse width modulation, width of a pulse train is varied according to the instantaneous width of the modulating signal.

9. In pulse position modulation, the position of a pulse train is varied according to that of the modulating signal.

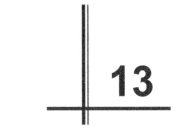

# 13

# Digital Pulse Modulation

All types of modulations discussed so for are analog modulations. In analog modulation the modulated parameter varies continuously and can take any value according to the range of the message. When the modulated wave is mixed with noise, there is no way for the receiver to detect the exact transmitted value of the signal.

An analog signal can be converted into digital signal by means of "sampling" and "quantizing". The process of analog to digital conversion is referred as digital pulse modulation.

The random noise can be virtually eliminated, this is the whole idea of digital modulation

## 13.1 ANALOG AND DIGITAL SIGNALS

An analog signal can be best illustrated by a sinewave. Note that like sine wave, an analog signal is continuous and its value at any instant can be anywhere within the range of its extremes (Fig. 13.1 *a*).

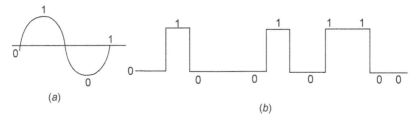

(*a*)

(*b*)

**Fig. 13.1**

The digital signal (Fig. *b*) is not a continuous representation of the original signal. Instead, the digital signal represents the data as a series of digits such as a number. This digital representation is considered as a code which, approximates the actual value.

## 13.2  ADVANTAGES AND DISADVANTAGES OF DIGITAL COMMUNICATION

(*a*) The use of digital communication has many *advantages* over analog communication, some are given below:

    (*i*)   Noise immunity of digital signals,

   (*ii*)   Digital hardware implementation is flexible,

  (*iii*)   Ease to multiplex several digital signals with more efficiency,

  (*iv*)   Reliable reproduction of digital message,

   (*v*)   Easy and economic storage of digital signals,

  (*vi*)   Privacy of information during transmission.

(*b*) The *disadvantage* of digital communication (or transmission) are as follows:

    (*i*)   The transmission (channel) band width required by the digital communication system is much more than the analog communication system.

   (*ii*)   The digital communication systems are more complex than the analog communication systems.

  (*iii*)   A precise time synchronization is required between transmitter and receiver in Digital System.

## 13.3  LOGIC SYSTEM

The digital communication employs logic circuits, which usually work on binary number system. A digital circuit has two states, the output is either low (0) or high (1). In positive logic system in general 0 represents *zero volts* and 1 represents +5 volts. Its reverse case is known as negative logic system (0, —5*V*).

### Binary Number System

A decimal number system (base 10) has ten digits 0, 1, 2, 3, 4, 5, 6, 7, 8, 9. The word binary means two. The binary number system uses only two digits, 0 and 1. Thus, a binary number is a string of zeros and ones. In binary system, the base is 2.

The abbreviation for binary digit is bit. The binary number 1111 has 4 bits, 110011 has 6 bits and 11001100 has 8 bits. A string of 8 bits is known as a **byte**. A byte is the basic unit of data in computers. In most of the computers, the data is processed in strings of 8 bits or some multiples, (*i.e.*, 16, 24, 32, etc.). The memory also stores data in strings of 8 bits or some multiples of 8 bits. Table 13.1 shows binary and decimal equivalence.

**Table 13.1** Binary and Decimal Equivalence.

| Binary | | | | Decimal |
|:---:|:---:|:---:|:---:|:---:|
| 0 | 0 | 0 | 0 | 0 |
| 0 | 0 | 0 | I | 1 |
| 0 | 0 | 1 | 0 | 2 |
| 0 | 0 | 1 | 1 | 3 |
| 0 | 1 | 0 | 0 | 4 |
| 0 | 1 | 0 | 1 | 5 |
| 0 | 1 | 1 | 0 | 6 |
| 0 | 1 | 1 | 1 | 7 |
| 1 | 0 | 0 | 0 | 8 |
| 1 | 0 | 0 | 1 | 9 |
| 1 | 0 | 1 | 0 | 10 |
| 1 | 0 | 1 | 1 | 11 |
| 1 | 1 | 0 | 0 | 12 |
| 1 | 1 | 0 | 1 | 13 |
| 1 | 1 | 1 | 0 | 14 |
| 1 | 1 | 1 | 1 | 15 |

The decimal number 435 can be written as

$$= (4 \times 10^2) + (3 \times 10^1) + (5 \times 10^0) \text{ in binary system}$$

Similarly binary number 1011 is

$$= (1 \times 2^3) + (0 \times 2^2) + (1 \times 2^1) + (1 \times 2^0)$$
$$= 8 + 2 + 1 = 11 \text{ in decimal system.}$$

(*i*) The method to convert a decimal number into a binary number is known as **double dabble**. This method involves successive division by 2 and recording the remainder (the remainder will always be 0 or 1). The division is stopped when we get a quotient of 0 with a remainder of 1. The remainder, when read upwards, gives the binary number.

(*ii*) A systematic way to convert a binary to a decimal number is as given below:

1.  Write the binary number.
2.  Write the weights: $2^2$, $2'$, $2^2$, $2^3$ ; *i.e.* 1, 2, 4, 8, etc., under the binary digit.
3.  Cross out any weight under 0.
4.  Add the remaining weights.

**Problem 13.1.** *Convert decimal number 14 into a binary number.*

**Solution:**

| 2 | 14 | |
|---|---|---|
| 2 | 7 | remainder 0 |
| 2 | 3 | remainder 1 |
| 2 | 1 | remainder 1 |
|   | 0 | remainder 1 |

The binary number is 1110. **Ans.**

**Problem 13.2.** *Convert 1010 into decimal.*

**Solution:**     1    0    1    0    Binary number

$\times$ 8 $\times$ 4 $\times$ 2 $\times$ 1   Add weights

8    $\phi$    2    $\phi$    Cross out weights under 0.

8 + 0 + 2 + 8  = 10

The decimal number is 10. **Ans.**

## Logic Gates

The elements required for performing logic functions or operations (addition, multiplication etc.) are called logic **gates**. These logic gates also work on binary number system.

Logic gates are the electronic circuits with many inputs but only one output. Logic gates are the most important building blocks of any digital system. They do all sort of logic (*i.e.*, Boolean) operations *i.e.*, Addition, multiplications etc.

**Truth table:** The truth table lists all possible combinations of inputs and the corresponding output of a logic gate.

**Table 13.1** Logic Diagrams, Functions and Truths Tables of few gates

| Sl. No. | Gate | Logic diagram | Function | Truth table | | |
|---|---|---|---|---|---|---|
| 1. | AND | A o—[ ]—o Y, B o | $Y = A$ and $B$ <br> $= A . B$ | Input | | Output |
| | | | | A | B | Y |
| | | | | 0 | 0 | 0 |
| | | | | 0 | 1 | 0 |
| | | | | 1 | 0 | 0 |
| 2. | OR | A o—[ ]—o Y, B o | $Y = A$ or $B$ <br> $= A + B$ | Input | | Output |
| | | | | A | B | Y |
| | | | | 0 | 0 | 0 |
| | | | | 0 | 1 | 1 |
| | | | | 1 | 0 | 1 |
| | | | | 1 | 1 | 1 |
| 3. | NOT (Inverter) | A o—[>o—o Y | $Y = NOT\,A$ <br> $= \overline{A}$ | Input | | Output |
| | | | | A | | Y |
| | | | | 1 | | 0 |
| | | | | 0 | | 1 |
| 4. | NAND | A o—[ ]o—o Y, B o | $Y = A\,NOT\,AND\,B$ <br> $= A\,NAND\,B$ <br> $= \overline{A.B}$ | Input | | Output |
| | | | | A | B | Y |
| | | | | 0 | 0 | 1 |
| | | | | 0 | 1 | 1 |
| | | | | 1 | 0 | 1 |
| | | | | 1 | 1 | 0 |
| 5. | NOR | A o—[ ]o—o Y, B o | $Y = A\,NOT\,OR\,B$ <br> $= A\,NOR\,B$ <br> $= \overline{A + B}$ | Input | | Output |
| | | | | A | B | Y |
| | | | | 0 | 0 | 1 |
| | | | | 0 | 1 | 0 |
| | | | | 1 | 0 | 0 |
| | | | | 1 | 1 | 0 |
| 6. | EX-OR | A o—[ ]o—o Y, B o | $Y = A\,EX\!-\!OR$ <br> $= A \otimes B$ <br> $= \overline{AB} + \overline{A}B$ | Input | | Output |
| | | | | A | B | Y |
| | | | | 0 | 0 | 0 |
| | | | | 0 | 1 | 1 |
| | | | | 1 | 0 | 1 |
| | | | | 1 | 1 | 0 |

## 13.4 PRINCIPLE OF DIGITAL COMMUNICATION

Figure 13.2 shows the functional elements and block diagram of a digital communication system. The source output may be an analog signal, such as voice or picture signal or music. If it is in analog form, it must be converted into digital pulses prior to transmission and converted back to analog form at the receiver end (or destination). This can be done by **input transducer** (A/ D converter) at the source and by the **output transducer** (D/A converter) at destination (or receiver). In a digital communication system, the information (or messages) produced by the source are converted into a sequence of binary digits using **source encoder**. The function of the source encoder is to convert the digital output of the source into a sequence of binary digits. This process is called as **data compression** (or source encoding). After source encoding, the sequence of binary digits (message sequence) is given to the **channel encoder**. The function of channel encoder is to introduce some rebundancy in the binary message sequence which is used at the destination (receiver) to overcome the interference and noise effects on the transmission of signal through the channel.

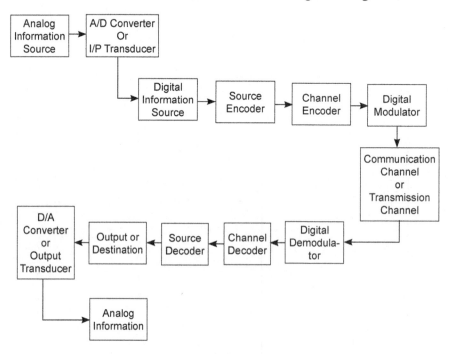

**Fig. 13.2**

The output of the channel encoder then goes to the **digital modulator** which converts the binary message (information) sequence into sinusoidal form. The **communication channel** is a medium used to send the signal from the transmitter to the receiver. During transmission through communication channel, the signal is affected by the noise such as man made noise, thermal noise, atmosphere noise etc. This noise affected signal is processed by the **digital demodulator** at the receiver end and converts this sinusoidal signal into binary sequence. This binary sequence is passed to the channel decoder which recovers the original information or message sequence from the code used by the channel encoder at transmitter, and rebundancy contained in the received data sequence. It is not easy to recover the original message (or information sequence) at the source decoder output because the noise and disturbances are already mixed in the communication channel. Because of this, the signal recovered/regenerated by the source decoder is an approximation of the original information its output (or destination). Then this digital message sequence is converted into analog form by the D-A converter at the receiver end.

## 13.5 COMMUNICATED SPEED

A **Baud** is the unit of signalling speed of communication channel. It is named after the name of French scientist JME Baudat, whose 5 bit code was adopted by the French telegraph system in 1877. A "Baud" is defined as the no. of code elements per second. If each signal element represents one bit information, the baud is one bit per second.

Now-a-days in digital communication instead of Baud, **"Bit Per Second"** (BPS) is used as the unit.

At present in standard digital voice communication, amplitude of the voice signal is sampled at the rate of 8000 samples per second. An 8 bit value for the amplitude of the frequency is transmitted digitally. The encoding scheme is known as **Pulse Code Modulation** and requires a bandwidth of 64 kilo bits per second (8000 samples per second at 8 bit per sample).

Now, digital transmission is widely used and much of the world utilises digital transmission links operating at 2.048 mega bits per second.

## 13.6 QUANTIZING

In pulse modulation, the pulses result from sampling the modulating signal wave. In other words, the modulating wave is **sliced** into small units, the process is called **Quantizing** or **Quantization**. These quantum points are then converted into binary code, which represents the amplitude of the waveform at the point.

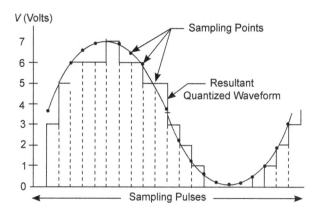

Fig. 13.3

In Fig. 13.3 the amplitude of modulating wave is 7 volts. Sampling points are taken at equal intervals and quantized waveform is obtained as shown in Fig. 13.4. Note that, if the modulating signal is not an exact value of the resulting binary code, distortion or error is introduced which produces a noise called **Quantizing noise**. The noise can be reduced by increasing sampling points, but this also increases the required bandwidth.

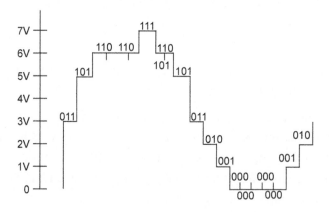

Fig. 13.4

The resulting pulse code wave forms are shown in the table below along with the binary code. In other words, each portion of the wave is converted into a binary number that represents the final wave.

| Quantizing level (Volts) | Binary Code | Pulse Code Wave |
|:---:|:---:|:---:|
| 0 | 000 | 0———————— |
| 1 | 001 | 0————⊓——— |

| Quantizing level (Volts) | Binary Code | Pulse Code Wave |
|:---:|:---:|:---:|
| 2 | 010 | |
| 3 | 011 | |
| 4 | 100 | |
| 5 | 101 | |
| 6 | 110 | |
| 7 | 111 | |

## 13.7 QUANTISING ERROR

A sampled signal exists only at discrete intervals but amplitude of the sampled signal depends on the amplitude of the signal. Each sample must be assigned a discrete measure of amplitude to produce a discrete representation of the signal. This is possible by comparing it with a scale having a finite no. of intervals and specifying it only by identifying the interval, in which it falls. This process is called **quantisation** and a **quantum** is defined as an interval on the scale.

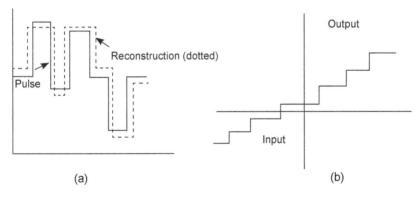

Fig. 13.5

The original signal is recovered from the sampled and quantised signal by generating each sample at time. A pulse equivalent to the mean amplitude is included at the appropriate interval.

A train of samples and its quantised reconstruction is shown in Fig. 13.5 (a). A quantised system transfer characteristic is stepped and is shown in Fig 13.5 (b). The step appears because of the fact that each discrete output step corresponds to a range of inputs. The difference between input and output is called a **quantised error.** The quantising error fluctuates over a range of **one quantum**. The smaller is the quantising error, the less is the effect of noise on reconstruction.

The Fig. 13.6 shows the linear and error components in quantization error.

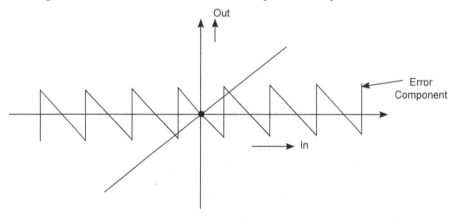

**Fig. 13.6**

## 13.8 TYPES OF DIGITAL PULSE MODULATIONS

The popular types of digital modulations are:

1. Pulse Code Modulation (PCM)

2. Differential PCM

3. Delta Modulation (DM)

4. Adaptive delta modulation (ADM).

We shall discuss these one by one:

## 13.9 PULSE CODE MODULATION (PCM)

This is a digital modulation process, in which signal to be transmitted is sampled at various instants. This results in a sequence of pulses each time, the signal is sampled depending upon magnitude of the signal at the sampling instant (Fig. 13.7). These pulses correspond to a certain code (usually Binary). These pulses modulate an RF carrier and are transmitted.

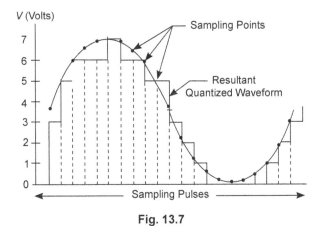

**Fig. 13.7**

At starting, a new code group, *reference pulses* are transmitted along the each code group. These reference pulses are different in amplitude or duration from the pulses carrying signal so that these can be easily **detected** at the receiver.

In simple words, amplitude of the modulating signal is converted into a digital code (which is generally binary number) at the transmitter. The process is similar to an analog to digital conversion where the amplitude of an analog signal is converted into a digital code.

At the receiver, the signal is decoded (say by a **computer**) and the original signal can be obtained.

## Block Diagram of PCM System

The **block diagram** for the PCM System has three parts (Fig. 13.8).

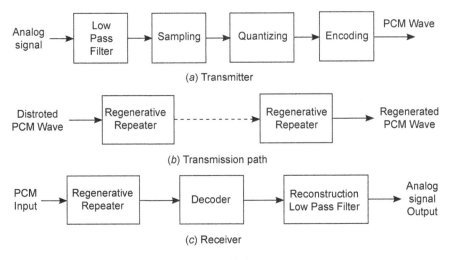

**Fig. 13.8**

(*a*) **Transmitter:** The essential operations in a PCM transmitter are: sampling, quantizing and encoding as shown in Fig. 13.8(*a*). The quantizing and encoding both are performed in the same circuit called Analog to Digital (A/D) converter. High frequency components are first attenuated by low pass filler.

(*b*) **Transmission Path:** At intermediate points along the transmission route from transmitter to receiver, **Regenerative Repeaters** are used to reconstruct (regenerate) the transmitted sequence to coded pulses to combat the effects of signal distortion and noise as shown in Fig. 13.8 (*b*).

(*c*) **Receiver:** At receiver, regeneration of impaired signals, decoding of the train of quantized pulses is carried out as shown in Fig. 13.8. (c). These operations are usually performed in the same circuit called Digital to Analog (D/A) converter.

**Note: Regenerative Repeater:** (Fig 13.9) The most important feature of a PCM system is its ability to control distortion and noise produced in the transmission of the PCM wave through the channel. This is accomplished by reproducing the PCM wave by means of a chain of regenerative repeaters located all over the route.

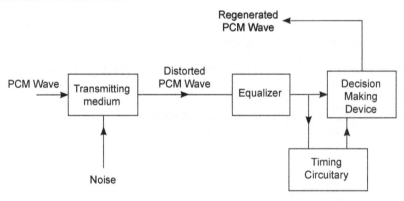

**Fig. 13.9**

A regenerative repeater performs the following functions:

(*i*) Reshaping the incoming pulses by means of **Equalizer** to compensate the effect of distortion in the transmitting channel.

(*ii*) The "Timing Circuitry" facilitates in sampling the **equalizer** pulses at the instants, where Signal-Noise ratio is maximum.

(*iii*) "The decision making device" is enabled, whenever the amplitude of the equalizer pulses plus noise exceeds a pre-determincd voltage level at the time of sampling.

## PCM Transmitter

A PCM transmitter is an Encoder, which converts a signal into a specified code. This is basically an Analog to Digital converter, which converts input analog signal to a Binary code. The encoders are electronic circuits, employing differential amplifiers, logic gates etc. The Fig. 13.10 shows a typical encoder circuit consisting of three differential amplifiers, one exclusive NOR gate and two AND gates.

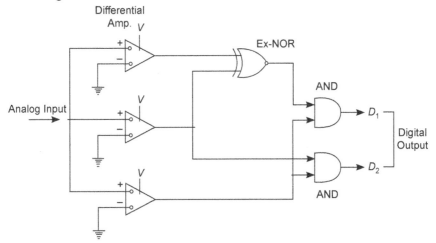

**Fig. 13.10**

## PCM Receiver

A PCM demodulator (Receiver) is just reverse of modulator, it is basically a **decoder**, which converts the coded signal into analog form. In other words it is a digital to analog convertor. A typical decoder may consist of analog switch, AND gate and shift register. The resulting waveform is then passed through a low pass filter and original signal is obtained.

## Advantages of PCM

(*i*) The major advantage of PCM is that it is much more immune to noise or interference. The intelligence gets distorted, when some characteristic of the intelligence is affected such as in PAM, PWM and PPM. As in PCM, no characteristic of signal is affected, this has a noise less reception.

(*ii*) The PCM permits repeating (or amplifying) the encoded signal without significant distortion being introduced.

(*iii*) The output SN ratio increases exponentially with bandwidth.

(*iv*) A PCM system designed for analog data can be readily adopted to other inputs such as digital data, thus promoting flexibility.

(*v*) Due to regeneration capacity, the PCM is beneficial for a system having many repeater stations.

## Applications of PCM

The applications of PCM are in the following fields:

(*i*) The PCM is used for multi channel communication **over wires** *i.e.*, long distance telephony.

(*ii*) The pulse code modulated signals may be used to modulate an RF carrier and can be transmitted as radio signals.

## Companding in PCM

In PCM, there are two major problems:

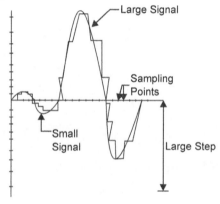

(*i*) The uniform step size in quantizing means weak signals will have more quantizing noise.

(*ii*) The signals with large dynamic range require many encoding bits, which may not be practically feasible.

The process to overcome these problems is known **Companding**. In this technique number of quantisation steps are increased for small signals and decreased for large signals. The Fig. 13.11 shows the technique.

**Fig. 13.11**

## Bandwidth of PCM

Refer the Single pulse shown in the Fig. 13.12 (*a*). The Fig. 13.12 (*b*) shows frequency spectrum of the pulse.

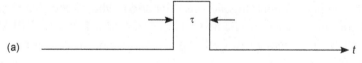

(a)

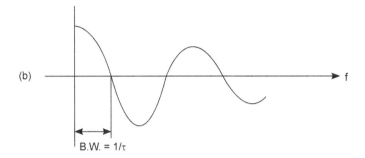

**Fig. 13.12**

The B.W. is equal to the distance of the first zero crossing *i.e.*,

$$\text{B.W.} = 1/\tau \text{ Hz} \qquad \qquad ...(i)$$

The PCM wave train is composed of sequence of 1s and 0s. Regardless of the sequence, the B.W. can not be greater than, that required for a single pulse, which has been expressed in eq. (*i*).

Suppose an analog signal is sampled at the "Nyquist rate", the time T between pulses is given by :

$T = \dfrac{1}{2f_m}$ where $f_m$ is the highest frequency in the signal. Expanding this to *n* signals (*i.e.*, multiplexing), the width of the sample pulse will be :

$$\tau_n = T/n = \frac{1}{2f_m.n}$$

If each pulse is digitized, each code word can last only for *T/n* seconds. For an *m* bit PCM word, the width of each bit pulse will be *T/ mn* (Where *m* is number of bits in the PCM word and *n* is number of signals). Using eq. (*i*), the PCM bandwidth will be given

$$\text{B.W. (PCM)} = \frac{1}{\tau} = \frac{nm}{T} \text{ Hz}$$

$$= m\left(\frac{w_m}{\pi}\right) \text{ Hz} \qquad \qquad ...(ii)$$

For the binary process, the number of levels *L* in the quantization process is equal to $2^m$.

Solving for *m* and substituting in eq. (*ii*), we get

$$\text{B.W. (PCM)} = \frac{n.w_m}{\pi} \log_2 (L) \text{ Hz} \quad \text{(For binary, the base is 2)}$$

$$= (2n.f_m).m. \ (w_m = 2\pi fm)$$

## 13.10  DIFFERENTIAL PCM [DPCM)

The differencial PCM differs from the conventional PCM in the respect that in this, only the relative amplitude of various samples is indicated and not the absolute magnitude. Each **word** in this system indicates difference in amplitude between present and the previous samples. The reason being that there are little variations from samples to sample and the transmission of the difference will need few bits and as well as smaller bandwidth.

The Fig. 13.13 shows transmitter and receiver for DPCM. The input to the Quantizer is between the **flat top** samples and **feedback signal** derived from the output of the quantizer. The estimated feedback signal is created by a "Predictor" circuit, which is basically a digital filter.

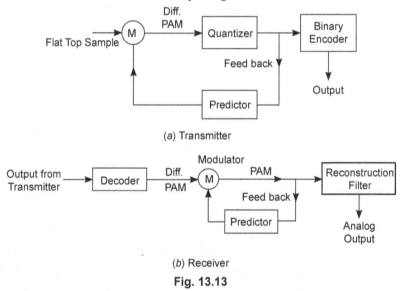

(a) Transmitter

(b) Receiver

**Fig. 13.13**

The incoming DPCM binary signal is decoded at the receiver to form a differential PAM (pulse amplitude modulated) signal. A predictor filter forms the estimated feedback signal. Now analog output is obtained from the "Reconstruction" filter.

## DPCM VS PCM

In case of voice signals, it is found that *signal to quantizing error* ratio (noise) of DPCM (diff. PCM) over conventional PCM is between 4-10 dB. For a constant signal to quantizing noise ratio, and assuming sampling rate of 8 kHz, the use of DPCM may provide a saving of 8–15 kilo bits per second or 1–2 bits per sample over conventional PCM. For monochrome TV signals, DPCM provides a *signal to quantizing noise ratio* of about 10 dB higher than conventional PCM. For a constant signal to quantizing noise ratio and assuming the sampling rate of 9 MHz, the DPCM provides a saving of 18 Mega bits per second or 2 bits per sample.

## 13.11 DELTA MODULATION (DM)

The Quantizing in DM (Fig. 13.14) is almost similar to differencial PCM. In this system, only one **bit** is transmitted per sample just to indicate whether the present sample is larger or smaller than the previous sample. The processes of coding and decoding are very simple, but this system cannot handle fast varying samples and therefore has also **not been much popular**.

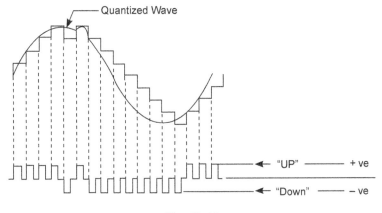

**Fig. 13.14**

Note that, a **digital regenerative repeater** can receive the signal like the receiver at the destination, identify the digits and regenerate the digits afresh. The regenerated digits will be almost identical with those at the first transmitter. Whereas in case of **analog signal repeater**, the noise power goes on accumulating at each repeater so that the total noise at the final receiver will be $n$ time ($n$ = no. of repeaters) as large as on the first repeater.

### DM Transmitter

The DM transmitter (Fig. 13.15) consists of a comparator, D/A converter, Binary up down counter and a $D$ type Flip Flop. The output of the counter goes to DA converter which has a staircase waveform that follows the modulating signal. The output of the comparator goes to the $D$ type Flip Flop, which gives the DM output.

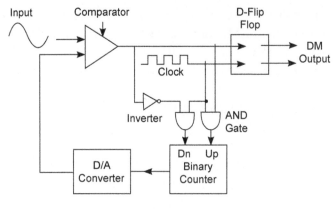

**Fig. 13.15**

## DM Receiver/Decoder (Figure 13.16)

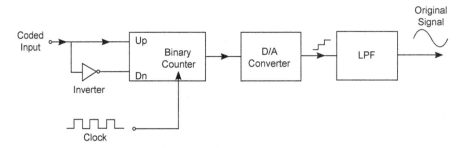

**Fig. 13.16**

The coded signal is given to the up/down binary counter. The output of the counter goes to D/A converter. The staircase output of the converter is passed through the low pass filter (LPF) to recover the original signal.

## Delta Modulation (DM) Vs Pulse Code Modulation (PCM)

(*a*) The two can be compared as below:

(*i*)  The DM has extreme simplicity of transmitter and receiver circuits than PCM.

(*ii*)  The DM system for the same bit transmission rate gives better performance over PCM.

(*iii*)  For speech transmission, DM gives better performance at lower bit rate. At higher bit rate, PCM has better performance.

(*iv*)  DM requires large bandwidth than PCM.

(*b*) The DM (delta modulation) system operating at 40 kB per second is equivalent to a conventional PCM system operating at a sample rate of 8 kB per second and 5 bits per sample. At lower bit rate, DM is better than conventional PCM but at higher bit rates, the PCM is better than DM. For telephone quality voice signals, 8 bit PCM is used but equivalent voice quality can be obtained with DM only by using bit rate much higher than 64 kB per second.

Also in DM, the SN ratio is increased by 9 dB by doubling the bit rate, whereas in conventional PCM we achieve 6 dB increase in SN ratio for each added bit.

**Advantages of DM:** From above, we can summarize advantages of DM as

(*i*) Reduction of bits to be transmitted

(*ii*) Simplified coding and decoding

(*iii*) The DM has almost noise free regeneration ability. The DM is therefore superior to PCM, particularly for long distance communication where use of regenerative repeaters becomes essential.

### Application of DM

The DM system is used in the following conditions:

(*i*) If it is necessary to reduce the bit rate below 40 kB per second and limited voice quality is acceptable.

(*ii*) If extreme circuit simplicity is more important and high bit rate is acceptable.

## 13.12 ADAPTIVE DELTA MODULATION (ADM)

The Delta modulation suffers from "slope overload", when the original signal has large variations in frequency. To rectify this problem, some type of signal compression is necessary. For this, the suitable method may be the adaption of steps size according to the level of the input signal derivative. This technique is called **Adaptive Delta Modulation (ADM)**.

While using ADM, a long string of 1 s and 0's indicates that a staircase voltage is continuously following the original signal. In other words, it is continuously adding/subtracting step voltages to form the output. More frequent the bit changing from 0 to 1, the more closely the staircase voltage is following the original waveform.

Here, the delta modulator is made adaptive wherein, the variable step size increases during an increase in amplitude of the sample values and decreases during decrease in amplitude of the sample values. In this way, the step size is adapted to the level of the input signal.

The Fig. 13.17 shows variation of step size in the adaptive delta modulator.

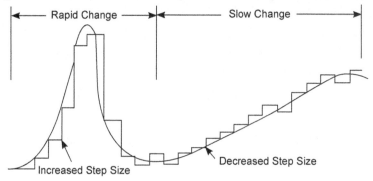

**Fig. 13.17**

Adaptive delta modulation is the modified form of the delta modulation and used when the step size is not fixed. The step size will become progressively larger when the slope overload occurs. As the method adjusts (adapts) the step size, it is called Adaptive DM. The difference between DM and ADM is Digital processor in case of ADM instead of UP/Down converter in case of DM. The digital processor which has an accumulator and a step is generated at each active clock edge which resets the accumulator. The step in ADM is always a multiple of basic step.

The Fig. 13.18 shows complete ADM transmission system.

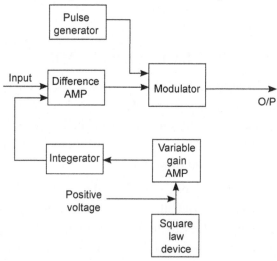

**Fig. 13.18**

## SUMMARY

1. The process of analog to digital conversion is referred as digital modulation. This is done by means of "sampling" and "Quantizing".

2. Digital modulation is noise free and privacy of information is possible.

3. Binary code is usually adopted for digital transmission.

4. The speed of communication is given in "Bauds" or Bits per second, the signalling speed may be of the order of 2.3 mega bits per second.

5. The examples of digital modulation are: pulse code modulation, differential modulation and delta modulation.

6. The steps of PCM are: sampling, Quantizing and encoding. At transmitting end, the analog signal is converted into digital (binary) code and transmitted. At the receiving end the digital signal is decoded into analog signal again.

7. In PCM, the modulating signal is "sliced" into small units, the process is called quantizing. The quantum points are then converted into binary code, which represents the amplitude of the wave at that point.

8. In PCM, the no. of quantisation steps are increased for small signals and decreased for large signals. The process is called "Companding". This is done to reduce quantising noise.

9. In DPCM, each word indicates difference in amplitude between present and the previous sample. In this way DPCM differs to the conventional PCM.

10. The Delta modulation is almost similar to the DPCM. In this, only one bit per sample is transmitted.

# Digital Carrier Modulation

In Digital pulse modulation, the digital signal to be sent is converted into analog signal and sent over copper wires. At reception, the received signal is converted into original digital signal. But if the digital signal is to be sent through space, some modulation technique is to be adopted. This chapter describes various digital carrier modulation techniques.

## 14.1 DIGITAL CARRIER MODULATION

Digital signals such as telegraphy and teleprinter signals are often required to be transmitted over long distance by means of radio waves. So some type of modulation is to be used. For this purpose Amplitude, Frequency and phase modulations are commonly used.

Because of the two level nature of the carrier, in this case,

- (*i*) The amplitude modulation is known as **Amplitude Shift Key** (ASK) or Binary ASK (BASK).

- (*ii*) The frequency modulation is known as **Frequency Shift Key** (FSK) or Binary FSK (BFSK).

- (*iii*) The phase modulation is known as **Phase Shift Key** (PSK) or Binary PSK (BPSK). The Fig. 14.1 shows waveforms for ASK, FSK and PSK signals for comparison.

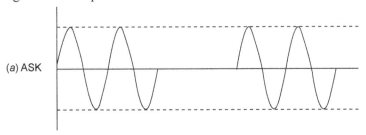

(*a*) ASK

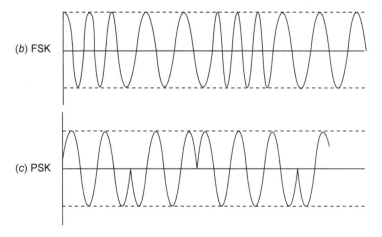

**Fig. 14.1**

**Note:** An ASK is analogous (similar) to AM, a FSK to FM and a PSK to PM.

Table 14.1   ASK Vs. FSK Vs. PSK

| ASK | FSK | PSK |
|---|---|---|
| The ASK also called "OFF Keying" is present for 1's carrier and absent for 0's carrier. | In FSK we have different frequencies for 0 and 1. | There is a difference of 180° in phase for each 0 and 1 communication. |

Table 14.2   ASK Vs. PSK

| ASK | PSK |
|---|---|
| 1. The carrier is switched ON and OFF. | The carrier is switched between levels + A and –A. |
| 2. It has an impulse at carrier frequency. | It does not have an impulse at the carrier frequency. |
| 3. The ASK is a linear modulation scheme. | The PSK is generally a non linear modulation scheme. |
| 4. Its performance is poor as compared to PSK. | Its performance is better than ASK. |
| 5. The average power is half of the PSK signal. | The average power of PSK signal is twice of the ASK signal. |

## 14.2   POWER SPECTRAL DENSITY (PSD)

The periodic waveforms are also called as "power waveforms". Another type of power waveform encountered in communication is "random waveform", for which the values may not be predicted. One of the examples of the random waveforms is "noise waveform", other examples are: ASK, FSK and PSK waveforms. These signals carry finite power. The Fourier methods applied to these random power waveforms result in a "power spectral density". **The PSD**

**is a curve, which shows the energy distribution as a continuous function of frequency**. A typical PSD curve has been shown in Fig. 14.2.

The units for PSD are Watt per hertz (or Joules). The area under the curve has units of Joules × hertz; which is dimensionally equivalent to Watts; But as Watts are the units of power, therefore total area under the curve gives average signal power; this means

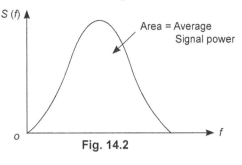

Fig. 14.2

the total average power is equal to the product of PSD (Watt/hertz) and the

Bandwidth (in Hz); $\left[\dfrac{\text{Watts}}{\text{Hz}} \times \text{Hz} = \text{Watts}\right]$

Now we will study ASK, FSK and PSK one by one:

## 14.3  AMPLITUDE SHIFT KEYING (ASK OR BASK)

In (Binary) ASK System, two binary (1,0) symbols are represented by two different amplitudes $(A_c, 0)$ of the carrier frequency $f_c$. Binary symbol '1' is represented by the presence of a constant amplitude of carrier for $T$ second where as other binary symbol '0' is represented by the absence of the carrier for $T$ second. This signal can be generated simply by turning on and off the carrier of a sinusoidal oscillator for the prescribed periods shown in modulating pulse train. That is why, this is also called as on-off keying (OOK). See Fig. 14.3.

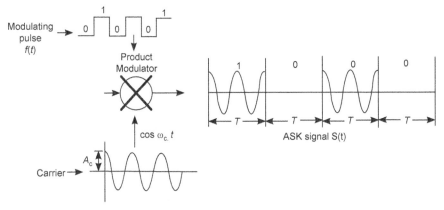

Fig. 14.3

Let the sinusoidal carrier be represented by

$$v_c(t) = A_c \cos(\omega_c t) = A_c \cos(2\pi f_c t) \qquad \qquad ...(i)$$

The binary amplitude shift Keying (BASK) signal can be expressed as $S(t)$, as shown in Fig. ($b$).

Then Binary ASK wave $S(t) = b(t). A_c \cos(\omega_c t) = A_c \cos(2\pi f_c t)$          ...($ii$)

$$b(t) = 1 \text{ for mark signal (binary 1)}$$

$$= 0 \text{ for space signal (binary 0)}$$

For a binary data represented by (10110110), a typical binary ASK waveform is shown in Fig. 14.4.

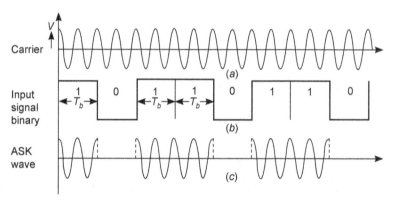

**Fig. 14.4**

## Generation and Reception of ASK (BASK) Signal

($a$) **Generation of ASK Signal:** By applying two inputs: binary data and the sinusoidal carrier wave to the balanced modulator, ASK signal can be generated. The output of the Balanced modulator is the ASK signal. The Fig. 14.5 ($a$) shows how the ASK wave is generated. A shift in the baseband signal spectrum is caused by the modulation [Fig. 14.5 ($b$).]

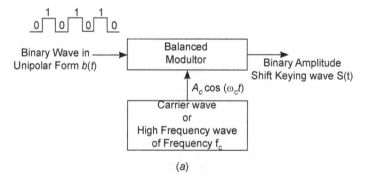

($a$)

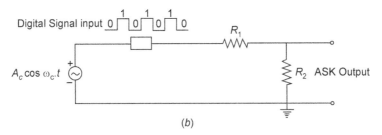

(b)

**Fig. 14.5**

(b) **Demodulation/Reception of binary ASK:** The demodulation of binary ASK can be obtained with the help of coherent demodulator/detector as shown in Fig. 14.6.

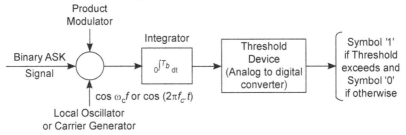

**Fig. 14.6**

Coherent detector comprises of integrator and threshold device (Analog to digital converter) that follow the product modulator. At the input of the product modulator, we apply two inputs, one the incoming ASK signal and other is a sinusoidal carrier '$A_c \cos \omega_c t$' where $\omega_c = 2\pi f_c$. The second input is sinusoidal carrier which is generated locally, of frequency $f_c$, with the help of local oscillator. The output of the product modulator is given to the integrator which operates on the output of the multiplier for successive bit intervals '$T_b$' and also perform a low pass filtering action. The integrator output then is given to the threshold device (Analog to digital converter) as input. The function of threshold device (ADC) is that, it compares the output of integrator with a preset threshold.

It makes a decision in favour of symbol 1, if the threshold is exceeded and in favour of symbol 0 otherwise. It has been assumed in the above discussed method for ASK detection that the locally generated carrier is in perfect synchronisation with transmitted carrier means that the phase and frequency of locally generated carrier is same as that of the carrier used in the transmitter. The coherent (or synchronous) detection involves the use of linear operation.

(*c*) **Power Spectral Density for ASK**. (ASK spectrum): The power spectral density [Average power per hertz (W/Hz)] of an ASK signal is same as that of baseband on-off signal ($f_b$) but only difference is that its power spectral density is shifted in the frequency domain by $\pm f_c$ (carrier frequency) as shown in Fig. 14.7 (*a*) and (*b*).$f_c \gg f_b$ where $f_b = 1/T$.

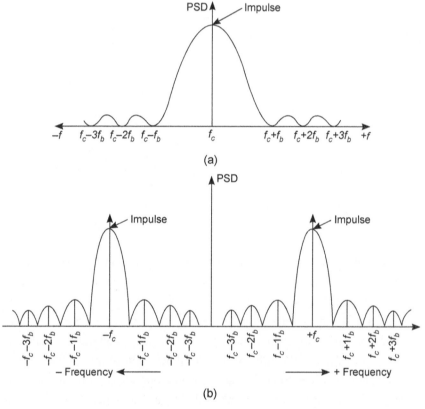

Fig. 14.7

In the Power spectral density curve of ASK, the two impulses occur at carrier frequency $\pm f_c$. The Fig. 14.7 (*b*) shows that ASK signal spectrum has an infinite bandwidth. For practical purposes, the bandwidth of $S(t)$ is often defined as the bandwidth of an ideal bandpass filter centred at carrier frequency $f_c$ whose output contains about 95% of the total average power of the content of binary ASK signal $\underline{S}(t)$. It can be shown that bandwidth of $S(t)$ or ASK signal is approximately $3f_b$ Hz and can be reduced by using smoothed versions of the binary pulse wave form $b(t)$ instead of rectangular pulse waveforms.

## Derivation of ASK Signal

Let $V(t)$ be the binary signal to be transmitted.

    1 is represented by $V(t) = +V$

    0 is represented by $V(t) = -V$

The ASK signal obtained can be expressed as :

$$V_{ASK}(t) = \mu(t).\, A \cos \omega_0 t.$$

Where                     $\mu(t) = 1$ for $V(t) = +V$

                            $\mu(t) = 0$ for $V(t) = -V$

$A \cos \omega_0 t$ is the carrier signal.

It can be seen that the carrier signal is ON, when $V(t) = +V$ and it is OFF, when $V(t) = -V$. So the ASK system is also called as **ON OFF KEYING** (OOK).

The Fig. 14.8 (b) shows ASK signal wave shape.

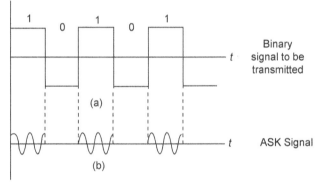

**Fig. 14.8**

**Note:**

1. In case of carrier signal, the generator should only be capable of formulating one of the two distinct AM wave segments for ON OFF keying. For generation, we need a simple switch and oscillator. The Fig 14.9 shows a simple ASK modulator.

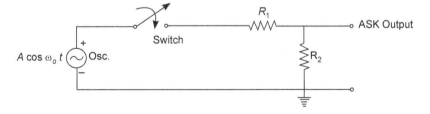

**Fig. 14.9**

2. The ASK does not give satisfactory results, **hence this system is rarely used**.

## Coherent and Non-Coherent ASK Receivers

In an ideal coherent detection of Binary ASK signals, we assumed that the binary ASK signal available at the receiver is an exact replica (or copy) of the orignal signal.

At Receiver, for local generation of an in-phase coherent carrier necessary for ideal coherent detection, is only possible by using very stable oscillators at both the transmitter and receiver whose cost is excessive. Due to this drawback, Non-coherent detection technique is used which do not require a phase-coherent local oscillator. It also simplifies greatly the design consideration needed in synchronous (or coherent) detection.

The Fig. 14.10 (*a*) shows coherent detector of BASK signal Fig. 14.10 (*b*) shows non-coherent detector for comparison. The non coherent technique involves some form of rectification and low pass filtering at the receiver. That is, in non-coherent detection, Binary ASK signal can be demodulated using simple envelope detector.

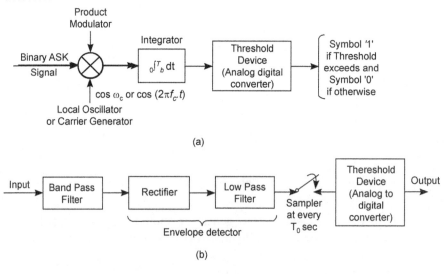

(a)

(b)

**Fig. 14.10**

The error probability of a non coherent Binary ASK receiver will always be higher than that of coherent Binary ASK receiver operating at same signalling speed and same signal power. In Fig. 14.10, $T_b$ = bit transmission period.

## 14.4 FREQUENCY SHIFT KEYING (FSK/BFSK)

The frequency shift keying is the oldest and simplest method of modulation used in modems. Two sinusoidal wave frequencies are used to represent binary symbols Os and Is. For example, a frequency of 1070 Hz in data communication is used to represent binary '0' and a frequency 1270 Hz is used to represent binary '1', for transmission of binary data as shown. These two frequency tones are well within the 300 to 3400 Hz bandwidth associated with telephone system.

The Fig. 14.11 (*a*) shows sinusoidal carrier wave, the Fig. 14.11 (*b*) shows Binary modulating signal and Fig. 14.10 (*c*) shows FSK signal.

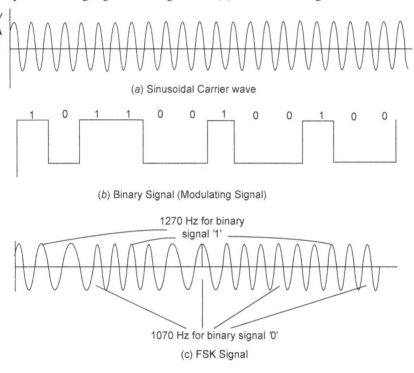

(*a*) Sinusoidal Carrier wave

(*b*) Binary Signal (Modulating Signal)

1270 Hz for binary signal '1'

1070 Hz for binary signal '0'

(c) FSK Signal

**Fig. 14.11**

## Two Tone Modulation Technique for BFSK

(*a*) In conventional/standard Frequency Modulation (F.M.), the frequency of the high frequency wave (or carrier wave) is varied in accordance with the instantaneous value of modulating signal, provided amplitude & phase of the carrier remains constant throughout, but in BFSK (Binary Frequency Shift Keying), the modulating signal is binary in nature *i.e.*, it has two discrete voltage levels rather than a continuously changing analog value. The frequency of carrier signal in BFSK is varied to represent binary value 1 or 0. See Fig 14.12 (*a*, *b*, *c*). The Fig. 14.12 (*d*) shows FSK generation. The FSK can be considered as consisting of two pulses sinusoidal waves of frequencies $f_{c1}$ and $f_{c2}$.

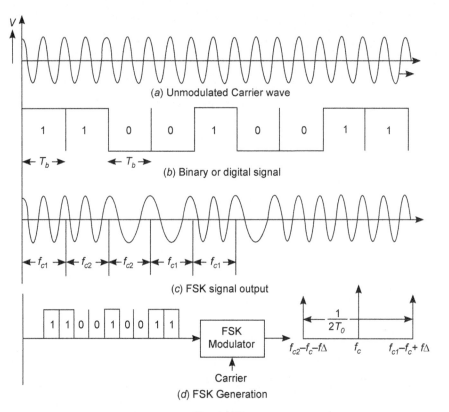

**Fig. 14.12**

In FSK, two sinusoidal carrier waves of same amplitude and phase, but of different frequencies $f_{c1}$, and $f_{c2}$ are used to represent binary symbols 1 and 0. The binary FSK wave S(t) can be represented as

$$S(t) = b(t) A_c \cos (2\pi f_{c1} t) \text{ for Binary Symbol 1} \quad ...(i)$$

$$= b (t) A_c \cos (2\pi f_{c2} t) \text{ for Binary Symbol 0}$$

The frequency of transmitted signal is high for 1 and low for 0.

From Fig. 14.12, the maximum **frequency** deviation is

$$\Delta_f = \frac{f_{c2} - f_{c1}}{2} = \frac{1}{4T_b} \text{ Hz} \qquad ...(ii)$$

where $T_b$ is the **bit transmission period** and $f_{c2}$ and $f_{c1}$ are expressed as

$$f_{c2} = f_c - \Delta_f = -\frac{1}{4T_b} \qquad ...(iii)$$

$$f_{c1} = f_c + \Delta_f = -\frac{1}{4T_b} \qquad ...(iv)$$

Using equations (iii) and (iv), we have BFSK signal:

$$F_{FSK}(t) = A_c \cos [2 (f_c \pm \Delta f)t] \qquad ...(v)$$

Putting the value of $\Delta f$ from eq. (*ii*)

$$F_{\text{FSK}}(t) = A_c \cos\left[2\left(f_c + \frac{1}{4T_b}\right)t\right]$$

(*b*) The Power Spectral Density of binary FSK signal is the sum of two ASK spectra at frequencies $f_{c1}$, and $f_{c2}$ as shown in Fig. 14.13.

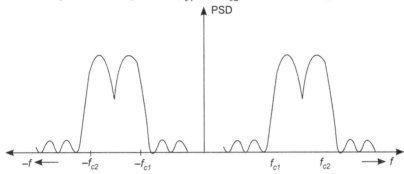

**Fig. 14.13**

(*c*) **Advantages of BFSK over BASK**

Frequency Shift Keying is used much more than ASK, because of the following advantages:

1. Noise immunity in FSK is more than ASK.

2. Improved SNR ratio that helps in reducing the effect of Interference.

3. Coding is inexpensive, simple and easy hardware implementation.

4. The FSK receiver can amplify the received signal without AGC (Automatic gain controller) which is required in FM receiver because Frequency Shift Keying (FSK) is similar to standard frequency modulator.

## Coherent (Synchronous) Detection of BFSK Signal

(*a*) The FSK signal can be demodulated using coherent or synchronous detector as shown in Fig. 14.14. This type of detector has two correlators. Both the correlators are tuned to two different carrier frequencies, $f_{c1}$ and $f_{c2}$ to represent binary symbol 1 and 0. Each correlator consists of a multiplier which is followed by an integrator. The binary FSK signal is first applied to multiplier of the correlators. Two carriers of frequency $f_{c1}$, and $f_{c2}$ are applied to the inputs of two multipliers 1 and 2 respectively as shown in Fig. 14.14.

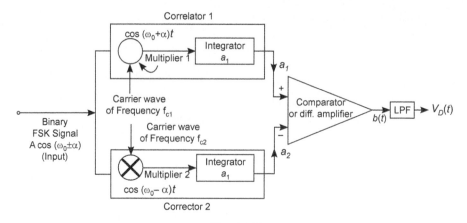

**Fig. 14.14**

The multiplied output of multiplier '1' is given to integrator '1' and multiplied output of multiplier '2' is given to the integrator '2'. The output of two integrators ($a_1$ and $a_2$) are then passed to the comparator. A comparator is a circuit that accepts two input signals and generates a binary output, which is at one level or other depending on which input is larger. Thus at the comparator, output $b(t)$ will be produced. Passing through LPF, we get $V_D(t)$. This type of the digital communication receiver system is also known as **correlation receiver**.

(b) The **disadvantage** of the synchronous or coherent method is that two synchronous local **carriers** of angular frequency $(\omega_0 + \alpha)\, t$ and $(\omega_0 - \alpha)\, t$ are required.

- **Derivation of FSK signal :** When $V_{FSK}(t) = A \cos(\omega_0 + \alpha)\, t$, the inputs to the comparator are:

$$a_1 = A \cos(\omega_0 + \alpha)\, t. \cos(\omega_0 - \alpha)\, t$$

$$= A \left[ \frac{\cos 2(\omega_0 + \alpha)t + 1}{2} \right]$$

and

$$a_2 = A \cos(\omega_0 + \alpha)\, t. \cos(\omega_0 - \alpha)\, t$$

$$= A \left[ \frac{\cos 2\omega_0 t + \cos 2\alpha.t}{2} \right]$$

Hence, output of the comparator is

$$V_0(t) = A \left[ \frac{\cos 2(\omega_0 + \alpha)t + 1}{2} \right] - A \left[ \frac{\cos 2\omega_0 t + \cos 2\alpha.t}{2} \right]$$

$$= \frac{A}{2} + \frac{A}{2} \left[ \cos(\omega_0 + \alpha)\, t - \cos \omega_0 t - \cos 2\alpha t \right] \quad ...(i)$$

Similarly it can be shown that when

$$FSK\ (t)\ =\ A\cos\ (\omega_0 - \alpha)\ t,$$

The output of the comparator is

$$V_D\ (t)\ =\ \frac{-A}{2} + \frac{A}{2}\ [\cos 2\omega_0 t + \cos 2\alpha.t - \cos 2\ (\omega_0 - \alpha)t]$$

$$...(ii)$$

The low pass filter (LPF) separates the dc terms from eq. ($i$) and ($ii$).

Thus, if $V_D\ (t)$ is $\left(-\dfrac{A}{2}\right)$ : output is 1

If $V_D\ (t)$ is $\left(-\dfrac{A}{2}\right)$ : output is 0.

## 14.5 PHASE SHIFT KEYING (PSK/BPSK)

($a$) In Binary Phase shift Keying system, the sinusoidal carrier wave of fixed frequency $f_c$ and fixed amplitude $A_c$ is used to represent binary symbols 0 and 1 except that the phase of carrier for one symbol (1) differs by a phase angle of 180° to the phase of carrier for other symbol (0).

Let the unmodulated carrier be represented by equation

$$c(t)\ =\ A_c \cos\ (2\pi f_c t)$$

Then the binary PSK signal S ($t$) can be written as

$$S(t)\ =\ b\ (t)\ A_c \cos\ (2\pi f_c t)\ \text{for binary symbol 1}$$

$$=\ b\ (t)\ A_c \cos\ (2\pi f_c t + \pi)\ \text{for binary symbol 0}$$

where $b(t)$ is binary data (modulating signal) as shown in the Fig. 14.15.

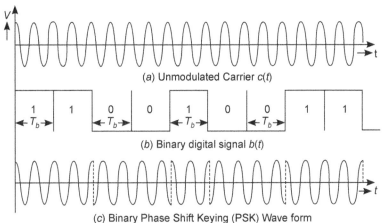

(a) Unmodulated Carrier c(t)

(b) Binary digital signal b(t)

(c) Binary Phase Shift Keying (PSK) Wave form

**Fig. 14.15**

Here, it may be noted that unlike ASK transmission, the PSK transmission is **polar**. Polarity changes in the **binary signal** $b(t)$ are used to produce 180° changes in the carrier phase. This may be achieved by using *double sidebands suppressed carrier modulation* (DSBSC), with binary signal as a polar **NRZ** waveform. The carrier amplitude is multiplied by $\pm 1$, pulsed waveform. When the binary signal $b(t)$ is $+ 1$, the sinusoidal carrier is unchanged, and when b(t) is $- 1$, the carrier is changed in phase by 180°. Binary phase shift keying is also known as **Phase reversal keying** (PRK).

(*b*) Since the PSK signal may be obtained by the product of binary data $b(t)$ (which is in polar form) and the sinusoidal carrier, the **power spectral density** of BPSK (PSK spectrum) shown in Fig. 14.16 is same as that of the polar baseband signal shifted to $\pm f_c$ (carrier frequency). The shape| of power spectral density of the BPSK signal and BASK signals are similar. The only difference is that the BPSK does not have impulse at carrier frequency.

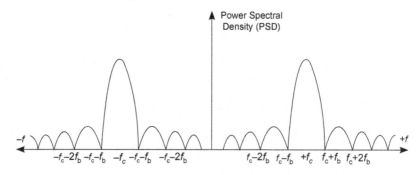

**Fig. 14.16**

The bandwidth requirements for both BPSK & BASK signals is also same. The most significant difference between the two modulations is that BASK is a *linear modulation* technique, whereas BPSK is a *non linear modulation* technique.

(*c*) The advantage of the BPSK modulation is that it has a superior performance over BASK in noisy environment.

## Generation of PSK

Binary phase-shift keying may be achieved by using the binary Polar NRZ signal to multiply the carrier generated locally as shown in the Fig. 14.17.

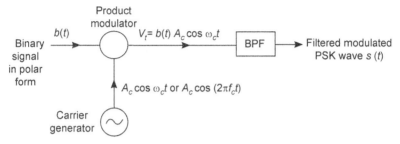

**Fig. 14.17**

For a binary signal b(t), the modulated wave may be written as:

$$v(t) = b(t) A_c \cos 2\pi f_c t$$

When $b(t) = +1$, $\qquad v(t) = A_c 2\pi f_c t,$

When $b(t) = -1$, then $v(t) = -A_c \cos 2\pi f_c t$, which is equivalent to $A_c \cos (2\pi f_c t \pm 180°)$. Band pass filtering (BPF) of the modulated wave may be used instead of base band filtering to limit the radiated spectrum. The generation technique of BASK and BPSK is same. The only difference is that the incoming binary data $b(t)$ in BPSK should be in polar form. BPSK may be achieved through the use of *double side band suppressed carrier* modulation (DSBSC).

## Derivation of PSK Signal Expression

A binary signal $V(t)$ is used to generate a waveform:

$$V_{PSK}(t) = A \cos [\omega_0 t + \phi t] \qquad \qquad ...(i)$$

The two values of $\phi(t)$ are chosen to represent 1 or 0. Normally for $V(t) = +V$ (1), $\phi(t)$ is taken as 0 and for $V(t) = -V(0)$, $\phi(t)$ is chosen as 1.

$$V(t) = \pm V.\phi(t)$$

The eq. (i) can be written as

$$V_{PSK}(t) = \frac{V(t)}{V}. A \cos \omega_0 t = A \cos \omega_0.t$$

Thus $\qquad V_{PSK}(t) = +A \cos \omega_0 t$ for $V(t) = +V(1)$

$\qquad\qquad V_{PSK}(t) = -A \cos \omega_0 t$ for $V(t) = -V(0)$

## Coherent Detection of PSK

At the coherent receiver, the received modulated carrier wave or BPSK signal will undergo further band pass filtering as shown in Fig. 14.18 (*a*) to complete the **raised-cosine response** and to limit input noise. Then the output of the bandpass filter $v'(t) = b'(t) A_c \cos \omega_c t = b'(t) A_c \cos (2 \pi f_c t)$ is passed into another multiplier circuit, where it is multiplied by replica of the carrier wave $\cos \omega_c t$ ($\cos 2\pi f_c t$). Then at O/P of multiplier is $b'(t) Ac^2 \cos^2 2\pi f_c t$ and can be expanded as $A_c^2 b'(t) (0.5 + 0.5 \cos 2\omega_c t)$. The second harmonic component of the carrier is removed by using lowpass filter, leaving the low frequency output, which is $0.5 A_c b'(t)$, where $b'(t)$ is the filtered version of the input binary $b(t)$. The Fig. 14.18 (*b*) shows the complete block diagram.

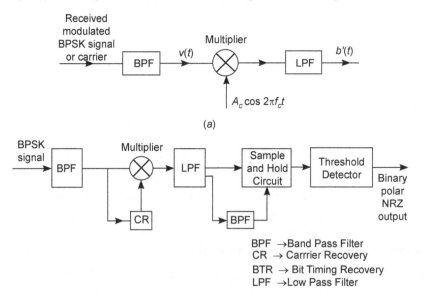

(a)

BPF →Band Pass Filter
CR → Carrrier Recovery
BTR → Bit Timing Recovery
LPF →Low Pass Filter

**Fig. 14.18**

The coherent detection necessitates recovering the unmodulated carrier phase information from the incoming modulated carrier wave, and this is achieved in the carrier recovery (CR) section as shown.

To avoid **inter symbol interference**, sampling must be carried out at the bit rate and at the peak of the output pulses. This requires sample and hold circuit to be accurately synchronized to the bit rate which necessitates a bit timing recovery (BTR) section.

The thermal noise present at the receiver will also be phase modulated. The demodulated waveform v' (t) along with noise is given to threshold detector which reproduces a *noise free output*.

## 14.6 TYPES OF PSKS

The PSK may be of two types:

(*i*) DPSK

(*ii*) QPSK

These are discussed below:

## 14.7 DIFFERENTIAL PHASE SHIFT KEYING (DPSK)

This is a type of Phase Shift Keying, in which phase of carrier is changed only if the current data bit differs from the previous one. A reference bit must be sent at the start of message. The DPSK is used to eliminate the need of the phase synchronization of coherent receiver (with PSK), with the help of a differential encoding system. The digital information content of the binary data is encoded in terms of signal transition *i.e.*, symbol '0' may be used to represent transition in given binary sequence and symbol ' 1 ' to represent no transition.

### Modulation of DPSK

The DPSK modulation technique is shown in the Fig. 14.19.

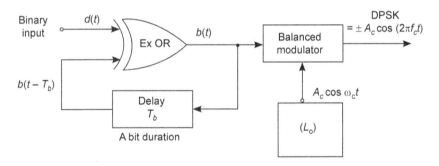

**Fig. 14.19**

The data stream to be transmitted $d(t)$ is applied to one input of an Exclusive OR gate. The second input to the same gate as its own output delayed by a bit $(T_b)$.

The output of X-OR gate is applied to the balanced modulator as one input, other input is a sinusoidal carrier generated by a local oscillator $(L_o)$ of fixed amplitude and frequency.

**Detection/demodulation of DPSK:** A method of receiving the data bit stream from the DPSK signal is shown in the Fig. 14.20. Here the received signal and the same delayed by the bit time $T_b$ are applied to the multiplier. The multiplier output is applied to a bit synchronizer and integrator.

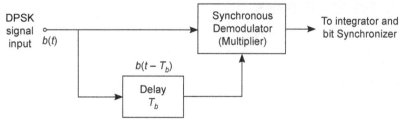

**Fig. 14.20**

The transmitted data bit $d(t)$ can readily be determined. If $d(t) = 0$ there was no phase change and $b(t) = b(t - T_b)$ both being $+1$V or both being $-1$V. In the case $b(t).b(t - T_b) = 1$. Then $d(t) = 1$, there was a phase change.

The table will make clear the differential data decoding.

**Table 14.3**   Differential data decoding.

| Shifted Differentially encoded data | 0 | 1 | 0 | 0 | 1 | 0 | 0 | 1 | 0 | 0 |
|---|---|---|---|---|---|---|---|---|---|---|
| Phase of Shifted DPSK | $\pi$ | 0 | $\pi$ | $\pi$ | 0 | $\pi$ | $\pi$ | 0 | $\pi$ | $\pi$ |
| Detected binary | 0 | 0 | 1 | 0 | 0 | 1 | 0 | 0 | 1 | 1 |

## 14.8  QUADRATURE PHASE SHIFT KEYING (QPSK)

In Quadrature Phase-shift Keying, the binary data is first converted into two bit symbols which are then used to phase modulate the carrier. The $2^2 = 4$ combinations are possible from a binary alphabet (logical l's and 0's), the phase of the carrier can be shifted to one of the four states (00, 01,10 and 11) (Fig. 14.21).

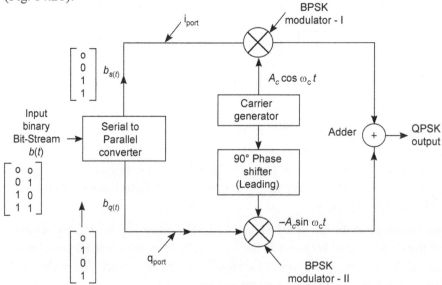

**Fig. 14.21**

The input binary bit stream $b(t)$ is converted in serial to parallel converter into

two binary streams. The serial to parallel converted bit $\begin{bmatrix} 0 \\ 0 \\ 1 \\ 1 \end{bmatrix}$ to the $i$th port,

and the same time, serial to parallel converted bit $\begin{bmatrix} 0 \\ 1 \\ 0 \\ 1 \end{bmatrix}$ to the $q$th port. In the

process, each bit duration is doubled and so the bit rates at $i$ and $q$ outputs are half that of the input bit rate.

The $b_i(t)$ bit stream is combined with a carrier $A_c \cos \omega_c t$ in a BPSK modulator, while $b_q(t)$ is combined with (90° carrier phase shift) a carrier $-A_c \sin \omega_c t$, also in another BPSK modulator.

The output of the two BPSK modulators are added to give QPSK wave. The QPSK modulator states are shown in the table.

**Table 14.4** Truth table of QPSK Wave

| $b_i(t)$ | $b_q(t)$ | QPSK Output |
|----------|----------|-------------|
| 0 | 0 | −135° |
| 0 | 1 | −45° |
| 1 | 0 | +135° |
| 1 | 1 | +45° |

The corresponding QPSK modulation phasor diagram of the truth table is shown in Fig. 14.22.

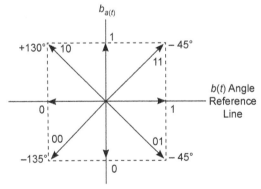

**Fig. 14.22**

## Detection of QPSK Signal

Detection of QPSK signal can be done with the help of circuit shown in the Fig. 14.23. With the incoming carrier represented as $b_i(t) A_c \cos \omega_c t - A_c b_q$

($t$) sin $\omega_c t$, after low pass filtering, the output of the two BPSK modulators are 0.5 $b_i(t)$ and 0.5 $b_q$ ($t$). These two outputs are combined in parallel to serial converter and the desired binary output signal $b(t)$ is obtained.

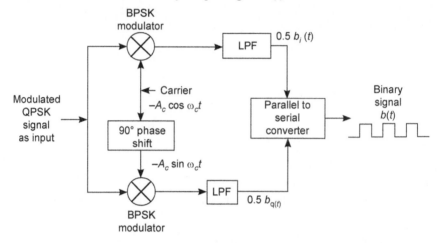

**Fig. 14.23**

## 14.9 MULTIPLEXING

This is a process used for transmitting telephonic messages or signals. It is a process of transmission of more than one signals together and simultaneously through the same line. This increases the handling capacity of the line.

Multiplexing may be of following types:

1. Time division multiplexing

2. Frequency division multiplexing

1. **Time Division Multiplexing (TDM):** The pulses are generally narrow in width and separation between them is larger. This fact is utilized in this multiplexing. The space in between two pulses can be utilized by other signals. One line is assigned to channels turn by turn. In low speed TDM, rotating switches (mechanical in nature) are used in transmitter as well as in receiver synchronized with one another. Number of channels are fed to the transmitter switch which are separated by the receiver switch. In high speed TDM, electronic switches replace the mechanical switches. If the speech wave is sampled at a frequency greater than the highest frequency present in the speech, we get an output sample wave as shown (Fig. 14.24) in which the speech signal is present and same may be obtained at the receiver.

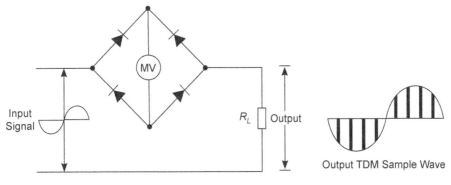

**Fig. 14.24**

For generating such a wave at the transmitter, a Multivibrator (MV) producing 10 kHz carrier wave with a bridge rectifier circuit is used. Across the load $R_L$ the sample wave is obtained. At the receiver, the same circuit detects the sample wave across its load, provided both the circuits (at transmitter as well as at receiver) work in synchronism. If two persons are provided with such a circuit they can hear each other. Moreover, a *number of signals can pass* through the same transmission line.

2. **Frequency Division Multiplexing (FDM):** This is also known as **earner telephone system**. In this system, carrier frequency is modulated by the speech signal, and as a modulated wave, we get original carrier along with side bands. The side bands which carry the signal are sent. At the receiver, the original speech signal is

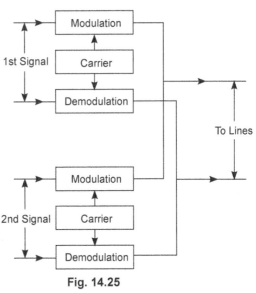

**Fig. 14.25**

detected or demodulated. By using different carrier frequencies we get different side bands of different frequencies and therefore *any number of signals* can be transmitted simultaneously through the same line. Similarly the signals can be received at the receiver by using different filter circuits.

The Fig. 14.25 shows simplified diagram of an electronic telephone exchange using FDM. Only two signals have been shown but any number of signals can be sent through a pair of lines. In this system, each channel is assigned a carrier frequency which is modulated by the channel signal. The modulated carriers then travel through the line simultaneously.

## 14.10  TRANSMISSION AND RECEPTION OF TDM

The TDM is the process of utilizing the time scale for simultaneous transmission of more than one intelligence signals on the same carrier.

In a pulse communication, if the unmodulated pulse train is having a *low duty* cycle, the interspace between the pulses can be utilized for transmission of another intelligence. A synchronizing pulse may be added for each group of pulses (consisting of one pulse from each signal) so that there is no problem in separating the signals at the receiver. The Fig. 14.26 shows transmission of two signals by TDM.

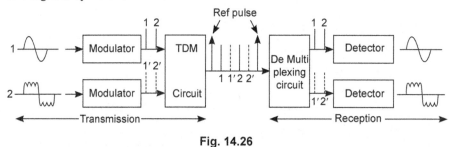

**Fig. 14.26**

At the receiver, these signals can be separated by proper synchronous detector. With the help of this method, we recover the various signals in the time domain.

## 14.11  TRANSMISSION AND RECEPTION OF FDM (FIG. 14.27)

The process of utilizing frequency scale for simultaneous transmission of more than one signal in the same carrier is called FDM. If two signals having same frequency modulate a carrier simultaneously, they will interact. Therefore to avoid interaction the signals first modulate same carrier (different for different signals) called "sub carrier" then the modulated signals which differ in frequency, modulate a common RF carrier, without any interaction. The output is a composite carrier signal, which is transmitted through an antenna.

At the receiver, this composite carrier signal is split into the individual carriers. These carriers are passed through detectors and original signals are obtained.

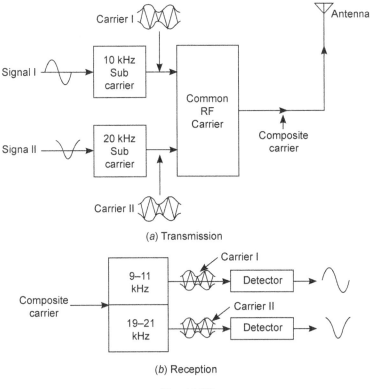

(a) Transmission

(b) Reception

**Fig. 14.27**

## 14.12 TDM vs FDM

1. **Time division multiplexing (TDM):** In this, complete channel width is allotted to one user for a fixed time slot. This technique is suitable for digital signals as these signals are transmitted intermittently and the time between two successive signals can be utilized for other signals. The TDM suffers from inter symbol interference (I.S.I).

   In case of TDM (Time division multiplexing), the signals are mixed (jumbled) in frequency domain, but separated in "time domain" such that the frequency spectra of the various signals occupy the same frequency and their wave shape identity is maintained.

2. **Frequency division multiplexing (FDM):** The signals are separately modulated and transmitted. Any type of modulation can be used, however SSB (Single Side Band) modulation is most widely used. At the receiver, the signals are separated by band pass filters and then demodulated. The FDM is used in telephony, telemetry and TV communications. The FDM suffers from the problem of **"cross talk"**.

Note that a signal is completely specified either by "Frequency domain" or time domain. In FDM (Frequency division multiplexing), all the signals to be transmitted are continuous and are mixed (jumbled) in the "Time domain" but separated (maintain there identity) in frequency domain, as spectra of the various modulated signals occupy different bands in frequency domain, *i.e.* their frequency spectra identity is maintained and can be separated by filters.

The difference between TDM and FDM can be represented graphically on a communication space, which is used to transmit information.

The Fig. 14.28 (*a*) shows that for TDM, each signal occupies a distinct time interval, not occupied by any other signal, but spectra of the signals have components in the same frequency interval.

The Fig. 14.28 (*b*) shows that for FDM, each signal is present on the channel all the time and all signals are mixed, but each of the signal occupies a finite and distinct frequency interval not occupied by any other channel.

**Note** The BW requirement for sending number of channels either by TDM/ FDM is same. If $n$ signals each of band limited to $f_m$ are multiplexed by TDM/FDM using PAM or AM/SSB techniques, the BW required is $nf_m \ H_z$ and in case, PAM or AM techniques are used, the B.W. required is $2nf_m$ Hz.

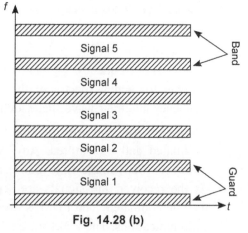

**Fig. 14.28 (a)**

**Fig. 14.28 (b)**

**Table 14.5   The TDM and FDM are Compared in Tabular Form.**

| S. No. | TDM | FDM |
|--------|-----|-----|
| 1. | The signals which are to be multiplexed can occupy the entire bandwidth but they are isolated in time domain. | The signals which are to be multiplexed are added in the time domain. But they occupy different slots in the frequency domain. |
| 2. | TDM is preferred for digital signals. | FDM is preferred for analog signals. |
| 3. | Synchronization is required. | Synchronization is not required. |
| 4. | The TDM circuitry is simple | The FDM requires a complex circuitry at the transmitter and receiver. |
| 5. | The TDM suffers from the problem of inter symbol interference (I.S.I). | The FDM suffers from the problem of cross talk due to imperfect band pass filters. |
| 6. | Due to fading, only a few channels are affected. | Due to wideband fading in the transmission medium, all FDM channels are affected. |
| 7. | Due to slow narrow band fading, all the TDM channels may get wiped out. | Due to slow narrow band fading taking place in transmission channel, only a single channel may be affected in FDM. |

## 14.13 TDM IS SUPERIOR TO FDM

The TDM is supposed to be superior in the following aspects:

(*i*) TDM instrumentation is simpler, whereas FDM requires modulators, filters and demodulators. But TDM synchronisation is slightly more demanding than that of FDM with suppressed carrier Modulation.

(*ii*) TDM is invulnerable to the usual sources of FDM inter channel crosstalk. In fact, there is no crosstalk in TDM, if the pulses are completely isolated and non-overlapping. (Since message, information or modulating signal separation is achieved by decommutation, or gating in time, rather by filtering).

(*iii*) In order to reduce crosstalk, the TDM system is provided with guard time between pulses, analogous to the guard band in FDM. Thus, a practical TDM system will have both guard times and guard bands, the former to suppress crosstalk, the latter to facilitate message reconstruction with practical filters.

(*iv*) One more point regarding the bandwidth requirement for TDM system to multiplexing '$n$' signals, each band limited to '$f_m$' requires a bandwidth of '$nf_m$' Hz and if modulated by the carrier, the bandwidth becomes '$2nf_m$' Hz. Now if the '$n$' signals are multiplexed in FDM, using a single side band (SSB) technique, the bandwidth is $nf_m$. Thus TDM system using PAM (Pulse Amplitude modulation) signals has the same bandwidth as AM-SSB systems. Similarly if AM-DSB technique is used for sending 'n' signals, the bandwidth is $2nf_m$, which is the same as required for PAM/AM systems.

TDM systems are being used more commonly in long distance telephone communications. The TDM and FDM technique ares together used in radio telemetry. Note that Radio telemetry is used for measurements on distant objects.

## 14.14 HOW THE PRINCIPLE OF TDM IS DIFFERENT TO FDM?

(a) When very short pulses are used, there is a considerable amount of unallocated time between the pulses. Take for example, the case of "pulse position modulation" (PPM), in which each pulse is 1 μs long, and maximum time deviation corresponding to peak modulation is ± 4.5 μs. Each pulse must then be allocated approximately 10 μs, whereas if the sampling rate is 10,000 times per second there are 100 micro seconds available per second. The remaining 90 μs (100 μs – 10 μs) can be utilised by other pulses, transmitting other signals.

This gives the possibility of "time multiplexing", in which successive interval of time are assigned to different channels. Thus in the case, described above, in each 100 μs period, we can simultaneously transmit 8 signals assigned successive 12 μs time interval (10 μs for the pulse + 2 μs to provide protection against adjacent channels) and send an extra long synchronising pulse during the remaining 4 μs. [(12 μs × 8) + 4 μs = 100 μs],

(b) The method for transmitting several messages on one channel by dividing the time domain slots, one slot for each message in TDM. In TDM (time division multiplexing), use is made of the fact that narrow pulses with enough space between them are sent in any of the form of pulse modulation so the space can be used by signals from other sources.

The information contents of any band limited signal are completely specified by its signal samples spaced at least $\dfrac{1}{2f_m}$ seconds apart, where $f_m$ is the maximum frequency of the signal, that is time duration (T) between the samples should be less than $\dfrac{1}{2f_m}$ second.

*i.e.*
$$T \le \frac{1}{2f_m} \text{ or } \frac{1}{f_s} \le \frac{1}{2f_m} \ (f_s = \text{Sampling frequency})$$

or
$$f_s > 2f_m$$

where $f_s \left( = \dfrac{1}{T} \right)$ is called **sampling frequency**. The time interval in between

signal samples is utilized by the samples of other signals.

The complete signals can be reconstructed from the knowledge of signal at discrete intervals. The channel is occupied only at these instants and conveys no signal for the rest of the time. During this vacant time, we may transmit the samples of other signals and accommodate the samples of other several signals on this channel. At the receiver side, these samples can be separated by proper synchronous detector. With the help of this method, we recover the various signals in the time domain.

## 14.15  TELEPHONE MODULATION SYSTEM

Speech produces much more complex modulation pattern as compared to other signals but the principle of amplitude modulation in telephony remains the same.

By having a number of carrier frequencies, telephonic conversations can be "stacked" one above the other in term of frequencies and therefore take place without interfering with one another. Upto 300 simultaneous conversations may be carried out in this way over a single pair of cable.

An illustration of such a telephonic modulation system in given in Fig. 14.29.

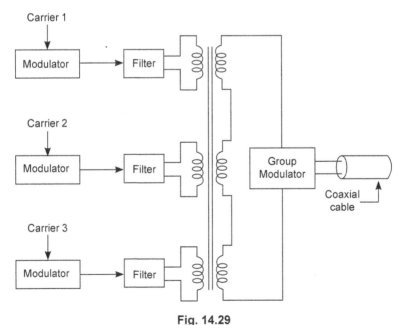

**Fig. 14.29**

## 14.16  TRANSMISSION MEDIUM CHARACTERISTICS FOR DIGITAL COMMUNICATION

The transmission medium is characterised by the following factors:

1. Bandwidth             2. Signalling Speed

3. Noise                 4. Distortion

5. Cross Talk.

1. **Bandwidth:** In digital communication, the message is in the form of pulses. The level of the pulse varies at a determined rate. The transmission medium should provide bandwidth for transmission at these pulse rate variation. The bandwidth can also be determined by the time (duration) allotted for a bit. Smaller is the bit duration, more is the required bandwidth. Most of the data transmissions utilise telephone channel, which has a bandwidth of 4 kHz.

2. **Signalling Speed:** It is also known as Baud rate. It is equal to the no. of symbols transmitted per second. Note that "Baud rate" is different from the "Bit rate". However, in two level transmission, both are equal but when more than two levels are used, "bit rate" is more than "baud rate".

3. **Noise:** In communication, the most common type of noise encountered is thermal noise. The value of signal is determined by sampling the signal at the rate of $2B$, where $B$ is the bandwidth. The noise affects the process of sampling, as the noise may get intercepted as a data bit.

4. **Distortion:** When the output waveform coming out of a channel is not an exact replica of the input waveform, **Distortion** is said to have occured. In communication "phase delay distortion" has the most significance, which occurs when signals of different frequency are passed through the same channel.

   The phase delay distortion can be reduced by "equalization" of the channel. For this "equalizer filters" are used which compensate for the "delay" of the channel.

5. **Cross Talk:** In a communication system in which more than one signals are transmitted simultaneously crosstalk is experienced. It is also experienced when two wires carrying signals run adjacent to each other. The problem becomes severe when length of the wires is large.

## 14.17 MODULATION USED IN VARIOUS COMMUNICATIONS

The type of modulation used in various communications are described below:

1. **Radio broadcasting:** It is done on long, medium and short wave bands in the frequency ranging from 20 Hz to 30 MHz. The modulation used is *amplitude modulation*. This signal travels through Ionosphere.

   The FM sound broadcasting is also in practice and it is done at VHF bands, *i.e.*, from 20 to 250 MHz. The short range of transmission, smaller wavelength, wider bandwidth and higher quality are the few good characteristics of FM transmission.

2. **TV broadcasting:** The TV broadcasting is done in VHP and UHF band of frequencies. For picture signal, AM and for sound signal FM systems are used.

3. **Satellite broadcasting:** A satellite is a transmitter installed in space. The part of earth in its range can receive signals sent from the satellite. Signals up to 30 GHz (1 GHz = $10^9$ Hz) are used in this system. For individual reception, a dish antenna of 1 m diameter may be required.

4. **Communication satellites:** A satellite is installed in space at a height of about 40,000 km. For reception, parabolic dishes of diameter 10 to 20 m are used at each station. The earth transmitter sends signals to the satellite which reflects it back to earth at another carrier frequency. The TV and telephone signals are transmitted by this method.

5. **Telephony, telex, telegraphy:** For transmission of these signals, *fixed radio links* at frequency up to 30 MHz are used. Now such links have been replaced by microwave or satellite links which give more reliability, FM is invariably used.

6. **Mobile Systems:** For radio communications between mobile units, *e.g.*, marine, ships, aircraft etc., amplitude modulation is used; but above 30 MHz, FM is used.

## SUMMARY

1. The digital signals are a string of 1 and 0. If they are to be transmitted over copper wires they can be transmitted directly at two voltage levels +ve and – ve. But if they are to be transmitted through space using antenna, some form of modulation technique is required, the modulation techniques used are:

   (*i*)    Amplitude shift keying (ASK) (analogous to amplitude modulation)

   (*ii*)   Frequency shift keying (FSK) (analogous to frequency modulation)

   (*iii*)  Phase shift keying (PSK) (analogous to phase modulation)

Here "modulation" means to superimpose the "binary waveform" over a carrier.

2. Power spectral density is a curve which shows energy distribution of a waveform (e.g. ASK, FSK, PSK) with respect to frequency.

3. Phase shift keying may be

   (*i*)   Differential phase shift keying (DPSK)

   (*ii*)   Quadrature phase shift keying (QPSK)

4. The receivers may be coherent or non coherent.

5. The multiplexing is the process of sending more than one signal simultaneously. This may be done by Time Division Multiplexing (TDM) or Frequency Division Multiplexing (FDM).

6. The characteristics of transmitting medium are: Bandwidth, signalling speed, noise, distortion and crosstalk.

□□□

# Antennas

In earlier pages, we have discussed transmission and reception of radio waves. The *transmitters* transmit or propagate and receivers receive, what has been propagated. Thus in order to couple the transmitters and receivers with the space some sort of *interface* is essential. A structure must be provided which is capable to radiate electro magnetic waves from the transmitter and to receive the same in the receiver. This structure is known as *Antenna* or *Aerial*.

## 15.1 ANTENNA

An antenna or aerial may be defined as a "metallic structure often a wire or collection of wires/conductors which *converts h.f. currents into electromagnetic waves* and vice-versa.

Transmitting antenna and Receiving antenna behave identically *i.e.*, their behaviour is reciprocal. (Fig. 15.1)

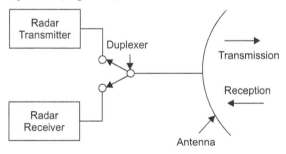

**Fig. 15.1**

The first type of conversion takes place, where an antenna is used for transmission, where it converts *h.f.* electrical energy into e.m. waves. The second conversion takes place at the receiving end where the receiving antenna converts the e.m. waves into electrical signal, which is applied at the input of the receiver.

## 15.2  ANTENNA RECIPROCITY

The phenomenon of using the same antenna for transmission as well as for reception is called **Antenna Reciprocity**.

In applications where transmission and reception are not simultaneous, single antenna can serve both the purposes equally well. The common example is **Radar** systems (Fig. 15.1) where during transmission, output of the transmitter is connected to the antenna, and during reception, the same antenna is connected to the input of the receiver. The switching over of connection is done by a device called **Duplexer**.

## 15.3  ANTENNA AS A TRANSMISSION LINE

A transmitting antenna or aerial can be considered as an open circuited transmission line. The space around the line may be considered as *load* with complete *mismatch* conditions. This line (antenna) radiates signal in the form of e.m. waves.

Consider an open circuited transmission line shown in Fig. 15.2 (*a*). It is seen that forward and reverse travelling waves combine to form a standing wave pattern on the line.

It is to be mentioned that all the transmitted energy is not reflected back by an open circuited line but a small portion of the electromagnetic energy escapes from the line and thus radiated. The radiated energy is very small, most of the energy is returned back to the source due to the follow reasons:

(*i*) A complete "mismatch" exists between source (line) and the load (space) and thus a very little energy is dissipated to the load.

(*ii*) Two wires are close together with opposite polarity, the radiation from one wire cancels the radiation from the other.

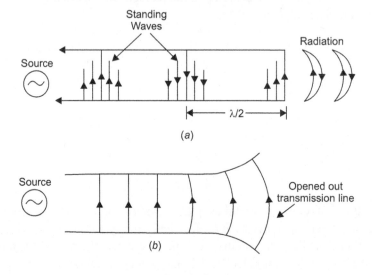

Fig. 15.2

If the lines are spread out as shown in Fig. 15.2 (*b*), due to the "enlargement" of the open circuit, the mutual cancellation of the radiation is reduced and also a better "*match*" is established of the line with the space, more energy is dissipated to the space *i.e.*, radiation increases.

Wavelength of antenna, $\lambda = \dfrac{\text{Velocity of e.m. waves}}{\text{Frequency of e.m. waves}} = \dfrac{3 \times 10^8 \text{ m/s}}{f}$

## 15.4 EVALUATION OF A DIPOLE–THE BASIC ANTENNA

The radiation efficiency of the system is further improved if the wires are bent so as to be in the same line, see Fig. 15.3. The electrical and magnetic field is now fully *coupled* with space, instead of being confined between the two wires. This is a *dipole antenna* which is the basic form of antenna today in use. When the total length of the two wires is *a half wave length,* the antenna is called a *half wave dipole.* By keeping half wave length ($\lambda/2$), even greater radiation occurs. The reason for this is that half wave dipole may be regarded as having the same impedance properties as a similar length of transmission line. Accordingly, we have the antenna behaving as a piece of "quarter wave" transmission line bent out and open circuited at the far end. This results as a high impedance at the far end of the antenna reflected as a low impedance at the end connected to the main transmission line. This means that a large current will flow at the input of the half wave dipole and efficient radiation will take place.

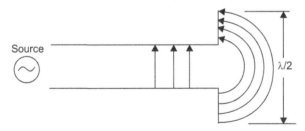

**Fig. 15.3**

## 15.5 TERMS RELATED TO ANTENNAS

The important terms related to antennas are defined below:

1. **Gain:** The gain of an antenna is not same in all directions, no doubt its radiation and the gain in one particular direction will be maximum. Accordingly an antenna have the following gains:

   (*a*) **Directive gain or directivity :** This is defined in a particular direction as the "ratio of the power density radiated in that direction by the antenna to the power density that would be radiated by an *isotropic* antenna."

Here we should note that an **isotropic antenna** may be assumed as an antenna which is "omini-directional", *i.e.*, which radiates (equally) in all directions. It is just an imaginary antenna used as a *reference*, through practically no such antenna exists.

Directive gain or directivity is a ratio of powers and its value (measured at some standard distance from antenna) is more than 1 in all cases. The value of this gain for **resonant antenna** is from 1.64 to 7.2 in free space and for **non-resonant antenna** it is from 3.2 to 17.5. The non resonant antenna has more directive gain than the resonant antenna of equal length. The directive gain is expressed in *decibel*.

(*b*)    **Power gain :** Power gain is "the ratio of the power that must be radiated by an isotropic antenna to develop a certain field strength at a certain given distance to the power fed to the antenna to develop the same field strength at the same distance in the direction of maximum radiations".

Thus this definition is similar to that of the "Directive gain", the only difference is that in case of "directive gain or directivity" the power radiated by the antenna is considered, while in the case of *power gain* "the power fed" to the antenna is taken. Thus the two terms are identical except that power gain also takes into account the antenna losses. Directivity is of theoretical, whereas power gain is of practical importance.

**Note:**

(*i*)    If $F_1$ is the field strength at a particular distance in the desired direction due to an antenna, and $F_2$ is the field strength at the distance and direction when the antenna is replaced by an isotropic antenna, the gain of the antenna = $F_1/F_2$.

(*ii*)    When antenna gain is referred in terms of power, it is called *power gain*.

Since Power $\alpha$ (voltage)$^2$, $\dfrac{P_1}{P_2} = \dfrac{F_1^2}{F_2^2} = \left[\dfrac{F_1}{F_2}\right]^2$

In decibels power gain is given as,

$$\text{P.G.} = 10 \log_{10} \dfrac{P_1}{P_2} = 10 \log_{10} \left[\dfrac{F_1}{F_2}\right]^2 = 20 \log \left[\dfrac{F_1}{F_2}\right]$$

(*iii*)    For maximum power transfer from transmitter to the antenna, the input impedance of the antenna must be equal to the impedance of the transmission line that is used for connecting the transmitter output to the antenna or the antenna to the receiver input.

2. **Antenna Resistance:** An antenna has two resistances:

   (*a*) ***Radiation resistance:*** Radiation resistance may be defined as the "ratio of power radiated by the antenna to the square of the currents $(P/I^2)$ at its feed point". It is an ac resistance like an equivalent resistance of a parallel tuned circuit. In simple words, we may say that this is the resistance, which if replaces the antenna, would dissipate same power as the antenna radiates.

   (*b*) ***Loss resistance :*** In addition to the energy radiated by an antenna, some energy is also dissipated as a result of **antenna and ground resistance**, corona and eddy currents induced in the antenna. This all lumped resistance is called as *loss* resistance.

   The value of both resistances can be found from "antenna hand-book".

3. **Bandwidth:** The bandwidth of an antenna is the frequency range over which the antenna will give satisfactory performance. In case of antennas, there are two bandwidths. One referring to the radiation pattern and the other to the input impedance. Everything being same, the radiation pattern bandwidth is equal to the difference between the frequencies, at which the received power falls *to half of the maximum.*

4. **Beamwidth:** Beamwidth of an antenna is the angular separation between its two 3 dB *down points* on the field strength radiation patterns. This is given in degrees. (See Fig 15.4). The term is used for narrow beam antennas.

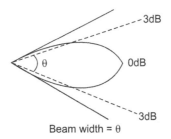

Beam width = θ

**Fig. 15.4**

5. **Law of reciprocity for antenna:** According to this law, the same antenna can be used as *transmitting* as well as *receiving* antenna.

## 15.6 OMNI DIRECTIONAL AND DIRECTIONAL ANTENNAS

(*a*) An Omni directional antenna is that which radiates and receives energy equally in all directions [Fig. 15.5(*a*)]

**Direction pattern** of an antenna is a polar plot, which gives us the strength of radiated/received signal by the antenna as a function of distance and angle. The pattern is affected by the shape and size of the antenna point of feed of RF signal etc. The 15.5 (*b*) and (*c*) show two antenna pattens.

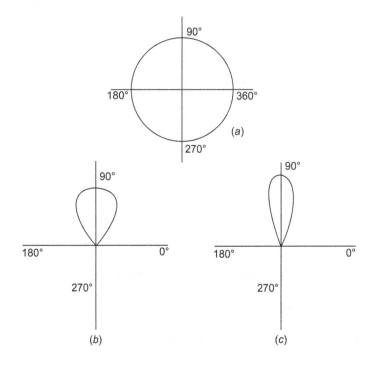

**Fig. 15.5**

(*b*) The *direction antenna* is that, which radiates in and receives from a particular direction. Figure 15.5 (b) is more directional.

## 15.7 LENGTH OF ANTENNA

The length of antenna should be some multiple of the wave length ($\lambda$) of the **applied RF energy**. This is, because the antenna must present itself as a **resonant circuit** to the RF signal so as maximise power in the antenna. In simple words, the antenna should behave as a resonant circuit at the frequency of operation. In most of the cases the selected *length of the antenna is half of the wave length of the RF signal applied*. This can be seen that the length of antenna is inversely proportional to the frequency.

## Illustration

(*a*)  At $f$ = 30 MHz = 30 × 10$^6$ Hz

Wave length,                    $\lambda = \dfrac{V}{f} = \dfrac{\text{Velocity of waves}}{\text{Frequency}}$

$$= \dfrac{3 \times 10^8 \text{ m/s}}{30 \times 10^6 \text{ Hz}} = 10 \text{ m}$$

[Velocity $V = 3 \times 10^8$ m/s]

Length of antenna $= \dfrac{\lambda}{2} = \dfrac{10}{2} = 5$ m

(b) At $f = 3000$ MHz $= 3000 \times 10^6$ Hz

$$\lambda = \dfrac{V}{f} = \dfrac{3 \times 10^8 \text{ m/s}}{3000 \times 10^6 \text{ Hz}} = 0.1 \text{ m} = 10 \text{ cm}$$

Length of antenna $= \dfrac{\lambda}{2} = \dfrac{10}{2} = 5$ cm.

## Effective Length of Antenna

The components of the line produce capacitive and inductive effects. The effective length of the antenna therefore increases. This is also called as *electric length*, it becomes necessary that length of the antenna is designed in such a way that the length is slightly less than the electrical length so as to allow compensation for the increase in length due to components of the line.

The effective or electrical length is 5% more than the physical length,

*e.g.*, if physical length is 100 m, the effective or electrical length is 105 m. In other words, a 100 m antenna behaves as if it is 105 m.

## 15.8 POLARISATION

This is the characteristic of the em wave that gives the direction of electrical component of wave w.r.t. ground. Accordingly there are three types of waves.

(i) *Horizontally polarised* waves have electric field component parallel to the ground. These waves are produced by horizontal antennas.

(ii) *Vertically polarised waves* have their electrical field component perpendicular to the ground. These waves are produced by vertical antennas.

(iii) *Elliptically polarised waves* have their electrical field component changing direction between horizontal and vertical.

The Fig. 15.6 shows linear, circular and Elliptical polarisation.

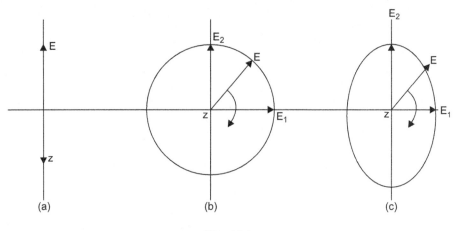

**Fig. 15.6**

At very high frequencies, for better reception, the polarisation of the received em wave should be same as that of the transmitted em wave. In other words, the receiving antenna should be vertical, if the transmitting antenna is vertical and so on.

The type of communication to be used is determined by the signal frequencies, the type of transmitting antenna, mode of propagation and polarisation etc. For communication in broadcast band, vertical polarisation is used.

## 15.9   HOW DOES AN ANTENNA RADIATE ENERGY?

When RF signal is applied to an antenna, the result is a current and voltage distribution. Now a current flowing through a conductor is surrounded by a magnetic field, whereas voltage produces an electric field.

If the RF signal applied to the antenna can be represented sinusoidally, at peak +ve point the electric field around the antenna is maximum and after that it starts decreasing. If frequency of RF signal is low, the field may collapse in the antenna, but if the frequency is very high, the field cannot collapse so fast and the result is that there is large electric field, even voltage or current across the antenna is zero.

Thus there is an electric field with no voltage. During the next (–ve) cycle when the field builds again, the previous field gets repelled by this new field. This is repeated again and again and a series of detached fields move outward from the antenna. The same phenomenon happens with magnetic field.

We know that according to the laws of em induction, a moving electric field produces a magnetic field and vice versa. The produced fields are in phase with the producer field and their direction is perpendicular to direction of propagation. The detached electrical field produces magnetic field and vice versa. These fields when added vectorially give one single wave which travels in a direction perpendicular to the electric and magnetic fields both of them at perpendicular to each other.

## 15.10 SKYWAVE COMMUNICATION

The transmitting antenna radiates signal energy in the form of em waves which travel through the layers of the *"ionosphere"*. The radiation changes with season and also with day and night. These waves propagated by antenna are called *Sky waves*.

The ionosphere consists of four layers: (Fig. 15.7)

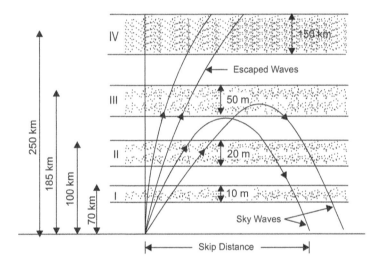

**Fig. 15.7**

   I. **First layer:** exists at a height of 70 km from grounds and is 10 km thick. It disappears at night and reflects RF signals.

  II. **Second layer :** exists at a height of about 100 km and is about 20 km thick. It also disappears at night.

 III. **Third layer :** exists at a height of about 185 km. Its thickness is about 50 km, but at night it mixes with fourth layer and thus disappears. It absorbs high frequencies.

 IV. **Fourth layer :** It exists at about 250 km with 150 km thickness which becomes 200 km at night.

Sky waves are radiated by transmitting antenna at various angles, (see Fig. 15.7).

The distance between transmitting antenna and point where sky wave is first received on earth after reflecting from ionosphere layers is known as *skip distance*. If the signal is strong it can be reflected back by the earth which is again reflected by ionosphere to the earth thus, through series of reflections, signal can be transmitted *around the world*. The angle above which sky wave no longer returns to earth and escapes into space is called *critical angle*. At high frequencies, ionosphere becomes less effective, at frequencies above 30 MHz, the signal does not return to the earth.

The maximum frequency that can be used for "sky wave communication" is known as *maximum usable frequency* (MUF). There is a different value of *muf* for each pair of points on the earth, its normal value being between 10-30 MHz.

## 15.11 TYPES OF ANTENNAS

The important antennas are described below:

### 1. Basic Antenna—A Simple Dipole [Figure 15.8 (a)]

A basic antenna is a metal rod having a length $\lambda/2$ at the frequency of operation. This is also called *half wave dipole*. Its two ends are at equal potential w.r.t. to its mid-point.

A dipole may be a "folded dipole" which consists of two $\lambda/2$ dipoles: one a continuous rod and other splitted at centre and both connected in parallel as shown in Fig. 15.8 (*b*). The transmission line is connected to the splitted rod.

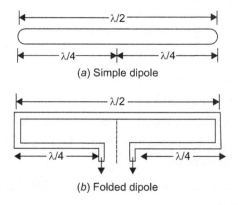

(a) Simple dipole

(b) Folded dipole

**Fig. 15.8**

## 2. Resonant Antenna

A resonant antenna corresponds to a resonant transmission line and the **dipole antennas** are the example of a resonant antenna. Such an antenna can be taken as an opened out transmission line, open circuited at the far end of the resonant length *i.e.*, a multiple of a quarter wavelength so that the length of the antenna is a multiple of half wavelength.

When length of the antenna is a full wave length, the polarity of current in one half of the antenna is opposite to that on the other half as shown Fig. 15.9. It is obvious that radiations at right angles from this antenna will be zero, because the field due to **one half** fully cancels the field due to **other half** of the antenna. The direction

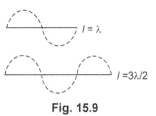

**Fig. 15.9**

of maximum radiations now for a full wave dipole is not at right angles but at about 55° from the antenna. As the length of the dipole increased to $3\lambda/2$, current distribution is changed as shown, and radiation occurs at right angles to the antenna but not maximum.

The Figure 15.10 shows radiation patterns of various resonant dipoles. Recall that the radiation pattern is a line joining points, *which have equal field intensity* due to this source. It is similar to "isothermal lines".

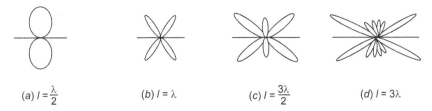

(a) $l = \dfrac{\lambda}{2}$      (b) $l = \lambda$      (c) $l = \dfrac{3\lambda}{2}$      (d) $l = 3\lambda$

**Fig. 15.10**

## 3. Non Resonant Antenna

This antenna is like a non resonant transmission line, on which there are no *standing waves*. In a correctly matched line all the transmitted power is dissipated in the terminating resistance. When antenna is terminated, about 2/3 of the forward power is radiated, the remainder is dissipated in the antenna and none is reflected back.

The non resonant antenna is unidirectional, since there are only forward travelling waves on it.

The Fig. 15.11 (*a*) shows layout and current distribution and (*b*) shows radiation pattern of a non resonant antenna. ]

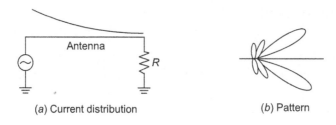

(a) Current distribution                    (b) Pattern

**Fig. 15.11**

These are of two types:

(a) **A non resonant (or a long wire) antenna:** Basically is a long wire of several wave lengths suspended at a height from the ground. No reflection of the transmitted wave takes place, about half of the energy is radiated into space. This antenna has low efficiency and requires more space to install. It is cheap and can be used for *point to point communication*. It has the advantage of increased signal pickup and share directivity along the wire itself.

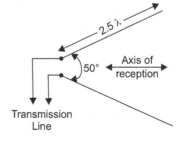

**Fig. 15.12**

The Fig. 15.12 shows long wire $V$ antenna; in which two long wires are bent to make a $V$ shape. For maximum gain the included angle should be minimum. The gain of this antenna is about 7 dB.

(b) **Long wire Rhombic Antenna:** It is a network of interconnected long antenna in the shape of a rhombus. The length and angles are to be designed properly to get the best results. It is a highly directional antenna and used for point to point communications with higher gain reception. Actually, it consists of "Two horizontal $V$ antennas". The gain of this antenna is about 10 dB. (See Fig. 15.13).

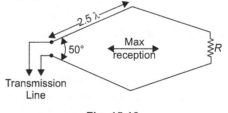

**Fig. 15.13**

## 4. Broadside Antenna

It is a network of number of dipoles of equal size spaced by $\lambda/2$ from each other. All the dipoles are fed from the same source. (See Fig. 15.14).

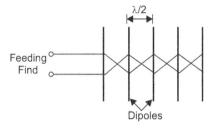

**Fig. 15.14**

## 5. Parasitic Arrays

This is the most popular antenna used for T.V. reception. When current flows in the receiving antenna, it radiates a part of the intercepted signal energy that may be wasted out. But, if a conductor approximately of λ/2 length is placed parallel to the main receiver element (half wave dipole) but not connected to it, this conductor will intercept some of the energy radiated out by dipole, therefore the lost signal is regained thus increasing the efficiency of the antenna. The conductor used for this purpose is called **parasitic element** as it is free and not connected to the dipole. A parasitic element placed behind the antenna is called **reflector** and if placed in front of the antenna is called a director. The dipole itself is the main antenna and called the **driven** element. This can be a straight or folded dipole.

A dipole with one or more parasitic elements is called as a parasitic array (*i.e.*, net work). This is the most popular TV receiver antenna as it is simple and cheap in construction, easily oriented and increases directivity and gain. The main **disadvantage** is that it is **unidirectional** *i.e.*, it can intercept signal from one direction only. The parasitic antennas may be of the following types:

(*a*) **Antenna with a dipole and reflector:** In this, only one reflector is placed behind the dipole at a distance of 0.2 λ. The length of the reflector is 5% more than the dipole. The spacing is designed carefully to get the required results. The gain of this antenna is about 5 dB. (See Fig. 15.15)

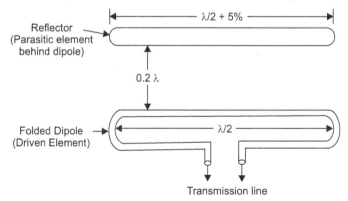

**Fig. 15.15**

(b) **Antenna with a dipole and director (Fig. 15.16):** The antenna has one dipole and one director placed infront of the dipole. The spacing of the director from dipole is about 0.15 $\lambda$. The length of the director is about 5% less than that of the dipole. As the dipole is $\lambda/2$ in length, the length of director is $\lambda/2 - 5\%$. The antenna with one parasitic element use "reflector" (instead of director) because director needs close spacing for the same gain which narrows the frequency response of the antenna.

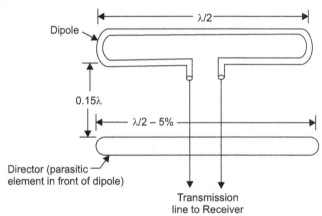

**Fig. 15.16**

(c) **Antenna with one reflector and one director:** Such antenna uses one reflector at the back and one director at the front. They operate at narrow frequencies. Such an antenna has a gain of about 7 dB. (See Fig. 15.17).

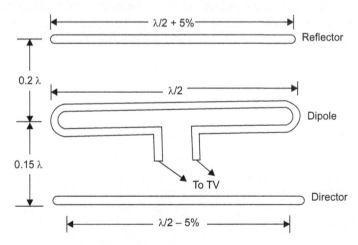

**Fig. 15.17**

(*d*) **Antenna with one reflector and more than one directors** *i.e.,* **yagi-uda Antenna:** This is a compact, high gain, narrow bandwidth network antenna used in low signal areas to cover several adjacent TV channels. The gain of yagi-uda antenna with 3 parasitic elements is about 10 dB. A high impedence folded dipole is used so that the antenna impedence can be about 300 Ω. We can increase the number of directors but generally one reflector is sufficient as additional reflectors do not improve performance of the antenna. Each director reduces its length by 5% of the previous, so that antenna is *tapered* in the direction of transmission. The elements get currents induced from the main element (folded dipole) or directly from the transmitted waves.

The Fig. 15.18 (*a*) shows symbolic Yagi-Uda antenna, (*b*) shows pattern and (*c*) shows its optical equivalence. Figure 15.18 (*d*) shows complete yagi-uda antenna.

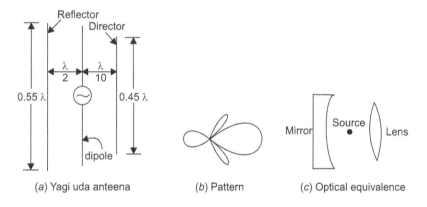

(*a*) Yagi uda anteena    (*b*) Pattern    (*c*) Optical equivalence

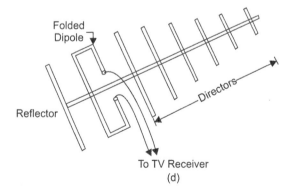

To TV Receiver
(d)

**Fig. 15.18**

(*e*) **Antenna with dipole and corner reflector:** As shown in Fig. 15.19, this has a reflector constructed as a corner conducting sheet behind the dipole. The reflector can be a solid metal sheet or a wire. The dipole insulated from the reflector is mounted along the line bisecting 90° angle. This antenna is useful for VHF bands.

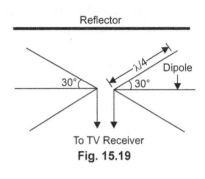

**Fig. 15.19**

## 15.12 UHF AND MICRO WAVE ANTENNAS

The transmitting and receiving antenna used at UHF (0.3 to 3.0 GHz) and micro waves (1–100 GHz) must be directive, because they are used in radars and such like purposes, where their function is *direction finding*, hence the need of a *directive* antenna is must. The other fields of their use is *micro wave communications* where a directional antenna helps a lot due to its "high gain".

The *yagi uda* antenna mostly used as a TV receiving antenna discussed previously is a VHF (30– 300 MHz) antenna. The VHF region is an "overlap" region, therefore VHF antennas can also be used in UHF region.

Here we will discuss some antennas used in UHF/micro wave regions:

### 1. Antenna With Parabolic Reflectors (Dishes)

A "parabola" may be defined as the "locus of a point which moves such that its distance from a straight line (*called directrix*) and from another point (*called Focus*) remains constant".

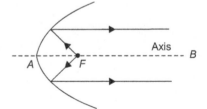

The Fig. 15.20 (*a*) shows a parabola whose focus is at *F* and axis is *AB*. If a source of radiation is placed at *F*, all waves coming from the source will be reflected by parabola as a result radiation will be very strong and concentrated along axis *AB*. The practical reflector is a three dimensional surface called *paraboloid or parabolic reflector or a micro wave dish*. According to the "principle of reciprocity" the properties

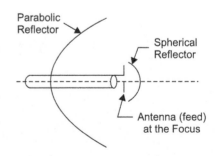

**Fig. 15.20**

of an antenna are independent whether it is used for transmission or reception purposes. The gain of such an antenna is also very high as the "dish" acts as a strong reflecting mirror and collect radiations from a large area and concentrates at the focal point. The *antenna feed* is placed at the focus of the parabolic reflector, the arrangement gives best results in transmission as well as reception. (Fig. 15.20*b*)

The radiations from the "feed" which are not reflected by the parabola, spread out in all directions and spoil the directivity of the antenna. For this, a spherical reflector can be put [(See Fig. 15.20 (*b*)] which will redirect all such radiations back to parabola. At the place of a reflector, a yagi"dipole" can also be used. The another way to deal with the problem is to use a "horn antenna' pointing at the main reflector.

## 2. Cut Parabolic Reflectors

A part (cut piece) of a parabola can also be used in place of full parabola. The only disadvantage is that the beam is not so directional as in the case of full parabolic reflector. In many applications like in "ship to ship" radar, directivity is immaterial. Figure 15.21 (*a*), (*b*), (*c*) show various shapes of cut parabolic reflectors.

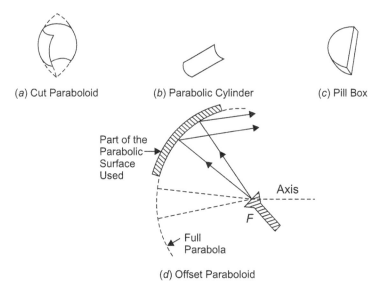

(*a*) Cut Paraboloid    (*b*) Parabolic Cylinder    (*c*) Pill Box

Part of the Parabolic Surface Used

Axis

F

Full Parabola

(*d*) Offset Paraboloid

**Fig. 15.21**

The another form of cut parabolic is an "offset paraboloid" Fig. (*d*) in which focus (F) is located outside the aperture. If an antenna feed placed at the focus, the reflected rays will pass above it, without any interference.

## Beam Width of Paraboloid Reflector

The directional pattern of an antenna using a paraboloid reflector has a main lobe in the direction of *AB* (See Fig. 15.22). If the feed antenna is non directional, the paraboloid produces a beam of radiations, whose beam width is given by

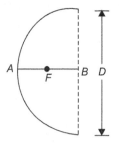

$$\phi = \frac{70\,\lambda}{D}$$

**Fig. 15.22**

Where                    $\phi$ = Beam   width   between half power points

$\lambda$ = Wavelength.

D = Mouth diameter.

and beam width between nulls is given by

$$\phi_0 = 2\phi$$

## 3. Horn Antennas (Fig. 15.23)

In the previous pages, we have explained that electromagnetic waves will travel from one point to another if suitably radiated. It has also been explained how it is possible to "guide" radio waves from one point to another by the use of "transmission lines". Now we introduce a new term *i.e.*, *"wave guide.* < = Any system of conductors (like transmission lines) for carrying electro magnetic waves can be called a wave guide, though in practice this name is used for specially constructed hollow metallic.pipes-rectangular or circular. As transmission lines are used at low frequencies, in the same way, wave guides are used at microwave frequencies but wave guides are preferred as they incur less loss.

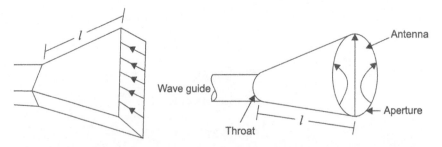

**Fig. 15.23**

As mouth of the transmission lines is opened out to get a dipole structure, similarly mouth of a wave guide is opened out to get a horn antenna.

Thus a horn antenna is nothing but a "wave guide terminated by a horn".

A horn antenna provides a better impedance match with the space around. Hence all the energy travelling in the wave guide is radiated as a result, directivity is improved, diffraction reduces, and a very little energy is returned back.

## 4. Lens Antennas

In a parabolic reflector, we have seen how principle of optics were utilised to receive or transmit a signal. Lens antenna also utilises the same principles and is very useful at 3 GHz or at even higher frequencies.

Figure 15.24 (*a, b*) shows the basic principle of a lens antenna. It works in the same manner as a glass lens used in optical instruments.

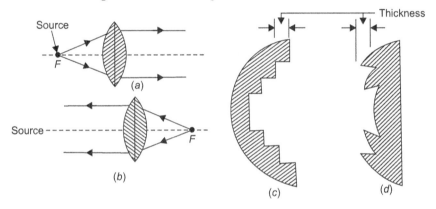

**Fig. 15.24**

A lens antenna also proves the principle of reciprocity of antennas. If a source is kept at the *focus* of a convex lens, the rays are diverged that gives us a *transmitting lens antennas*. (See Fig. *a*) On the other hand, if it is used as a receiving antenna, it will converge all the signal waves at its focus. (See Fig. *b*)

Lens antenna suffers from a serious problem that it requires excessive thickness at frequencies below 10 GHz. To remove this problem, the lens antenna in practice are *stepped* that cures the problem of greater thickness at the centre of the antenna, when used at lower micro wave frequencies.

Figure 15.24 (*c and d*) shows two stepped lens antennas.

## 15.13 SPECIAL ANTENNAS

These are wideband antennas *i.e.*, which can be operated over a wide range of frequencies. These antenna have special applications in TV broadcasting, radar and satellite communications. Some are described below:

## 1. Helical Antenna

It is a broadband VHF and UHF antenna. (Fig. 15.25).

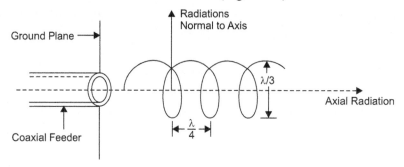

**Fig. 15.25**

It consists of a loosely wound helix backed up by a "ground plane" which is a *screen* made of "chicken wire". There are two modes of radiations *i.e.*, *towards* axis and *normal* to the axis. The helical antenna is used for transmission and reception of VHF signals through "ionosphere". It is widely used for satellite communications.

## 2. Discone Antenna

A discone antenna is a combination of a *disc* and a *cone* as shown (See Fig. 15.26). The disc acts as a "reflector" and the antenna has an enormous bandwidth. However, it is a low gain but "**all directional**' antenna. It is used as VHF and UHF receiving antenna at the airports, where it maintains communication with the aircrafts coming from all directions.

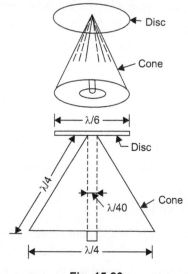

**Fig. 15.26**

## 3. Direction Finder (DF) Loop Antenna

The DF loop antenna is used for "direction finding" purposes and may be mounted on a portable radar receiver.

Basically, it is a single turn coil carrying RF currents. The loop of the wire is surrounded by magnetic field everywhere perpendicular to the loop. It acts as a dipole and its working is independent of its shape. The Fig. 15.27 shows a circular and a rectangular loop antenna.

No radiation is received normal to the plane of the loop, this makes it useful for direction finding purposes. Due to its being small and simple in size it is very much suitable to be mounted on portable radar receivers, and the output of the receiver may be connected to a meter and thus distance of the *target* can also be calculated.

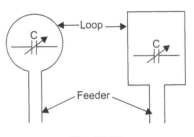

**Fig. 15.27**

Having tuned to the desired transmission, the loop is rotated till the received signal is *maximum*, the plane of the loop is now perpendicular to the direction of radiations. Loops are some times provided with several turns and also with a fertile core.

## 4. Mobile Antenna [Car Antenna]

Most of the medium frequency broadcast antennas fall in this category. This type of antenna is also known as "Marconi Antenna", and made up of a vertical mast, pole or rod, which becomes the main radiating conductor. It may be supported by insulated guy wires and is placed at a good electrical ground.

These antennas are used at VHF for mobile service because of their simple construction. The radiator length may be few feet (or few inches). For fixed antennas, a series of radial ground plane rods may be used (Figure 15.28 *a*) while for the mobile antenna [Fig. 15.28 (*b*)] the metal top of the vehicle may be used as ground.

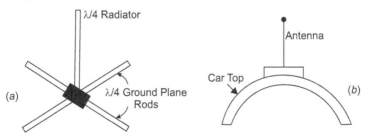

**Fig. 15.28**

## 5. Mobile Phone Antenna (Fig. 15.29)

This antenna is made for mobile phone signals. When the mobile signals are weak, *i.e.*, in remote areas, this antenna helps in listening the messages without disturbances.

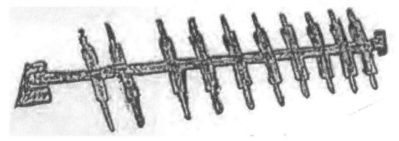

**Fig. 15.29**

## 6. Ferrite Rod Antenna (Fig. 15.30)

This antenna is made by winding a coil on a ferrite rod. The ferrite exhibits a property called "Ferrimagnetism". It has high permeability as well as high bulk resistivity. In other words, at high frequencies, we can make coils with high $Q$ factors. Moreover, a high length – diameter ratio of the rod gives desirable higher permeability. The size and positioning of the coil on the rod

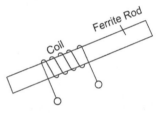

**Fig. 15.30**

are also deciding factors for permeability and $Q$ factor. For medium frequency band, about 175 turns coil is required.

This antenna is a modification of loop antenna.

## SUMMARY

1. An antenna or aerial is a metallic structure or wire, or collection of wires which converts h.f. currents into e.m. waves and vice versa.

2. The same antenna can be used for transmission as well as for reception.

3. An antenna may be considered as a transmission line with open circuited ends.

4. An omni directional antenna is that which radiates (and receives) energy equally in (and from) all directions.

5. The required length of antenna is inversely proportional to the frequency. The effective or electrical length is five percent more than the physical length.

6. The transmitting antenna radiates signal energy in the form of e.m. waves which travel through the layer of the Ionosphere. The waves propagated by antenna are called sky waves and the communication is called "skywave communication".

7. The antenna may be resonant antenna or non resonant antenna.

8. For receiving TV signals, yagi antenna is the most popular.

9. The parabolic or dish antenna are U.H.F. and microwave antenna.

10. Other antennas are Helical antenna, Lens antenna, Mobile antenna.

❑❑❑

# 16

# Television –
# Monochrome (T.V.)

Television is a very popular audio video device. It is now a basic need of every house. It is a source of entertainment and education.

## 16.1 TELEVISION

"Tele-Vision" means to "see at a distance". The visual information in the picture is converted into electrical signal for transmission. At receiver it is reconverted into its original form. In monochrome TV, the picture is reproduced in *black and white* (whereas in colour TV, it is reproduced in its original colours).

The TV was invented by J.L. Baird.

J.L. BAIRD

## 16.2 TV APPLICATIONS

Below important TV applications are described in brief:

(*a*) *Cable TV (CATV)*: When signal is to be sent to shorter distances, a cable is connected from transmitter to the receiver. Generally modulation is not required. The signals are distributed by co-axial cable to the consumers. (See Fig. 16.1)

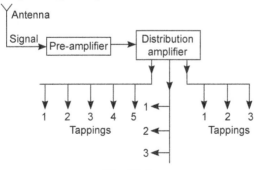

**Fig. 16.1**

(b) *Close circuit TV (CCTV)*: In this system, video signal output of camera is connected by cable directly to the monitors placed at distance. Thus picture is reproduced on each monitor. (A monitor is a video display device. It is a TV receiver without RF and IF circuits). The CCTV equipment is available for monochrome (B & W), as well as colour TV. A network of co-axial cable is used for connection. [See Fig. 16.2 (*a*), (*b*), (*c*)]

(*a*) Camera directly linked with one monitor

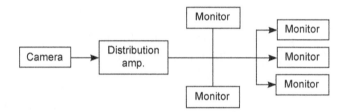

(*b*) Camera linked with several monitors

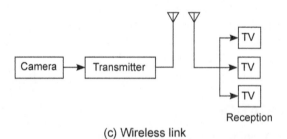

(c) Wireless link

**Fig. 16.2**

(c) *Picture phone:* This is a *"telephone plus TV"*. We can see as well as hear each other by this phone.

(d) *Fax (Facsimile)*: This is electronic transmission of visual information over telephone lines. It is also called *"slow scan TV"* as scanning in this case is slow. Only still picture can be transmitted on fax.

(e) *Satellite TV*: By installing a satellite in space we can transmit signals on TV receivers over large areas. Satellites are used as *relay stations* to provide world wide TV broadcasting.

The Fig. 16.3 shows block diagram of a typical satellite transmission. A high power satellite is installed at a height of about 36000 km. The TV programs from an earth station are transmitted to the satellite at 6 GHz (FM carrier) with the help of a dish antenna.

The downward transmission from the satellite is done at 850 MHz. The earth transmitting station may be equipped with a dish antenna and the receiving station may also have a dish antenna to receive back the signals reflected from the satellite.

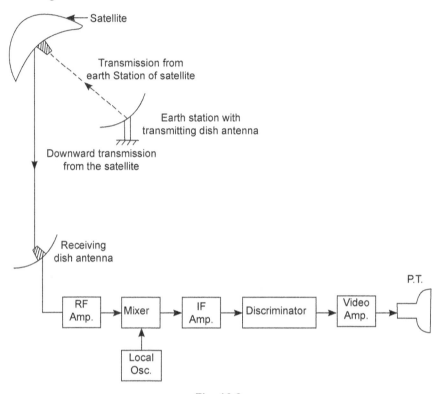

**Fig. 16.3**

(*f*) *TV Games*: The earlier TV games were using TTL (Transistor-Transistor Logic) circuits. Now with the development of *microprocessors*, more sophistication has been obtained. Basically all TV games are logic circuits.

A TV game has two important parts. The *first* part is the "game unit" (which further includes control, logic circuit and RF oscillator circuit) and *second* part is the TV receiver. The system has been shown through a block diagram in Fig. 16.4.

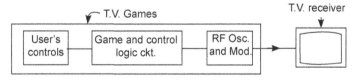

**Fig. 16.4**

## 16.3 BROADCASTING, TRANSMISSION AND RECEPTION OF MONOCHROME TV

(*i*) *"Broadcasting"* means to *"cast broad' i.e.,* to send a massage in all directions. The TV signal has audio as well video information. The transmitting antenna radiates TV signals in the form of electro magnetic waves that can be picked up by a TV receiver placed in the range. The amplitude modulation is used for the picture signal and frequency modulation for the sound signal.

(*ii*) *Transmission (Fig. 16.5a)*: In a TV studio, audio signal produced at mike is converted into corresponding electrical pulses and fed to amplifier. Before picture transmission, the picture signal (light energy) is converted through camera into electrical pulses. This video signal (electrical pulses) are given to a video amplifier. The both *i.e., sound as well as the picture* are sent to the transmitting antenna for transmission.

(*iii*) *Reception (Fig. 16.5b)*: Separate carrier waves are used for sound and picture signals but they are radiated by one transmitting antenna. At TV receiver also, the same antenna is used to pick up both signals. The signals in the receiver are amplified and separated. The picture signal is given to picture tube, which by its transducing action converts it into original picture produced in the camera. The sound signal is fed to the loudspeaker which gives the original sound produced before mike.

During transmission, to convert picture into video signal (electrical pulses), the camera *scans* the picture into horizontal and vertical lines (the picture is divided into 625 lines).

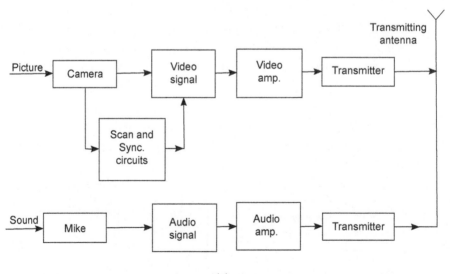

(*a*)

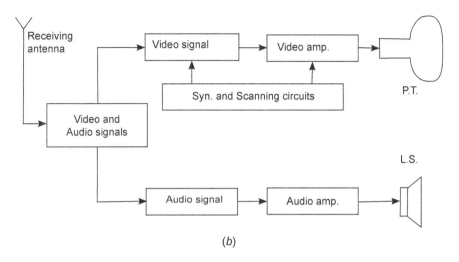

(b)

**Fig. 16.5**

This process is known as *scanning*. Similarly during reception, the picture tube re-assembles these lines to obtain the original picture. It is important that both the scanning processes taking place in camera as well as in the picture tube should be property *synchronized*. This is done by "scanning and synchronizing circuits".

## 16.4 TV CAMERA

Camera is the first and basic equipment in a TV. The input to a camera is the light from the picture or scene to be televized and output obtained from camera is the electrical pulses corresponding to the information contained in the picture.

The TV camera is just *analogous* to human eye. The basic principle of all TV cameras is based on the fact that each picture may be assumed to be composed of small elements with different light intensity. The camera picks up each element and by *transducing action* converts it into "electrical signal" proportional to its brightness. There is a photosensitive layer called *target or image plate* in each camera which performs this job. At the same time simultaneous pick up of this information is also necessary. For this purpose, there is an *electron gun* (which produces an electron beam) which *scans* the image plate at a fast speed. Thus *opto-electric conversion* as well as *pick up* of the signal takes place simultaneously and at a fast speed.

## 16.5 WORKING PRINCIPLE (FUNCTION) OF CAMERA

(*a*) The camera works on one of the two principles:

    (*i*) *Photo emessivity:* Certain metals emit electrons when light falls on them. The property is called photo emission. The emitted electrons are called *photo electrons* and the surface is called *photo cathode*. Generally, *Cesium* or *Bismuth* are used for making photo cathodes of cameras.

(*ii*) *Photo conductivity:* is the property, according to which a metal loses its resistance when light is incident upon. Generally *selenium* and *lead* make good photo conductive surfaces in cameras.

(*b*) A video camera performs the following functions (Fig. 16.6):

(*i*) If converts the picture signal into electrical pulses of equivalent brightness. The picture is composed of small elements of different brightness. There is a photosensitive plate called "Image plate or Target plate", which picks up each element of the picture and converts into electrical pulse of varying brightness.

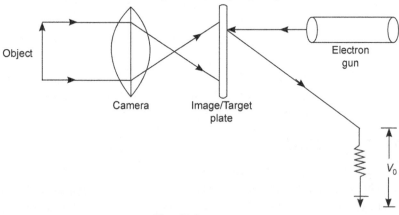

**Fig. 16.6**

(*ii*) The camera has also an "electron gun" which "scans" the image plate and picks up these pulses simultaneously at a very fast speed.

## 16.6 BASIC CONSTRUCTION OF CAMERA

The following are the components of cameras:

(*i*) *Target/Image Plate/Signal Plate*: Construction of target is different in different cameras, but the basic is same. The plate has a photo emissive/conductive coating which is *scanned* by the electron gun of the camera tube. By means of photo electric effect, the picture (visual) signal is converted into electric signal. The target is also called *signal plate or image plate*.

(*ii*) *Electron multiplier* (*Fig.* 16.7): Electron multipliers are like amplifiers which amplify the photo electric current obtained from picture. They work on the principle of *secondary emission*. The electron multiplier has a series of electrodes called *dynodes* each at a progressively higher positive potential. The Fig. 16.7 shows five such dynodes. From photo cathode, which is at zero potential, the primary electrons are made to bombard the first dynode which gives secondary electrons. These electrons bombard to each dynode as shown and finally we get a high anode current. The device is noise free, whereas the conventional amplifiers produce much noise.

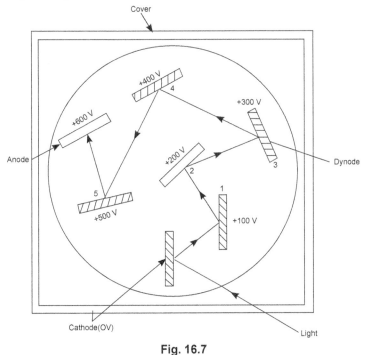

**Fig. 16.7**

## 16.7 RESOLUTION OF A CAMERA

By resolution of a camera we mean, how efficiently a camera can *scan* the picture element. If resolution of camera is high, we can see finer details of the picture *e.g.*, wrinkles on the face, or a *til/mole* on the forehead etc. Resolution is given in percentage.

## 16.8 PERSISTENCE OF VISION AND ASPECT RATIO

(*a*) Our eye can retain a picture only for 50 ms after the picture is removed. This property of eye is known as persistence of vision. Video systems' (TV etc.) work on this principle.

(*b*) It has been found that the scanned picture should have a rectangular format with width to height ratio of 4 : 3 called *Aspect ratio.* This ratio is most pleasing to eyes with least fatigue.

## 16.9 VARIOUS TV CAMERAS

The various TV cameras are:

(*i*) Image Orthicon                (*ii*) Vidicon                (*iii*) Plumbicon

These are discussed below:

## 16.10 IMAGE ORITHICON (I.O.) CAMERA

This camera lube can be divided into 3 sections—image section, scanning section and the multiplier section. (See Fig. 16.8)

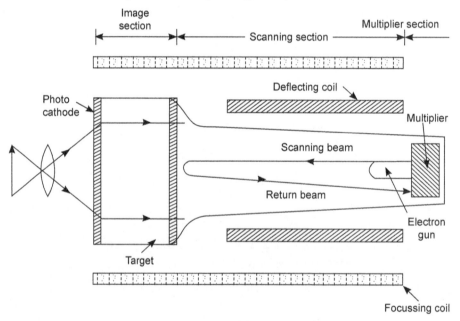

**Fig. 16.8**

## Working

Light from the scene is focussed on the photo cathode. This gives an electron image corresponding to the picture. Needless to tell, that the *cathode* is made of photo emissive material like cesium oxide. This electron image is accelerated to the *target* to produce secondary emission. The secondary emission produces on the target a pattern of positive charges corresponding to the picture, the *white* colour being most positive. The scanning beam from the electron gun neutralises these positive charges and the excess electrons of the beam return back to the gun.

As the beam *scans* the target, the electrons turning back provide a signal current in accordance with the picture. The signal current is maximum for the *black* colour of the picture. Now the signal current passes through the *multiplier* section, where it is amplified. The amplified current passes through the load and produces the camera signal output voltage. If signal current is assumed as 5 μA and load as 20 K, the output voltage is 5 μA × 20K = 0.1 V.

## 16.11 VIDICON CAMERA

Vidicon is a very small camera tube of about 8″ long and 1.5″ dia. It is also simple in construction as it has only *target plate* and the *gun*.

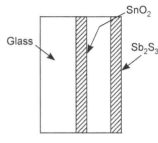

**Fig. 16.9**

(*a*) *The target* (See Fig. 16.9) has two layers, one layer is of transparent glass coated with some conducting material usually *Tin oxide*. The second layer is coated with thin photo conductive material like Antimony compound ($Sb_2S_3$). The Tin oxide acts as positive terminal for the target.

(*b*) *Electron Gun*: The electron gun has a heated cathode, a control grid and a focussing grid. The deflection of the beam produced by the gun is used for scanning, and the direction of the beam is controlled by *deflecting* coils. [See Fig. 16.10 (*a*)]

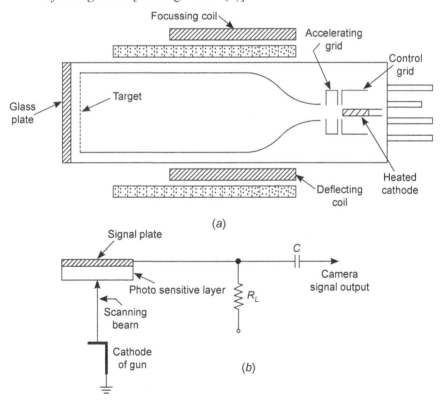

**Fig. 16.10**

## Working

With the optical image focussed on the target (or signal plate), it produces a *charge image* which is scanned by the electron gun. Each point in the charge image has a different positive potential on the side of the target facing the gun. The beam from the gun deposits its electrons on the photo layer surface of the target reducing the positive potential to zero. Excess electrons, are returned back but they are not used in vidicon, though they were used in the orithicon. T his change in potential of the target plate causes signal current to flow in the circuit. This has been shown in Fig. 16.10 (*b*) producing voltage across $R_L$. For black in the picture where photo layer is less positive than the *white*, the deposited electrons cause a small change in signal current, hence the signal current results from the change in the potential difference between the two surfaces of the photo layer.

### Characteristics of Vidicon Camera Tube

(*i*) *Dark current sensitivity*: When the camera lens is closed, very small signal current flows in the target circuit of the tube as the target is in total darkness. This small current is known as, *dark current*. The value of dark current in vidicon is about 20 nano amp. See Fig. 16.10 (*c*). The dark current affects the *sensitivity* of the tube. For higher value of dark current, sensitivity is more. A typical value of sensitivity of vidicon is 120 µA per lumen illumination on the target.

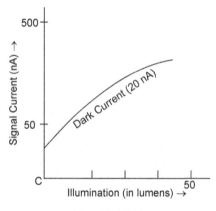

**Fig. 16.10** (*c*)

(*ii*) *Resolution*: As defined already resolution offers the smallest details of the picture that can be *resolved* by the camera. The resolution of vidicon is about 55% which can be said as satisfactory.

(*iii*) *Signal/Noise Ratio*: The S/N ratio of vidicon is about 47 dB. It is a quite high value.

#### Applications

Vidicon is compact in size and simple in operation. It is just a target and gun assembly. It is used in outdoor and indoor shootings with black and white TV.

## 16.12 PLUMBICON CAMERA TUBE

Plumbicon is similar to vidicon. The electron gun is also similar to that of vidicon but with a different target plate which is basically a PIN ($P$, $I$ and $N$ type) semi conductor diode. On one side of the plate, PBO ($P$ type semi conductor material) is deposited and on the other side a layer of $SnO_2$ ($N$ type material) is deposited—both separated again by a PbO layer which is an intrinsic (I type) material. Thus a *PIN* diode is formed. The $SnO_2$ coated side works as a signal plate. [See Fig. 16.11 (*a*)].

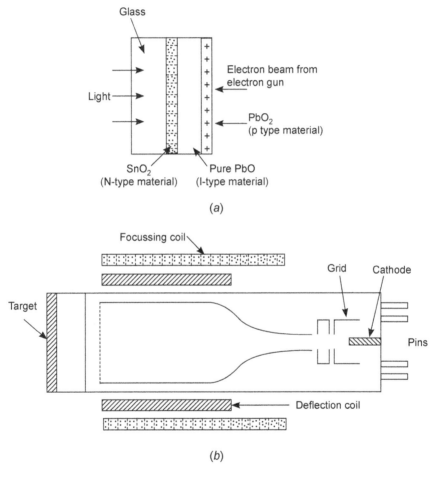

(*a*)

(*b*)

Fig. 16.11

*Construction of Plumbicon* [See Fig. 16.11 (*b*)]

The construction of plumbicon is just similar to the vidicon. It is also a target and gun assembly. It also has a cathode, that emits electron beam which is controlled by deflecting and focussing coils.

## Working (Fig. 16.12)

The $SnO_2$ side is connected to a supply of 40 V through a load $R_L$, across which output signal voltage is developed. When the electron beam *scans* the target, the signal current varies in accordance with the light of each part of the picture.

The working of plumbicon is also similar to the vidicon, with the difference that as in vidicon each element acts as a *leaky capacitor*, in plumbicon, it acts like a capacitor in series with light controlled diode, without light the diode is inverse biased and there is no output.

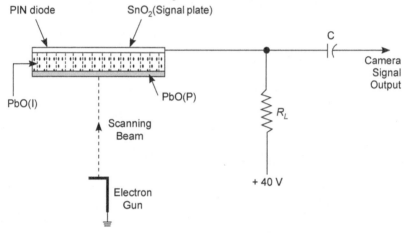

**Fig. 16.12**

When there is no light, the *PIN* diode is reverse biased, and a negligible current flows. When light falls on the PIN diode (target), it becomes forward biased in accordance to the intensity of light. The bias is the result of photo excitation of $P$ layer and intrinsic layer of the target, thus the target behaves as a capacitor in series with the *PIN* diode, and output is obtained.

### *Characteristics of plumbicon*

(*i*) *Dark current and sensitivity*: The dark current of plumbicon is nearly 1 nano amp. and sensitivity is better than vidicon. The value being 400 μA per lumen illumination of the target.

(*ii*) *Resolution*: The resolution of plumbicon is poor than vidicon. The typical value is 45%.

(*iii*) *S/N Ratio*: The SN ratio of plumbicon is higher than vidicon. The typical value is 52 dB.

**Application**

It is also simple in construction and working, therefore, used in studio as well as outdoor shootings specially with colour TV.

## 16.13 HUMAN EYE

Human eye is just like a colour camera. An eye is an important video system of human beings. Simplified diagram of the human eye is given in Fig. 16.13. The main parts of the human eye are; cornea (*C*), iris (*I*), pupil (*P*), ciliary muscles (*M*), convex lens (*L*), retina (*R*) and optic nerve (*N*).

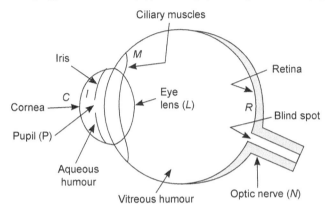

**Fig. 16.13**

The front part of the eye is called cornea. It is made of a transparent substance and it is *bulging outwards*. The light coming from objects enters the eye through cornea. Just behind the cornea is the iris (or *coloured diaphragm*). There is a hole in the middle of the iris which is called pupil of the eye. The pupil appears *black* because no light is reflected from it.

The eye-lens is a *convex lens* made of a transparent and flexible material like jelly made of *proteins*. The eye-lens is held in position by ciliary muscles. *The focal length* of eye-lens (and hence its converging power) can be changed by the action of ciliary muscles.

The screen on which the image is formed in the eye is called retina. The retina is behind the eye-lens and at the back part of the eye. The retina of an eye is just like the film in a camera. It should be noted that there is an *eye-lid* in front of the eye which is just like the shutter in a camera. When eye-lid is open, light can enter the eye but when eye-lid is closed, no light enters the eye. The space between cornea and eye-lens is filled with a viscous liquid called '*aqueous humour*'. And the space between eye-lens and retina is filled with another liquid called 'vitreous humour'.

*Working of the eye*: The light rays coming from the object kept in front of us enter the pupil of the eye and fall on the eye lens. The eye-lens is a convex lens, so it converges the light rays and *produces a real and inverted image of the object on the retina*. The image formed on the retina is conveyed to the brain by the optic nerve and gives rise to the sensation of vision. Actually, the retina has a large number of *light-sensitive* cells. When the image falls on the retina, these light-sensitive cells get activated and generate *electrical signal*. The retina sends these electrical signals to the brain through the optic nerve and gives rise to the sensation of vision. Although the image formed on the retina is inverted, our mind interprets the image as that of *an erect object. A small region of the retina where the optic nerve enters is insensitive to light and it is called blind spot.*

## 16.14 ACCOMMODATION POWER OF HUMAN EYE

A normal eye can see the distant objects as well as the nearby objects clearly. We shall now discuss how the eye is able to focus the objects lying at various distances. An eye can focus the images of the distant objects as well as the nearby objects on its retina by changing the *focal length of its lens*. The focal length of the eye-lens is changed by the action of ciliary muscles. The ciliary muscles can change *thickness of the eye-lens* and hence its focal length which, in turn, changes the converging power of the eye-lens. Let us see how it happens.

(*i*) When the eye is looking at a distant object (at infinity), the ciliary muscles of the eye are fully relaxed and the eye-lens is very *thin* (*or less convex*) in this position. Since the eye-lens is very thin, its focal length is maximum in this position and it can form the image of the distant object on the retina as shown in Fig. 16.14 (*a*). When the eye is looking at a distant object, the eye is said to be '*unaccommodated*' because it is in the relaxed state.

(*ii*) When the same eye has to see the nearby objects, the ciliary muscles become tense due to which the eye-lens becomes *thick* (or more convex) and its focal length decreases. Due to short focal length the converging power of *eye-lens increases* and it can focus the nearby objects on the retina. This is shown in Fig. 16.14 (*b*), in which an object *O* is near to the eye. It has been focussed by the eye-lens to form an image on the retina. When the eye-lens becomes more convex to focus the nearby objects, the eye is said to be '*accommodated*'. We can now say that: **The ability of an eye to focus the distant objects as well as the nearby objects on the retina by changing the focal length or converging power of its lens is called accommodation.**

The maximum "accommodation" of a normal eye is reached when the object is at a distance of about 25 cm from the eye. After this the ciliary muscles cannot make the eye-lens bulge more. So, an object placed at a distance of less than 25 cm cannot be seen clearly by a normal eye because all the power of accommodation of the eye has already been used. Thus, **a normal eye has a power of accommodation which enables objects as far as infinity and as close as 25 cm to be focussed on the retina.**

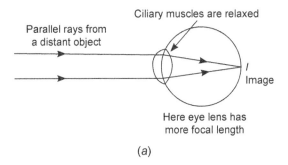

Ciliary muscles are relaxed

Parallel rays from a distant object

*I* Image

Here eye lens has more focal length

(a)

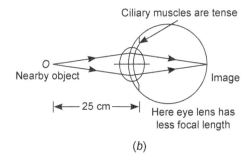

Ciliary muscles are tense

*O* Nearby object

Image

|←— 25 cm —→|

Here eye lens has less focal length

(b)

**Fig. 16.14**

## 16.15 COMPARISON BETWEEN HUMAN EYE AND THE CAMERA

The human eye works on the same principle as a camera. In the eye, a convex lens (called eye-lens) forms a real and inverted image of the retina whereas in a camera, the convex lens (called camera lens) forms a real and inverted image of the object on the photographic film. The main points of similarity and difference between the human eye and camera are given in Table 16.1.

Table 16.1: Human Eye vs. Camera

| The eye | The camera |
|---|---|
| **Points of similarity** | |
| 1. In an eye, the image is formed by a convex lens made of a transparent and flexible substance. | In a camera, the image is-formed by a convex lens made of glass. |
| 2. In the eye, a real and inverted image is formed on the retina. | In a camera, a real and inverted image is formed on the photographic film. |
| 3. Pupil in the iris of the eye controls the amount of light entering the eye. | The diaphragm controls the amount of light in a camera. |
| 4. The time of exposure in the eye is controlled by the eye-lids. | The time of exposure in a camera is controlled by a shutter. |

| The eye | The camera |
|---------|------------|
| **Points of difference** ||
| 1. The focal length of the eye lens can be changed by the action of ciliary muscles. | The focal length of a camera lens is fixed. It cannot be changed. |
| 2. The focussing in the eye is done by changing the focal length of the eye-lens. | The focussing in a camera is done by changing the distance between the lens and the film. |
| 3. The retina of the eye retains the image only upto $\frac{1}{20}$th a second after the object is removed. Thus, the image formed on the retina of an eye is not permanent. | The photographic film of a camera retains the image of the object permanently. |
| 4. Retina can be used again and again for forming the image. | A photographic film can be used only once for forming the image. |

## 16.16  PICTURE TUBE

The picture tube that provides a screen for a TV receiver is basically a cathode ray tube (CRT). It consists of an evacuated glass envelope. At its neck, there is an *electron gun* which supplies the electron beam. The inner surface of its face plate has a phosphor coating which produces light when the electron beam strikes. A monochrome (B & W) picture tube has one electron gun and a *continuous* phosphor coating that produces a picture in black and white. For colour picture tubes, the screen is formed of three different phosphors (red, green and blue). The neck of the colour picture tube may have one gun emitting three beams for the three phosphors. These 3 phosphors by combination can produce any colour; or the PT (picture tube) may have three guns each emitting different colour.

**Table 16.2:** Various screen phosphors for Picture Tubes

| S.No. | Phosphor colour | Applications |
|-------|-----------------|--------------|
| 1. | White | Monochrome picture tubes |
| 2. | Red, Green, Blue | Tricolour picture tubes |
| 3. | White, Yellow | Two layer screen |
| 4. | Green, ultra violet | Flying spot scanner |

## 16.17  MONOCHROME (B & W) PICTURE TUBE

The monochrome picture tube is used in B & W-TV. These tubes employ *electrostatic focussing* and *electromagnetic deflection*. The composite video signal (picture + sound) obtained is fed to cathode of the picture tube. The picture is constructed *bit by bit* but the same is perceived by the tube as a *complete* and *continuous* due to *persistence of vision*.

The important parts of a PT are described below (See Fig. 4.15)

(*i*) *Electron gun*: The electron gun that produces high velocity electron beam consists of an indirectly heated cathode (of tungsten) and 3 grids. These grids are control grid, accelerating grid and focussing grid and constitute the *electrostatic focussing* system of the tube.

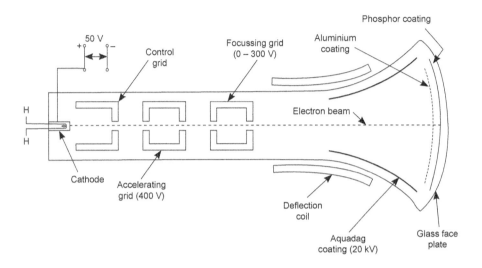

**Fig. 16.15**

(*ii*) *Focussing grids*: The grids control the movement of the beam through a smaller area and hence they 'focus' the beam at the screen. The control grid is maintained at a negative potential w.r.t. cathode. The accelerating and the focussing grids are maintained at positive potentials between (+200 to 600 V) w.r.t. cathode.

(*iii*) *Aquadag coating*: Starting from the half way into the neck to about 3 cm of the screen there is a special material coating called *aquadag*. This is a conducting coating generally of graphite. It is connected with a very high potential. Generally a 48 cm monochrome TV has an aquadag coating at 18 kV. Due to such a high positive potential, it increases the velocity of the electron beam to a very high value.

(*iv*) *Electromagnetic deflecting coils*: (See Fig. 16.15 and 16.16) Two pairs of coils are mounted outside and close to the neck of tube. These coils provide desired deflection (horizontal as well as vertical) to the beam.

These coils maintain a deflection angle between 55° to 110°. The deflection angle decreases with length of the tube. (See Fig. 16.17)

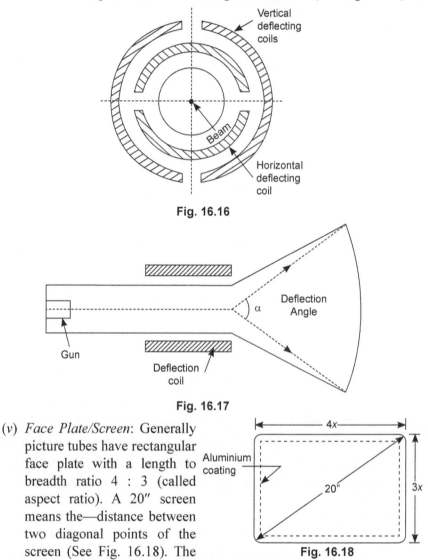

**Fig. 16.16**

**Fig. 16.17**

(v) *Face Plate/Screen*: Generally picture tubes have rectangular face plate with a length to breadth ratio 4 : 3 (called aspect ratio). A 20″ screen means the—distance between two diagonal points of the screen (See Fig. 16.18). The thickness of about 1 to 1.5 cm is sufficient to provide it the mechanical strength to withstand the air pressure on the evacuated glass envelope.

**Fig. 16.18**

(vi) *Aluminium coating*: An aluminium coating is provided at the back surface of face plate of the tube. The coating is very thin and high velocity beam can easily penetrate it to reach the phosphor screen.

About 50% of the light emitted is returned back into the tube after striking the screen. Another 20% is lost in internal reflections within the tube thus, only 30% light is utilised. The aluminium coating reflects back to the screen much of the light lost in the tube, thus it improves *brightness* on the screen.

## 16.18 SCANNING

In order to convert a picture into a *video signal* through a camera, the picture is divided into number of horizontal lines. The *picture tube* in turn re-assembles these lines into the original picture. These horizontal lines are produced by making the electron beam to *scan* the picture *line by line*. There may be 525 (or 625) lines per picture frame. The electron beam scans these lines horizontally as well as vertically. The process is known as *Scanning*.

The scanning in camera and scanning in the picture tube should be *synchronised*. This is essential to re-assemble the picture on the correct lines. This is known *Synchronisation*.

For both the purposes *scanning* and *synchronising circuits* are provided in TV transmitter as well as in the receiver.

The scanning is of two types:

(*a*) *Horizontal scanning*: For scanning, saw tooth currents (Fig. 16.19) are used. This current flows through horizontal deflection coils of the picture tube for horizontal scanning and through its vertical deflection coils for vertical scanning.

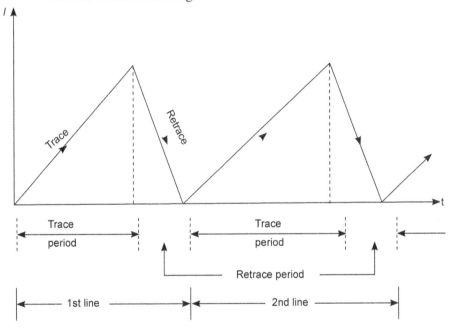

**Fig. 16.19**

The linear rise of the saw tooth wave deflects the beam across the screen with a continuous uniform motion for the *trace* from left to right. At peak the wave reverses and decreases rapidly to its initial value. This fast reversal produces *retrace* or *flyback*. (See Fig. 16.19 and 16.20)

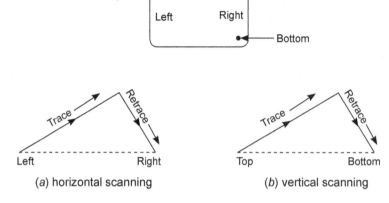

(a) horizontal scanning                    (b) vertical scanning

**Fig. 16.20**

The *start* of the horizontal trace is at the left edge of the raster and the *finish* is at its right edge. The flyback produces retrace back to the left edge. (See Fig. 16. 21)

(b) *Vertical scanning*: When made to flow through vertical deflecting coils of the P.T.; The saw tooth current moves the electron beam from top to bottom of the raster. While the electron beam is being deflected horizontally, the vertical deflection coils move the beam downward with uniform speed. Thus the beam produces complete horizontal lines

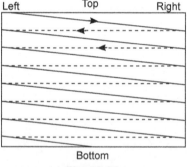

**Fig. 16.21**

one under the other. The trace part of the wave for vertical scanning deflects the beam to the bottom of the raster, then the rapid vertical retrace returns the beam to top. [See Fig. 16.22 (a), (b) and (c)].

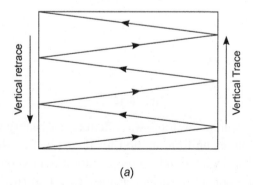

(a)

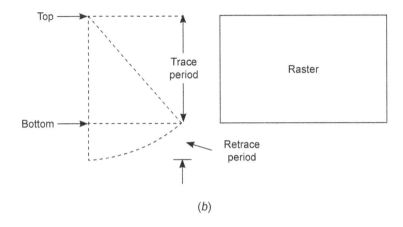

(b)

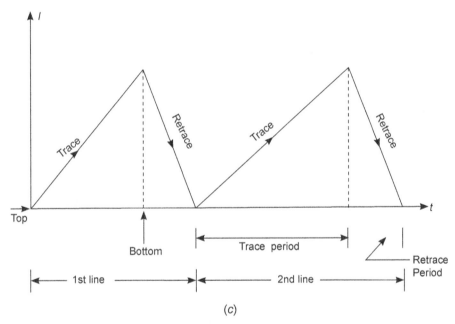

(c)

**Fig. 16.22**

## 16.19 SCANNING FREQUENCIES

Both trace and retrace are included in one cycle of the saw tooth wave. Since number of complete horizontal lines scanned in 1 second are 15,750 for horizontal deflection, the frequency of the saw tooth waves is 15750 cycles per second. For vertical scanning, the frequency of saw tooth wave is 60 cycles per second. As vertical scanning frequency is much lower than of the horizontal scanning frequency, many horizontal lines are scanned during one cycle of vertical scanning.

## 16.20  RETRACE TIME IN SCANNING

During retrace, the beam comes to its original point of start for the next horizontal or vertical line, therefore, this is a waste time, hence retrace time is kept as short as possible. For horizontal scanning, the retrace time is 10 per cent of the total time period of 63.5 µS for a complete line, its 10 per cent *i.e.*, 6.35 µS is the time for retrace or flyback.

The lower frequency vertical saw tooth waves usually have a retrace (flyback) time about 5% of one complete cycle, which comes to as 500 µS. It is note worthy that 500 µS is much more than 63.5 µS. Actually 500 µS includes approximately 8 lines.

## 16.21  NUMBER OF SCANNING LINES

How many scanning lines should be there in a picture to produce it effectively depends upon many factors. Greater the number of lines into which the picture is divided, better will be the resolution. It also very much depends upon the *resolving capacity* of human eye. There are other factors also which decide the total number of lines. For an ideal case the picture should be divided into 800 lines but above 500, the improvement is not significant. Morever with more number of scanning lines, the bandwidth also increases and this adds to the cost of the system. As a compromise between cost of the system and the quality of the picture, 625 lines have been fixed in India whereas in America this figure is 525 lines.

## 16.22  FLICKER IN SCANNING

Though scanning at the rate of 25 frames (pictures) per second produces illusion of continuity in TV system but when the screen is made alternately bright and dark in between two successive frames it causes a flicker of light which is very annoying to the viewer. To reduce the flicker, an improved method of scanning known as *interlaced* scanning is used.

## 16.23  INTERLACED SCANNING

In this scanning alternate lines are scanned, *e.g.*, all the *odd* lines from top to bottom of the frame are scanned first, leaving out the *even* lines. After this, a rapid vertical retrace moves the electron beam back to top of the frame and all the even lines on the frame which were omitted in the previous scanning are scanned from top to bottom.

This reduces flicker since the area of the screen is covered at a double rate. This is similar to reading alternate lines of the page of a book from top to bottom once and then going back to the top to read the remaining lines down the bottom.

This is illustrated in the Fig. 16.23: In this Fig.,

(*i*) Vertical retrace period has been assumed as zero.

(*ii*) Retrace lines not shown.

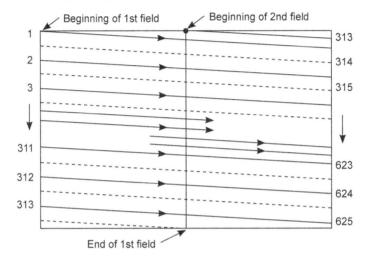

**Fig. 16.23**

In a 625 lines monochrome TV system, the lines of each picture (or frame) are divided into two sets of 625/2 = 312.5 lines, each set is called a *Field*. The each field is scanned alternately to cover the entire picture area. For this purpose, a horizontal sweep oscillator is used at a frequency of 15,625 cycles per second ($312.5 \times 50 = 15625$) to scan the same number of lines per frame ($15625/25 = 625$) but a vertical sweep oscillator is made to work at a frequency of 50 Hz. Since the electron beam is now deflected from top to bottom in half the time and the horizontal oscillator is working at 15625 Hz, only half of the total lines *i.e.*, 312.5 are scanned during each vertical sweep. The first field ends in half line, the second field starts from middle of the line on the top of the screen, the electron scanning beam is able to scan remaining 312.5 lines during its downward journey. In all, the electron beam scans total 625 lines at a rate of 15625 lines per second. Thus we are able to reduce the flicker effect without increasing the speed of scanning and thus no need to increase *band width* or cost of the system.

## 16.24 SCANNING PERIOD

The horizontal as well as vertical scanning currents are shown in the Fig. 16.24 (*a*) and (*b*) respectively. The nominal time for horizontal line = $10^6/1 5625 = 64$

μS out of which trace period is 52 μS and retrace period is 12 μS which is also called as *blanking period*.

Similarly nominal duration for vertical trace is 20 ms (1/50 = 20 ms). Out of this, 18.72 ms is spent to bring the electron beam from top to bottom and remaining 1.28 ms is spent by the beam to return back.

As horizontal and vertical sweep oscillators operate continuously, 1280 μS/64 μS = 20 lines are traced during each vertical trace interval. In this way, a total = 20 × 2 = 40 lines are lost per frame; as *blank lines* during the retrace of the two fields. In this way, the lines actually scanned per frame left = 625 – 40 = 585 lines.

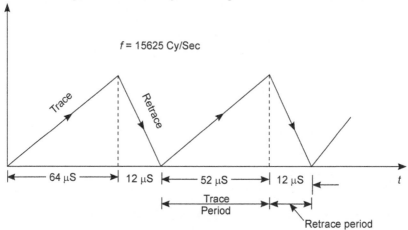

(*a*) Horizontal scanning currents

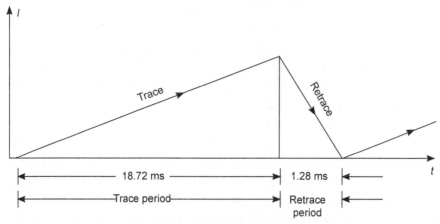

(*b*) Vertical scanning currents

**Fig. 16.24**

## 16.25 ACTUAL PROCESS OF INTERLACED SCANNING

The actual process of interlaced scanning has been illustrated in Fig. 16.25. To understand, start from point *A*. For line 1 [Fig. 16.25(*a*)] the electron beam sweeps across the frame with uniform velocity to cover all the picture elements contained in the line. At the end of the line, the beam retraces rapidly to the left side of the frame with more velocity to scan the next line. After line no. 1, the beam is at the left side ready to scan line 3, omitting line no. 2. This is the reason, that the process is also known as "*odd lines interlacing*". In this way, the beam scans all the *odd* lines reaching at point $B$ at the bottom of the frame. At $B$, vertical retrace begins and the beam comes back at top of the frame to scan the *even* lines. Now the beam moves from $B$ to $C$ traversing the whole number of horizontal lines.

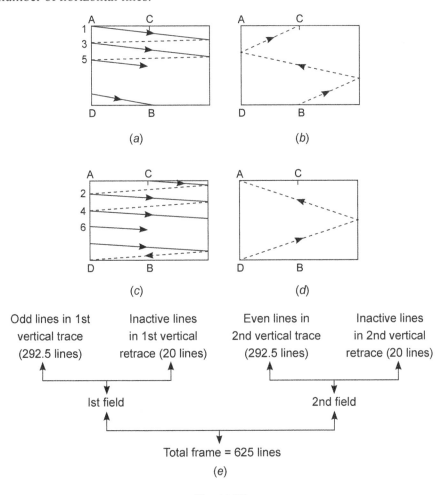

**Fig. 16.25**

The horizontal scanning during the second field begins with the beam at $C$ [Fig. 16.25 (c)]. This point is at the middle of horizontal line as the first field contains 312.5 lines. After scanning, a half line from $C$ the beam scans line no. 2 (even line) in the second field. The beam now scans all the even lines (2, 4, 6 ....) which were earlier skipped. The vertical motion is exactly same as in the previous field. Now the vertical retrace in second field starts at point

*D*. The beam is now returned to the top. The points *C*, *D* and *B* are only half a line apart.

Back at point *A*, the beam has just completed *two fields* or *one frame* and is ready to start the third field [Fig. 16.25(d)]. This process is continued at a rate of 50 times a second, which creates an illusion of *image continuity* as well as solves the problem of flicker.

The [Fig. 16.25(e)] explains the process through a line diagram.

## 16.26 SYNCHRONIZATION AND BLANKING PULSES

(*a*)  At the transmission end, camera scans the picture and the same process is carried out at the receiving end by picture tube. At the receiver the picture tube should reassemble the picture elements on each horizontal line with the same left-right position as the image at the camera. As the beam scans the successive lines in the picture tube vertically, the screen should show the same picture elements in the corresponding lines as at the camera. Therefore, a *horizontal synchronizing pulse* is transmitted for each horizontal line to keep the horizontal scanning synchronized and a *vertical synchronizing pulse* is transmitted for each field to synchronize the vertical scanning. Accordingly the horizontal synchronizing pulses should have a frequency of 15750 cy/s and vertical synchronizing pulses a frequency of 60 cy/s. The term "sync." is generally used as abbreviation for "synchronizing". The sync. pulses are transmitted as a part of the picture signal but are sent only during the *blanking period* when signal is to be transmitted.

The Fig. 16.26 shows the waveform of the sync. pulses for the purpose of synchronization. All pulses have the same amplitude but differ in pulse width or waveform. The sync, pulses include—3 horizontal pulses, 6 equalizing pulses, 6 additional equalising pulses and then 3 horizontal pulses. There are many additional horizontal pulses till *equalising* pulses occur again for beginning of the next field.

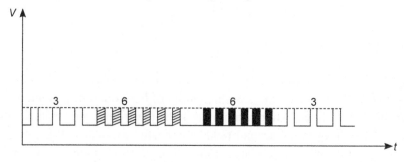

**Fig. 16.26**

The sync. pulses do not produce scanning. For scanning saw tooth pulses must be produced as explained earlier. However, the sync. pulses hold the picture on the screen in still position. If horizontal sync. is not provided, the picture *drifts* to the left or right on the screen and the picture is torn into diagonal segments. If vertical sync. is not provided, the picture appears *rolling up* and down on the screen.

(*b*)  *Blanking Pulses*: The purpose of blanking pulses is to make the retrace as black *i.e.*, invisible.

## 16.27 FREQUENCIES OF SCANNING, SYNCHRONISING AND BLANKING PULSES

The sync. and blanking pulses are of the same frequency as that of scanning. The values are shown in Table 16.3:

**Table 16.3** : Frequency of Various Pulses

| Particulars | Frequency (Hz) |
|---|---|
| 1. Horizontal scanning pulses | 15,750 |
| 2. Horizontal sync. pulses | 15,750 |
| 3. Horizontal blanking pulses | 15,750 |
| 4. Vertical scanning pulses | 60 |
| 5. Vertical sync. pulses | 60 |
| 6. Vertical blanking pulses | 60 |

## 16.28 HORIZONTAL AND VERTICAL BLANKING

In TV, "Blanking" means "going to black", as part of the video signal, the blanking voltage is at the black level. Video voltage at the black level cuts off the beam currents in the picture tube to black out the light from screen. The purpose of providing the "blanking pulses" is to make invisible the retraces of the scanning process. The horizontal blanking pulse at a frequency of 15750 Hz, blanks out the retrace from right to left for each line. The vertical blanking pulses at 60 Hz blank out the retrace from bottom to top for each field.

The time period of blanking pulses is 16% of the each horizontal line *i.e.*, = 16 per cent of 63.5 µS = 10.2 µS. In other words, retrace from right to left must be completed in 10.2 µS.

The time period of vertical blank pulses is 8 per cent of each vertical field. It comes equal to 8 per cent of 1/16 S = 0.0013 S. In other words, the vertical retrace must be completed within 0.0013 S.

A blanking pulse comes first to put the video signal at black level, then a sync. pulse comes to start the retrace. This sequence applies to blanking, horizontal and vertical retraces.

## 16.29 THE TV STANDARDS

We will discuss here about American and Indian standards:

(a) *American standard*: [Fig. 16.27 (a)] The band of frequencies assigned to a station for transmission of the signal is called a *channel*. Each TV station has a 6 MHz channel within specific bands for commercial broadcasting:

   (i) *Video modulation*: The 6 MHz bandwidth is needed for picture carrier signal. The carrier is amplitude modulated by the video signal.

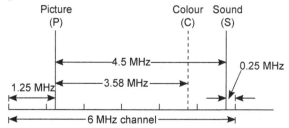

(a) Broadcasting channel (American standards)

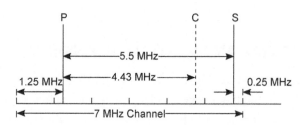

(b) Broadcasting channel (Indian standards)

**Fig. 16.27**

   (ii) *Chrominance modulation*: For colour broadcasting 3.58 MHz chrominance signal has the colour information. The colour signal (*C* signal) is combined with luminance signal (*Y* signal) to form one video signal, that modulates picture carrier wave for the transmission.

   (iii) *Sound signal*: In 6 MHz channel, sound carrier signal for the picture is also included. The sound carrier is a "frequency modulated" signal by audio frequencies between 50 Hz to 15 kHz.

(iv) *Carrier frequencies*: The Fig. 16.27 (a) shows how different carrier signals fit into the standard 6 MHz channel. The picture carrier frequency is always 1.25 MHz above lower end of the channel. At opposite end, the sound carrier frequency is 0.25 MHz below the higher end.

(b) *Indian standards:* The Fig. 16.27 (b) total channel width used in India is 7 MHz. The spacing between picture and sound is 5.5 MHz and between picture and colour signal is 4.43 MHz.

**Table 16.4:** Popular TV standards in the World

| S. No. | Particulars | India, Europe and Asian countries | America, Canada, Mexico and Japan | England | USSR | France |
|--------|-------------|------------------------------------|-----------------------------------|---------|------|--------|
| 1. | Lines per frame | 625 | 525 | 625 | 625 | 625 |
| 2. | Frame per second | 25 | 30 | 25 | 25 | 25 |
| .5. | Field frequency (Hz) | 50 Hz | 60 | 50 | 50 | 50 |
| 4. | Line frequency (Hz) | 15,625 | 15,750 | 15,625 | 15,625 | 15,625 |
| 5. | Channel BW (MHz) | 7 | 6 | 8 | 8 | 8 |
| 6. | Picture signal modulation | AM | AM | AM | AM | AM |
| 7. | Sound signal modulation | FM | FM | FM | FM | I'M |

## 16.30 COMPOSITE VIDEO SIGNAL

The video signal used in video system is not simple but it is a complex signal. It is a composite of the following signals:

(i) Camera signal—corresponding to the picture information.

(ii) Blanking signal—during retrace period to make it invisible.

(iii) Synchronizing signal—to synchronise the two scannings occurring in transmitter and receiver.

How this signal is formed is shown in Fig. 16.28

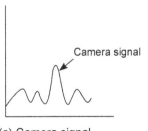

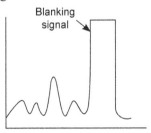

(a) Camera signal                    (b) Camera + blanking signal

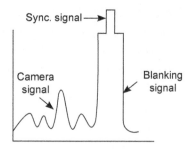

(c) Camera + blanking + sync. signal = Composite video signal

**Fig. 16.28**

*Video signal dimensions*: The Fig. 16.29 shows detailed dimensions of the composite video signal. The level of the video signal corresponding to the maximum whiteness to be handled is known as *peak while level*. This is about 10 per cent of the maximum value of the signal, while the black level is about 70 per cent. The sync. pulses are added at about 75 per cent level known as *black level*. In actual practice, these two levels are very close and are almost merged with each other.

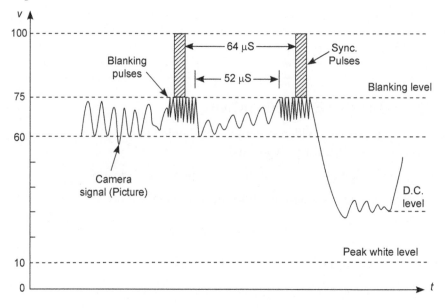

**Fig. 16.29**

In addition to continuous variation in amplitude for the picture elements, the video signal has a d.c., component corresponding to the average brightness of the scene. In its absence, grey picture on a black background will be same as a white picture on a grey background.

The video signal also contains blanking pulses which raises the signal amplitude slightly above the black level (75 per cent) so that the retrace is not visible. This signal contains horizontal as well as vertical blanking pulses. The frequency of horizontal blanking pulses is same as that of the horizontal scanning and frequency of the vertical blanking pulses is same as that of vertical scanning that is, 15625 Hz and 50 Hz respectively.

The *picture to sync. signal* ratio (P/S) is kept as 13:5 *i.e.*, 65 per cent of the maximum carrier amplitude is occupied by video signal and about 25 per cent by sync. pulses. This gives best results in reducing the noise level.

## 16.31 TV SIGNAL TRANSMISSION

In most of the TV transmission systems, the picture signal is *amplitude modulated* [Fig. 16.30 (*a*)] and the sound signal is frequency modulated [Fig. 16.30 (*b*)].

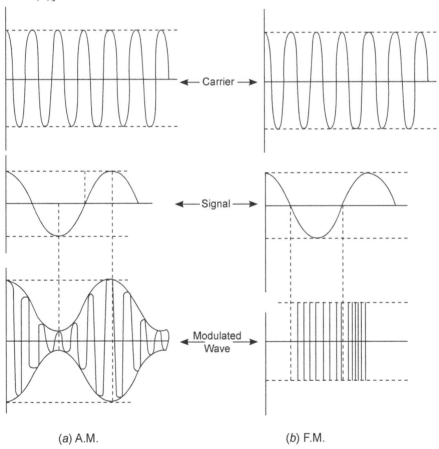

(*a*) A.M.  (*b*) F.M.

**Fig. 16.30**

**Need for modulation** has already been described. Just to remind that to transmit a signal directly (without modulation) a very very long antenna is required which is impractical. Further both picture and sound signals from different stations are concentrated within the same range of frequencies, therefore, both will be mixed up (if they are unmodulated) and it will be difficult to separate them in the receiver. Thus in order to make separation of "intelligence" (signal) from the different stations, it is necessary to keep them at different portions of the spectrum depending upon the carrier frequency assigned to each station.

(*a*) *Amplitude Modulation of Video Signal* (Fig. 16.31): Modulation of video signals is possible only by amplitude modulation because of their large bandwidth. The modulation can be of low level or high level. The high level amplitude modulation is done at final (power) stage. The intermediate frequency modulation is a low level modulation in which modulation of video signal is carried out at 38.9 MHz. The modulated IF is then changed to channel frequency by heterodyning.

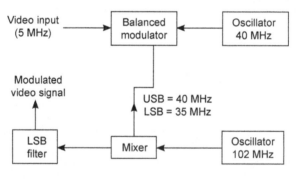

**Fig. 16.31**

To reduce the BW, the upper side band is transmitted fully and lower side band is transmitted partially. This is called *vestigial side band (VSB) transmission*. The LSB filter is used to allow only the desired frequency range.

(*b*) *Frequency Modulation of Sound Signal*: The output of all microphones terminate on the sound panel in the control room. Each microphone output is amplified before being fed to mixers. *Pre-emphasis* and *de-emphasis* are done to improve the quality of the sound. The modulated audio signal is transferred to the assigned channel sound carrier frequency by the use of multipliers.

## 16.32 PREFERENCE OF MODULATION FOR TRANSMISSION

(*a*) *Preference of FM over AM for Sound Transmission*: Because of crowding in *medium* and *short wave* bands in radio transmission, only 5 kHz frequency is used for signal as maximum. This is done to limit the BW to 10 kHz as a maximum. This enables to accommodate large number of radio stations.

Further FM provides almost noise free and high fidelity output. The maximum apace it needs is 200 kHz on each side of the picture carrier. As 7 MHz is allotted to the BW, it has no problem to accommodate the sound signal.

Hence FM is preferred for sound transmission in TV channels due to

(*i*) It reduces noise.

(*ii*) It increases, efficiency of transmission.

(*iii*) There is no inter-channel or inter-signal interference.

(*b*) *Preference of AM over FM for picture transmission*: The distortion which is produced due to interference between signals is more objectionable in FM then it is in AM, because frequency of the FM signal changes continuously. If FM is used for picture, it will produce the *shimmering effect* and the picture will not be steady. Further, the BW requirement in AM is lesser as compared in FM and also in FM, complexity of the circuitry increases. This is the reason, that for picture, AM is preferred.

Now we discuss various TV transmission.

## 16.33  TV TRANSMISSION

Here we discuss three techniques used for transmitting TV signals.

1. DSB transmission

2. SSB transmission

3. VSB transmission

(*a*) *Double Side Band (DSB) Transmission:* In a 625 line TV transmission in which frequency components ranging from 0 to 5 MHz are present, a *double side band* AM transmission will occupy a total BW of about 10 MHz. In addition, a *slope* of 0.5 MHz is to be added on both sides. Thus the total BW becomes 11 MHz. Further, each TV channel has an FM *sound signal* which is situated just out side the upper limit of the picture signal. A 0.25 MHz *guard* band is also added to this signal, which makes the BW equal to 11.25 MHz. (See Fig. 16.32)

Thus in a *double side band* (DSB) transmission, the BW is very large and would limit the number of channels for a particular station. In order to reduce this BW, single side band transmission is preferred.

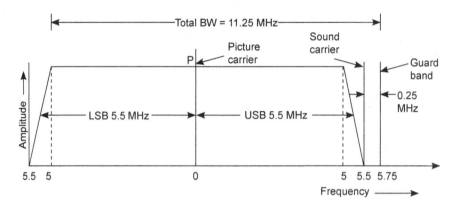

**Fig. 16.32**

(b) *Single Side Band (SSB) Transmission:* As discussed earlier, the *carrier does not contain signal.* The signal is contained in side bands only. But the transmission of carrier along with the side bands is necessary to make transmission and reception simple and inexpensive.

However, the two side bands are of the same amplitude, therefore, transmission of only one side band (with carrier) is sufficient to convey the total information and it saves a BW of 5 MHz per channel. No doubt magnitude of the detected signal in the receiver will be just half as only one side band is transmitted but it can be amplified by increasing the number of stages of the amplifiers in the receiver to the required value to compensate for the loss. But it will give great advantage as it is going to save a BW of 5 MHz, which in turn will increase the number of channels in a station. Generally lower side band is filtered out and only upper side band is transmitted. *This technique is not used* and VSB transmission is rather preferred.

(c) *Vestigial Side Band (VSB) Transmission (CCIR System B)* (See Fig. 16.33): Practically it is not possible to filter out LSB completely as it has been seen that the lower frequencies contain the most important information of the picture and filtering out the LSB (Lower Side Band) as a whole gives rise to distortions. Therefore, as a compromise full USB (Upper Side Band) and some part (Vestigial part) of the LSB (frequencies up 0.75 MHz) are transmitted. This is known as VSB transmission which is .practically done in 625 lines TV system. Now the composition of the SSB bandwidth comes to be as 7 MHz as shown.

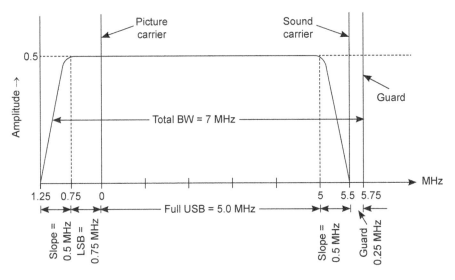

**Fig. 16.33**

## 16.34 TV RECEIVERS

Monochrome as well as colour TV receivers may be of the following 3 types:

(*i*) *All tube receivers*: The type mainly applies to monochrome and few old colour TV receivers. All the functions are carried out by electron tubes. A colour receiver may employ more number of tubes as compared to monochrome receive.

(*ii*) *Solid state receivers*: In this type, all the stages use solid state devices, except the picture tube (P.T.). The devices include semi conductor diodes, transistors and IC etc.

(*iii*) *Hybrid receivers*: This is a combination of the above two types. The deflection circuits use power tubes, while the signal circuits use solid state devices.

(*iv*) *LCD/LED receivers*: Recently, these have captured the market, the picture tube has been replaced by LCD/LED respectively.

In fact, a TV receiver is a combination of *AM receiver* (for the picture signal) and *FM receiver* for sound signal. In addition, it also contains other circuits for scanning, synchronising etc. Here we shall discuss monochrome TV circuitry as all the monochrome circuitry is needed in a colour receiver. The colour TV receiver is just a monochrome receiver *plus* colours.

## 16.35 MONOCHROME TV RECEIVER

The important sections/stages of a monochrome TV receivers are discussed as under: (See Fig. 16.34) (*i*) *Antenna*: The picture and sound signals are intercepted by antenna of the receiver. A wire connects the antenna to the input of the receiver. Twin lead is generally used. There are various antennas used but the most popular is "yagi-uda" which gives good output in fringe areas. If the signal is weak a *booster* can be used and, if signal is strong or if the transmitting station is near, either no antenna is required or a telescopic indoor antenna may be sufficient. The twin lead from antenna to the receiver has an approximate impedance of 300 ohm. A folded dipole has also the same impedance at its resonant frequency, hence both provide a good *impedance matching*.

(*ii*) *VHF Tuner*: The antenna input provides r.f. picture and sound signals for the r.f. amplifier stage. The amplified output of amplifier is given to the mixer. Also output of a local oscillator is given to mixer to heterodyne with signals as in the case of radio. When the oscillator frequency is set for the particular channel to be tuned, the signals are converted into intermediate frequency.

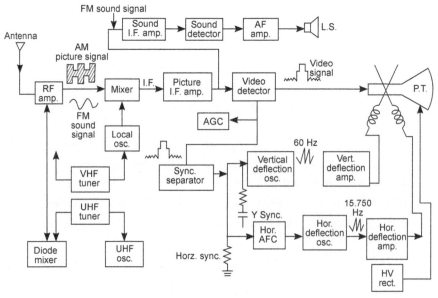

**Fig. 16.34**

There are two I.F. outputs from the mixer—one for the picture and other for the sound. The standard value for I.F. in a 625 line TV system is 38.9 MHz for picture and 33.4 MHz for sound.

The tuner selects the channel to be received by converting its picture and sound R.F. carrier frequencies into intermediate frequencies which can be amplified in I.F. amplifiers. The station selector is a gang switch that changes value of capacitor of the tuned circuits of the R.F. amplifier, mixer and local oscillator simultaneously. (See Fig. 16.35)

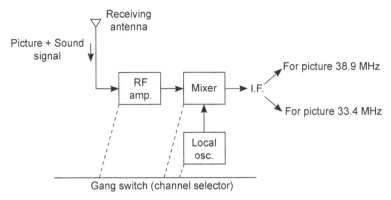

**Fig. 16.35**

(*iii*) *UHF Tuner*: When channel selector is set to UHF tuner position, the antenna input is obtained from a separate UHF antenna. This heterodynes the UHF input to the intermediate frequencies. These frequencies are amplified by required number of amplifier stages.

(*iv*) *I.F. Amplifiers*: These amplify output of the mixer in number of stages to get sufficient voltage for the video detector. The gain of the I.F. amplifier is controlled by an AGC.

(*v*) *Video detector*: The modulated I.F. picture signal is rectified and filtered in the detector to detect (recover) the amplitude modulated picture signal which is required for driving the picture tube.

(*vi*) *Sync. separator*: A sync. separator is basically a *clipper* circuit that separates the sync. pulses from the camera signal contained in the composite video signal. Since there are sync. pulses for horizontal as well as vertical scannings, output of the sync. separator is divided into two parts.

(*vii*) *Deflection circuits*: These include the vertical and horizontal oscillators for vertical and horizontal scannings. The deflection circuits produce the required scanning currents. The deflection oscillators basically are *A-stable multivibrators*, which do not need any external triggering pulse for operation.

(*viii*) *L.T. (Low Tension) supply*: This is needed for tubes and transistors. The dc output upto 280V is needed to vaccum tube amplifers. A dc supply upto 90 V is needed for rectifier and other circuits. The heater of the vacuum tubes may need dc supply of 6.3 V.

(*ix*) *EHT (Extra High Tension) supply*: The high voltage supply needed to rectifiers is given from horizontal amplifiers. The H.V supply is also needed to picture tube for its suitable operation. Approximate value of EHT supply for anode of the picture tube is 15—18 kV. This is needed for sufficient brightness. A voltage of 6 to 9 kV is developed across primary winding of horizontal output transformer. This is stepped up by an auto transformer to 15–18 kV (See Fig. 16.36).

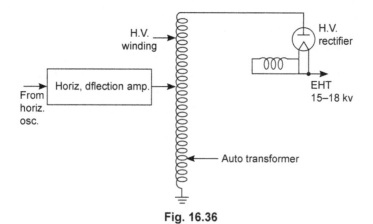

**Fig. 16.36**

(x) *AGC*: It stands for "Automatic Gain Control": It helps in getting constant amplitude of the video signal for different carriers. For the picture signal, AGC provides an automatic control of contrast in the reproduced picture. An AGC circuit has been shown in Fig. 16.37.

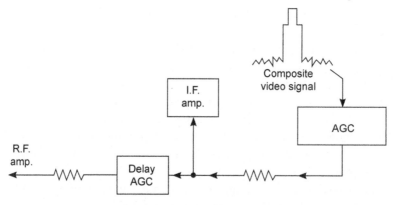

**Fig. 16.37**

(xi) *Video amplifier*: Before giving composite video signal to picture tube, it is amplified in video amplifier sufficiently. It may have number of stages as per requirement. The amount of video signal required for the picture tube is about 100 V for strong contrast. More strong video signal means more contrast.

The *blanking pulses* in composite video signal drive the grid voltage of the picture tube to cut off, blanking out the retraces. The function of *sync. pulses* is to drive the grid more negative than cut off. The Fig. 16.38 shows last stage of the TV receiver.

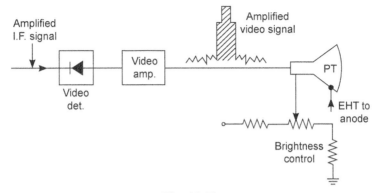

**Fig. 16.38.**

## SUMMARY

1. The "Television" means to see at a distance.

2. The few TV applications are: cable TV, close circuit TV, Picture phone, TV games etc.

3. The TV camera converts picture into electrical signal. The popular black and white TV camera tubes are: Orthicon, Vidicon and Plumbicon.

4. The B and W cameras are used in different configurations.

5. The picture tube converts the electrical signal into picture signal. It may be a monochrome picture tube or a colour picture lube.

6. The picture is "scanned" by a camera before converting it into electrical signal. Usually "Interlaced Scanning" is used; in which the picture is divided into 625 lines and is then scanned line by line.

7. The Vestigial side band (VSB) transmission is generally used for TV signals. In this technique, full upper side band and some part of lower side band (called Vestigial) are transmitted.

8. Few blocks in monochrome and colour TV are common.

9. In a TV, sound is given to microphone which converts it into electric waves, these waves are processed and transmitted into space through an antenna. The picture is given to a camera, which is also converted into electrical waves which are processed and transmitted into space through the same antenna.

   At the receiver, sound and picture both are detected separately. Sound after processing, is given to loudspeaker, which converts it into the original sound and the picture signal after processing is given to the picture tube, which converts it into original picture.

10. *The important TV processes are*: scanning, synchronization and blanking.

11. The picture signal along with blanking and synchronizing signals is called "composite video signal".

12. In TV, the sound signal is frequency modulated and the picture signal is amplitude modulated.

13. The TV receiver may be monochrome type (Black and white) or colour TV, solid state TV, LCD/LED TV, hybrid TV, etc.

❏❏❏

# 17

# Colour Televisions

The colour television produces transmitted information in the original coloured form. All stages described in the monochrome TV are used in a colour TV. In addition, few stages, which are required for treatment of colours are also included.

## 17.1 COLOUR TELEVISION

A colour is nothing but "a monochrome picture on a white raster with colours added". The required information is in chrominance (or $C$ signal) broadcasted along with monochrome signal. If we turn down the colour control on the receiver to eliminate the colour signal, what we get will be a black & white picture. With $C$ signal; the picture is reproduced in natural colours. All colours can be produced as a combination of red, green and blue colours, which are known as **Primary Colours**.

## 17.2 PRIMARY, SECONDARY AND COMPLEMENTARY COLOURS

(*a*) The colours which cannot be produced by mixing other colours are called **primary** colours *e.g.,* Red, Green and Blue are called primary colours. Sometimes Yellow is also taken as primary colour. There are two types of primary colors.

   (*i*) *Additive Primary Colours*: The red, green and blue are called **Additive Primaries** and are used when coloured light sources are blended to produce the required colour.

    (*ii*) *Subtractive Primary Colours*: The red, blue and yellow are called **Subtractive primaries** and are used when a picture on print is viewed by reflected light from a white source.

(*b*) The colours, which can be produced or obtained by mixing primary colours are called **secondary colours.**

(*c*) The two colours, which on mixing give white colour are called **complementary colours** to each other *e.g.,* yellow is a complementary to blue, magenta is complementary to green and cyan is complementary to red.

## 17.3  ADDITIVE AND SUBTRACTIVE MIXING OF COLOURS

The mixing of different colours to get new colour is known as **additive mixing** of colours.

When we mix pigments, this is known as **subtractive mixing** of colours. *e.g.* the additive mixing of red, green and blue gives white colour, while mixing them subtractively gives black colour. The Fig 17.1 shows additive mixing of colours:

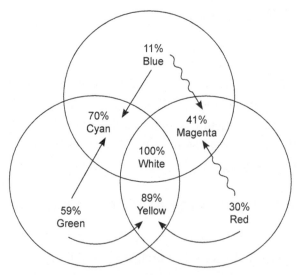

**Fig. 17.1**

*e.g.,*          Red + Green  = Yellow        $R + G = Y$

                 Red + Blue  = Magenta     $R + B = M$

The Fig. 17.2 shows subtractive mixing of colours *e.g.,* White – Green – Blue = Red ($W – G – B = R$) and White – Green = Magenta ($W – E = M$) etc.

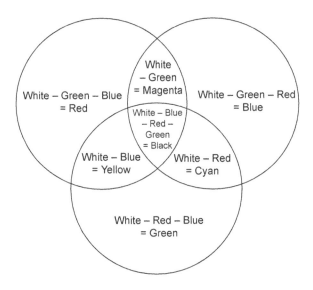

**Fig. 17.2**

As told earlier, the two colours which on mixing produce white colour are called **complementary colours**. This can be shown that

$$Red + Cyan = White$$

$$Green + Magenta = White \text{ etc.}$$

So, (Red, Cyan), (Green, Magenta) are pairs of complementary colours.

In colour TV system red, green and blue are used as Primary colours, By mixing these colours in required proportion any colour can be obtained.

## 17.4 TRISTIMULUS VALUES

The primary colour components values in a secondary colour are called **Tristimulus values.** The natural white has 30% red, 59% green and 11% blue. So one lumen of white light has 0.30 lumen of red + 0.59 lumen of green + 0.11 lumen of blue.

## 17.5 TRICHROMATIC VALUES (OR COEFFICIENTS) AND COLOUR TRIANGLE

The Tristimulus values (0.30, 0.59 and 0.11) are not very convenient. So we use Trichromatic ($T$) units or ($T$) Coefficients for obtaining secondary colours. The Trichromatic units of white colour consist of 1/3 of green, red and blue. These coefficients are represented as $x$, $y$ and $z$. Note that $x + y + z = 1$, so the $T$ coefficients of $x$, $y$ and $z$ can be represented by a triangle called *colour triangle*, as shown in Fig. 17.3.

The red is shown on $x$ axis, green is shown on $y$ axis, the third axis $z$ is perpendicular to $xy$ plane and shows blue. Since $x + y + z = 1$, the third axis need not to be shown.

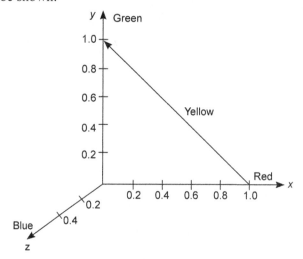

**Fig. 17.3**

The coordinates of the colours are as given below:

| | | | |
|---|---|---|---|
| Magenta: | $x = 0.5$, | $y = 0$, | $z = 0.5$ |
| White: | $x = 0.333$, | $y = 0.333$ | $z = 0.333$ |
| Yellow: | $x = 0.5$, | $y = 0.5$, | $z = 0$ |
| Cyan: | $x = 0$, | $y = 0.5$ | $z = 0.5$ |

Note that: $x$ = red, $y$ = green, $z$ = blue

See that in each case, $x + y + z = 1$

## 17.6 COLOUR CIRCLE

The colour triangle can be converted into colour circle where red, green and blue are radial vectors spaced at 120° from each other. The distance from centre to the circumference represents **Saturation (colour concentration).** The vector *reference axis* is R (Red). The colours are represented by the phase angle w.r.t. this axis (Fig. 17.4).

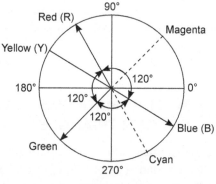

**Fig. 17.4**

If one end of a vector represents one primary colour, the other end represents complementary colour. Yellow is complementary to blue, magenta is complementary to green and cyan is complementary to red.

## 17.7 TYPES OF COLOUR VIDEO SIGNALS

The important types of colour video signals are three: Red, Green and Blue ($R$, $G$, and $B$), as a TV system starts with $R$, $G$, $B$ at the camera and finishes at $R$, $G$, $B$ in picture tube. However, colour mixtures are used for *coding* and *decoding*, because mixing of two colours can have all the *colour* (Chrominance) information of the three colours, allowing the third signal to be $Y$ (luminance) signal.

The important colour video signals and their combinations is given below.

   **1.** $I$ **Signal** $= 0.60\ R - 0.28\ G - 0.32\ B$.

   It is a combination of 60 per cent red, 28 per cent green and 32 per cent blue video signals. The minus sign indicates the addition of video voltages of negative polarity *e.g.*, $-0.32\ B$ means 32 per cent of the total blue video signal, but with inverted polarity, which reproduces blue.

   **Similarly**

   **2.** $Q$ **signal** $= 0.21\ R - 0.52\ G + 0.31\ B$.

   **3.** $B - Y$ **signal** $= -0.30\ R - 0.59\ G + 0.89\ B$.

   **4.** $R - Y$ **signal** $= 0.70\ R - 0.59\ G - 0.11\ B$.

   **5.** $G - Y$ signal $= -0.30\ R - 41\ G - 0.11\ B$.

## 17.8 CHROMINANCE AND LUMINANCE SIGNALS

   (*i*) *Chrominance or Colour Signal*: It is denoted by $C$. The colour TV system starts and ends with primary colours *i.e.*, red, green and blue colour signals. A colour camera tube had different camera tubes for red, green and blue colour. The screen of the colour picture tube has red, green and blue colour phosphorus. All the three camera tubes are used to produce the colour signal. Now the colour signals arc not directly transmitted or received. At transmitter, light of different colours is converted to different video signal voltages.

   (*ii*) *Luminance Signal*: It is denoted by '$Y$'. Luminance ($Y$) is the amount of light intensity perceived by the eye as brightness. Different colours have shades of luminance. Luminance signal contains the brightness variations for all the colour information in the picture. In colour televisions, it is obtained by mixing three colours red, green and blue in proportions of 0.3%, 0.59% and 0.11%. Thus,

$$Y = 0.30\ R + 0.59\ G + 0.11\ B$$

The above proportions of colours are selected depending upon the sensitivity of eyes to these colours.

## 17.9  COLOUR TV RECEIVERS-IMPORTANT TERMS

Any colour has 3 characteristics: First is *hue or tint,* which we generally call the "colour", the second is *saturation* and the third is *luminance*. The saturation describes the *concentration* of the colour, while the luminance indicates brightness or the "shade". The important terms are described below:

(*i*) *White*: Actually white light can be considered as a mixture of red, green and blue (primary colours) in right proportions. A white colour for colour TV is a mixture of 30 per cent *red*, 59 per cent *green* and 10 per cent *blue*. The percentage of the *luminance* signal is based on sensitivity of the eyes for different colours.

(*ii*) *Hue (or Tint)*: The colour itself is known as its hue *e.g.*, green leaves has a green *hue* and a red flower has a red hue.

(*iii*) *Saturation*: Saturated colours are deep and strong. Weak colours have no saturation. By saturation we mean how a colour can be diluted by white. When a saturated (strong) red colour is mixed with white, we get *pink* colour, which can be called as "desaturated" red.

(*iv*) *Brightness*: Each colour produces a certain amount of brightness. It is determined by the amount of light contained by it.

"Brightness" is different from "Saturation", one can be changed keeping the other constant. The colour TV has two different controls: Saturation and Brightness. When saturation control is operated, amount of white light contained in the colour is changed, when the Brightness control is operated white as well as coloured lights are changed.

So brightness is the measure of white and colour lights, whereas saturation is the measure of white light only.

(*v*) *Luminance*: By luminance, we mean amount of light intensity or brightness sensed by our eyes. In B & W picture, the lighter parts are more luminous than its darker parts. Some colours appear brighter than others. The luminance indicates how a particular colour will look in black and white reproduction. For example, a monochrome picture will show yellow colour as white, light blue colour as grey and dark red colour as black. Luminance signal is written as *Y*-signal. In brief, luminance means brightness.

(*vi*) *Chrominance*: The term is used to indicate *hue* as well as *saturation* of a colour. The chrominance signal includes all the colour information except the *brightness*. Chrominance and luminance together give complete information about the pricture. Chrominance is abbreviated as *chroma* and is also written as "*C*-signal" . As colour TV is concerned, chrominance means 3.58 MHz modulated sub carrier before modulation, and after demodulation it is the colour information in red, green and blue signals.

## 17.10  VISIBILITY CURVE

All types of radiations do not produce sensation of light on human eye. The radiations between wave lengths of 4000 Å and 7000 Å (Angstrom, 1 Å = $10^{-10}$ m) only produce sensation. The human eye is the most sensitive to radiations of 5500 Å, though it differs from man to man and from age to age. The curve between sensitivity of human eye and the wave length is shown in Fig. 17.5.

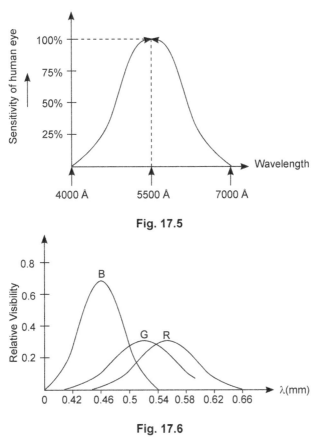

**Fig. 17.5**

**Fig. 17.6**

The colour corresponding to 5500 Å is **yellowish green** which is the most suitable colour tor all purposes.

The retina of the human eye consists of large number of minute cells called rods and cones, each of which is connected to the brain by optic nerve. These cells are sensitive to the quantity and colour of light falling on them and transmit the colour information to the brain.

The 'cones' and 'rods' are light sensing cells. 'Rods' help us to know about 'light' and 'darkness' while the 'cones' give an idea of colours. The 'cones' have special attraction for red and yellow wave lengths. This is why we like these colours.

The retina has 3 types of cones. The action of light on one kind of cones produces sensation of *red* colour (*R*), on a second kind, the sensation of *green* colour (*G*) and on third sensation of *Blue* colur (*B*). This has been shown in Fig. 17.6, which shows sensitivity of human eye as a function of light wave length for the three kinds of cones.

## 17.11 SUB CARRIER AND MULTIPLEXING

1. *Sub Carrier*: This is a carrier which modulates other high frequency carriers. In colour receiver the colour information modulates 3.58 MHz sub carrier.

2. *Multiplexing*: In this, one carrier is used to modulate two or more signals. In colour TV, 3.58 MHz *C*-signal is multiplexed with Y signal and then both modulate the main picture carrier.

## 17.12 COMPATIBILITY

Black and white television was established well before colour television. After the invention of colour TV, it was necessary that existing system (black and white) should be modified. In other words colour system must be compatible with the existing black and white TV system, which needs following requirements:

1. The colour TV signal must produce normal black and white picture on a monochrome receiver without much modification in the receiver.

2. The colour TV must be able to produce black and white picture from normal monochrome transmission.

   So to make the system compatible, following changes were needed:

   (*i*)   The colour signal should occupy same bandwidth as monochrome signal.

   (*ii*)  The location and spacing of picture and sound carrier frequencies should be same.

   (*iii*) The colour signal should have the same (brightness) information as would a monochrome signal, transmitting the same scene.

   (*iv*)  The composite colour signal should contain colour information together with synchronising signal.

   (*v*)   This extra colour information should not affect the picture reproduced on the screen of a monochrome receiver.

   (*vi*)  The same deflection frequencies and sync. signals should be used as were used for monochrome transmission and reception.

## 17.13 POPULAR TV SYSTEMS

The popular TV systems are:

   (*a*) NTSC system     (*b*) SESCAM system     (*c*) PAL system

These are discussed below one by one:

## 17.14 NTSC SYSTEM

It stands for "National Television System Committee system". In this system, the two colour difference signals created by subtraction of red and blue from the total signal are transmitted in quadrature (with one quarter cycle behind). The signals are then added together to get the chrominance signal.

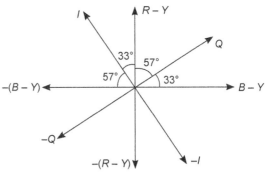

**Fig. 17.7**

The **Fig. 17.7** shows phase relationship in NTSC transmitter system.

## Transmitter

The **Fig. 17.8** shows NTSC transmitter.

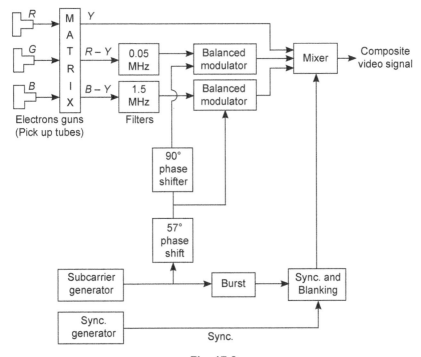

**Fig. 17.8**

The three primary signals generated by three electron guns or pick up tubes ($R$, $G$, $B$) are amplified and applied to a **Matrix** where these are combined algebraically to produce one **luminance** signal ($Y$) and two **chrominance** signals ($R$–$Y$, $B$–$Y$).

The chrominance spectras are clipped by **Filters**. The chrominance signals are modulated on the subcarrier. It accepts all the signals that make up a composite video signal. The sub carrier generator and the sync. generator must be linked in a suitable manner; so that the sub carrier may be an odd harmonic of half the frame (line) scanning frequency.

A colour subcarrier burst signal is inserted in the composite video signal during the line blanking interval. The frequency and phase of the burst should be equal to the chrominance subcarrier at the transmitter.

### Receiver

The **Fig. 17.9** shows NTSC receiver. After amplification and detection of R.F. and and I.F. signals, the composite signal is applied to the Filters, demodulators, amplifiers and through matrix, it is given to the picture tube.

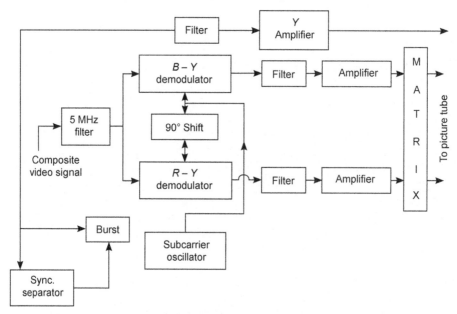

**Fig. 17.9**

### Limitations of NTSC System:

(*i*) The difference of phase between the sub carrier and local oscillator produces incorrect hues.

(*ii*) The cross talk between demodulator outputs at the receiver causes colour distortions.

## 17.15 THE SECAM (SEQUENTIAL) SYSTEM

It stands for " Sequential Colour a memory' system The system has a different mode of transmission. The colour difference signals are not arranged a quarter of cycle apart but are kept separate by transmitting them on alternate lines of the picture. Delay lines inside the receiver hold up one set of signals so that they can be recombined to form a picture from alternate lines of the signal.

The system avoids the problem of phase error. In SECAM, the two chrominance signals ($R$-$Y$ and $B$-$Y$) are transmitted on alternate lines in sequence. Though half the colour information is lost but, the human eye cannot distinguish this. The principle of SECAM has been shown in the **Fig. 17.10**.

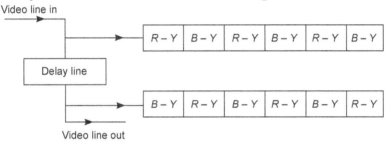

**Fig. 17.10**

The delay time is exactly one line scan interval, which is about 64 μs. For first time, the signal is taken from input (Video line) to the delay line and second time from output of the delay line.

## Transmitter

The **Fig. 17.11** shows block diagram of SECAM **Transmitter**. The colour difference signals ($R$-$Y$ and $B$-$Y$) are band limited to 1.5 MHz by filters. The two signals are applied to FM modulator alternately by using an **electronic switch**. The switch operates at line frequencies. The colour sub carrier is, therefore, modulated alternately. The channel B.W. is 8 MHz.

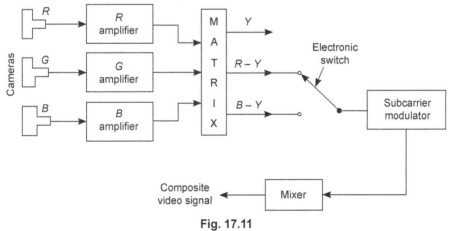

**Fig. 17.11**

To obtain the composite signal, the modulated subcarrier is mixed with luminance, Sync. and blanking signals.

## Receiver

The **Fig. 17.12** shows the SECAM receiver. In the receiver, the colour difference signals existing at input and output from the delay line are routed by an electronic switch to the two inputs of a MATRIX which drives the third colour difference signal ($G$-$Y$). With the chrominance signal transmitted sequentially (on alternate line scans), the delay line enables the three chrominance signals to be generated in receiver at the same time. This is the reason, that the SECAM system is also called **Sequential Simultaneous**.

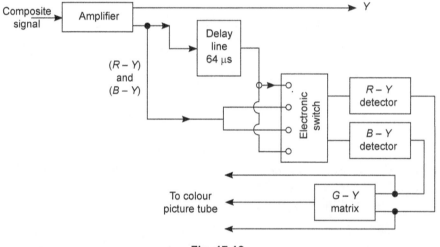

**Fig. 17.12**

**Advantages of SECAM**

(*i*) As FM is used, the SECAM receiver is free from phase distortion.

(*ii*) There is no **cross talk**, as the signals are not present simultaneously.

(*iii*) The SECAM receiver is simpler and cheaper than the NTSC and PAL receivers.

**Disadvantages of SECAM**

(*i*) The vertical resolution of SECAM system is inferior.

(*ii*) The colour is more saturated during fade to black. During fade, the pink colour changes to red.

## 17.16  PAL SYSTEM

PAL stands for "Phase Alternate Line" system and is adopted in India. It is an improved NTSC system and stands in between NTSC and SECAM systems.

In this system, the signal is transmitted in the same way but difference lies in reception. In receiver, the information is delayed at every line *e.g.*, if a certain line of the picture signal has a strong green signal, the receiver reverses the polarity of alternate lines, it assures that the next line contains very low green signal. The input sent to the picture tube is the average of the delayed line and corrected line; thus eliminating the error.

## Transmitter

The Fig. 17.13 shows PAL transmitter. Camera tube converts light into video signals having three primary colours. These colour signals are converted into luminance signal $Y$ and colour difference signals $B-Y$ and $R-Y$.

In this system, $B-Y$ modulates the sub-carrier in phase and $R-Y$ modulates the subcarrier with phase = +90° on one line and –9 0 ° on the next line and so on. So, the phase of the subcarrier changes are automatically corrected.

The phase shifter stage uses an electronic switch to change the phase from + 90 degree on one line to – 90 degree on the next line.

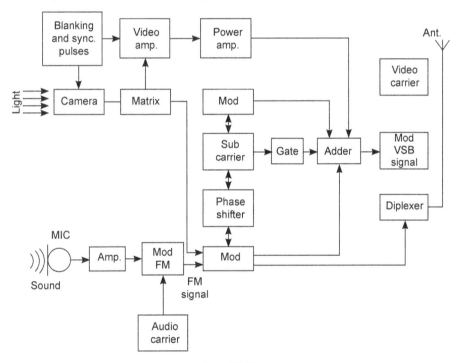

**Fig. 17.13**

The encoded signal is called chroma signal. The $Y$ signal along-with its control signal is added to the chroma signal by an adder. This colour video signal (CVS) modulates the main video carrier. Audio signal is converted into electrical signal by the microphone. After amplification is made, the audio carrier is frequency modulated, which is located within the 7 MHz channel width.

Both modulated signals (Video modulated signal and audio modulated signal) are sent to the transmitting antenna with the help of a diplexer circuit. The transmitting antenna transmits the signal into space.

## Receiver:

The Fig. 17.14 shows PAL receiver. Here, the receiving antenna picks up the signal and feeds it to the tuner stage. The tuner contains a low noise RF amplifier for amplification. The signal is converted into intermediate frequency signal by superheterodyning process. Conversion of RF and IF results in better selectivity, higher gain and better stability.

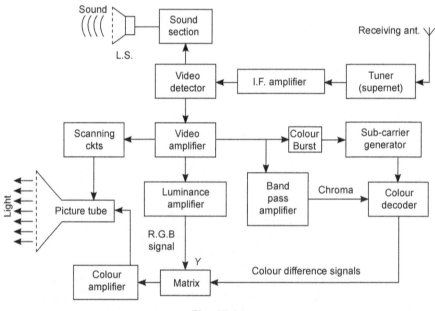

**Fig. 17.14**

The IF signal is amplified by three or four stage of amplification and then goes to video detector.

In the output of video detector, we get:

    I.  Intercarrier frequency signal (5.5 MHz)

   II.  Chroma signal

  III.  Colour burst (CB) signal

  IV.  Luminance (Y) signal

   V.  Sync. pulses.

Now sound IF is amplified and is fed to the FM detector to recover audio signal. After amplification it is fed to the loudspeaker which converts it into sound signal.

Chroma signal is decoded to retrieve original colour difference signals. The decoding is done by synchronous demodulation process. Luminance ($Y$) signal is added to the colour difference signals to get $R$, $G$ and $B$ signals in the same proportions of amplitude as in the output of the camera tubes.

$R$, $G$ and $B$ signals are fed to a colour picture tube which produces three beams of electrons. These beams are focussed on three vertical stripes of red, green and blue phosphors, respectively, through vertical slits.

The closely spaced phosphors produce red, green and blue light intensity in the same proportions as contained in the original light and our eyes integrate these colours to give the resultant colour of the original scene. Deflection of the beams in horizontal and vertical directions is achieved by horizontal and vertical deflection coils.

## 17.17 COLOUR TV CAMERA

Basically, colour TV camera consists of (Fig. 17.15)

- Diachromatic Mirrors
- Trimming Filters
- Camera Filters
- Video pre-amplifiers
- Resistive networks

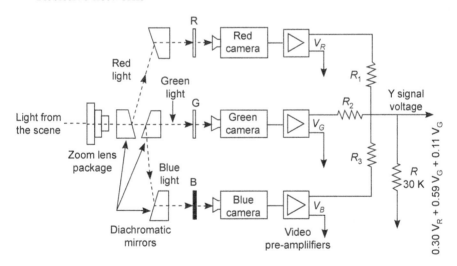

**Fig. 17.15**

Light from the scene is processed by the lens system. Diachromatic mirrors are specially designed which allow a particular colour to pass through and reject all other colours. So these are also a type of colour filters. Diachromatic mirrors are designed to pass one wavelength and reject other wavelengths. So the light is filtered into three colours red, green and blue. These colours also pass through colour filters (R, G, B) which finely tune each colour and provide highly precise colours. Now these colour signals are converted into corresponding electrical signals by camera tubes. So in this way three colour video signals are generated. As the magnitude of video signals is less, so these are amplified by video pre-amplifiers. Now these amplified video signals are passed through resistive network. All the resistors chosen are of different values. These are chosen to produce 30% Red, 59% Green and 11% Blue colour signal.

The output of colour TV camera is $0.30\ V_R + 0.59\ V_G + 0.11\ V_B$.

A colour TV camera is basically an arrangements of three black and white camera tubes. In case of three tube configuration, light falling on the target is split into three primary colours: $R$, $G$ and $B$. (Fig. 17.16)

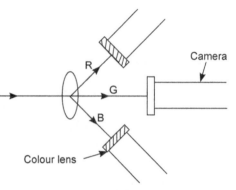

**Fig. 17.16**

Each tube now processes a particular colour. The two tubes or single tube arrangement can also work.

## 17.18 COLOUR PICTURE TUBES

The coloured picture tubes are provided in colour television.

The screen of the colour TV has three *i.e.*, red, green and blue phosphors. These are three primary colours and their combination can produce any desired colour.

There are 3 electron beams to excite these colours. A metal plate called *shadow mask* is also provided which has holes to allow the beam to reach the screen and the electrons that do not have the required angle are blocked. It is important to mention that each beam excites its respective phosphors on the screen. The screen phosphors produce colours, and *the electron beams have no colour of their own*. If only red gun is operated the whole screen will be red similarly by operating green gun, the screen can be made all green and so on.

In normal operation when the three guns are operated, they excite their respective phosphors. We get a picture on screen by *superimposition* of the three primary colours. For example, *yellow* colour is obtained by combination of red and green. *White* colour is obtained by proper combination of red, green and blue and black in the picture is obtained when all the three beams are off.

**Types of Colour Picture Tubes**

According to configuration of guns and the phosphors, three types of colour picture tubes are popular:

1. Delta gun tube (Delta tube) [Fig. 17.17 (*a*)]
2. Guns in Line tube (GIL tube) [Fig. (*b*)|
3. Single gun tube (S.G. tube) [Fig. (*c*)]

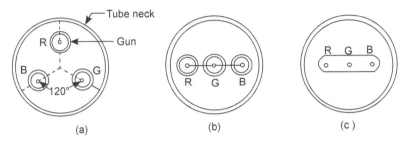

(a)                    (b)                    (c )

**Fig. 17.17**

*The basic construction of all the above tubes is same*, the difference lies only in their guns configuration. In delta tube, the guns are arranged in a *triangular* (delta) form, *i.e.* 120° apart, in GIL tube the 3 guns are arranged in a horizontal line and in single gun tube, as the name suggest, there is only one gun having 3 cathodes (*R, G, B*) in a line.

The delta gun tube is widely used and has been discussed below:

## 17.19 DELTA GUN COLOUR PICTURE LUBE

This is the most widely used colour tube.

The Fig. 17.18 shows delta P.T. with 3 electron guns. There are 3 separate cathodes and 3 separate control grids. Each gun has a separate screen grid, focus grid and accelerating grid.

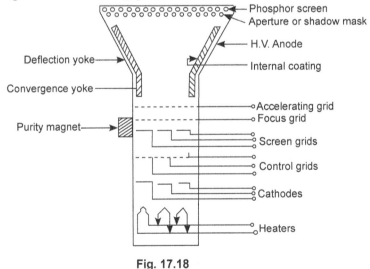

**Fig. 17.18**

At the neck of tube there are following components: (See Fig. 17.19)

(*i*) *Deflection yoke*: Its horizontal and vertical deflecting coils deflect the three beams.

(*ii*) *Convergence yoke*: This yoke possesses individual adjustment for *R, G* and *B* beams to make them *converge* through the openings of the *shadow mask*. For each beam there is a permanent magnet and coil.

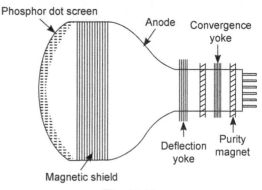

**Fig. 17.19**

(*iii*) *Purity magnet*: This magnet adjusts the three beams to produce pure red, green and blue colours.

*A colour P.T. is different from monochrome P.T. in the following ways:*

1. It has 3 guns, which provide three electron beams, one each of the three primary colours.

2. The screen of the picture tube is coated with three types of phosphor colours. When light from the three guns is incident on the phosphors, they separately emit red, blue and green lights.

3. The phosphors are embedded on the screen in triangular dot pattern (Triads). Each triad has a group of three phosphor dots.

4. In colour picture tube, a shadow mask is mounted, about 1/2 inch behind the screen.

## 17.20 COLOUR SCREENS

(*a*) Delta gun tube uses a screen with *R, G* and *B* phosphors in dots, forming *triads*. With this screen, shadow or aperature mask has holes opposite the triads through which beam passes and excites the trios. [See Fig. 17.20 (*a*)']

(*b*) GIL type tubes use screen phosphors in vertical strips in *R, G* and *B* colours, the green being in the centre. With the stripped screen, the mask has vertical slot for the beam. [See Fig. 17.20 (*b*)]

(*c*) Single beam type tube uses a vertical stripped three colour phosphor screen and opposite to it is *grill* shaped aperture mask. [See Fig. 17.20 (*c*)]

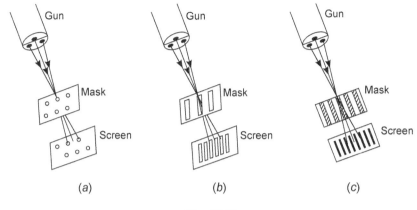

**Fig. 17.20**

Whatever may be shape of screen and the mask, only 20% of the beam is allowed by mask to reach the screen. This is the reason, why colour P.T. needs higher value of anode voltage, cathode current and heater power as compared to monochrome tubes.

## 17.21 TRANSMISSION (BROADCAST) OF COLOUR TV SIGNALS

The various methods of transmitting broadcasting TV signals are described:

1. *By Co-axial and Fibre Optic Cables*: Co-axial and fibre optic cables are used to carry TV signals for long distance or to areas, which have difficulty in receiving the signal through air. The TV network sends programs to their affiliated stations through co-axial cables. The cable TV system use co-axial or fibre optic cables to carry signals to the individual TV receiver.

2. *By Microwaves*: These are also electromagnetic waves. A row of towers at about 50 km span carry program to the affiliated stations through these waves. The equipment in a tower automatically receives, amplifies and then retransmits the microwave signals to the next tower.

3. *By Satellites*: The satellite is used between stations where cable or microwave towers cannot be built. The satellite receives coded signal from an earth station, amplifies and sends to another earth station. The two stations may be thousands of kilometers apart.

4. *By air (through space)*: This is the most popular method used for broadcasting (transmission) colour TV signals.

The transmitter at the broadcasting station amplifies the signal so that it has enough power to reach large distances. The transmitter increases the frequency of both the audio and video signals through "modulation". For this purpose, high frequency electromagnetic waves called "carrier waves" are generated by the transmitter. The video signal varies the amplitude of the carrier waves to carry video part of the TV signal. The process is called *"amplitude modulation"*. The video signal is then amplified to a power of $10^3$ to $10^6$ watts.

Another carrier waves modulate the audio part of the TV signal. The process is called *frequency modulation*. The readers know that a TV signal is a composite signal comprising of audio and video parts.

The transmitter then combines the modulated video and audio parts to form the composite TV signal. This signal is transmitted into space.

The Fig. 17.21 shows the process of transmission through air. The colour TV transmission begins with a TV camera, in which a mirror or prism system breaks light from the scene to be transmitted into 3 primary colours: red, blue and green. At the same time, the camera tubes change these colour signals

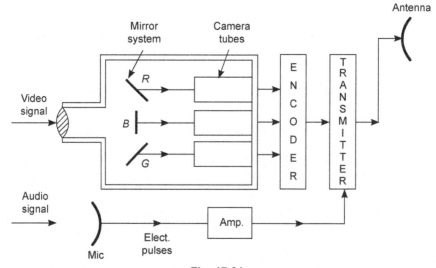

**Fig. 17.21**

into electrical pulses. These colour signals then go to "encoder" which produces a compatible colour signal and sends it to the transmitter. A microphone on the other side changes the audio (sound) signal into electrical pulses. The electrical pulses obtained from microphone are amplified and are also fed to the transmitter. The transmitter combines the audio and video signals for broadcasting from the transmitting antenna.

## 17.22 RECEPTION OF COLOUR TV SIGNALS (FIG. 17.22)

The TV signals can be received by a TV receiver through a receiving antenna. The best reception is achieved when the antenna is pointed towards the desired transmitting station. Some antennas can be rotated by remote control in the required direction.

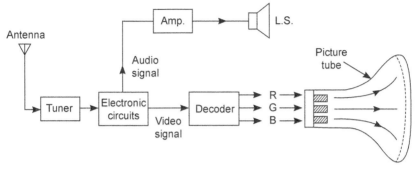

**Fig. 17.22**

The TV tuner selects only the signal, which is desirable. Few TV receivers have two tuners—one for selecting VHF channels and other for UHF channels. The electronic circuits in the receiver *separates* the received composite TV signal into audio and video signals. The audio signal is amplified and is changed into sound by a loud speaker. A *decoder* changes the video signal again into primary colour signals. The three electron guns of the picture tube—one for each primary colour scan the screen. The screen is covered with red, blue and green dots. When struck by a beam from the guns, the coloured dots glow and blend together in the viewer's mind to produce the original colour image.

## 17.23 COLOUR T.V. RECEIVER

All stages described in monochrome TV receivers are used in a colour TV receiver. In addition, some stages which are required for treatment of colours are also included in a colour receiver.

A colour picture is actually a "monochrome picture on a white raster with colours added". The required colour information is in the chrominance (*C* signal) broadcast along with monochrome signal. If we turn down the colour control at the receiver to eliminate the colour signal what we get will be a *black and white* picture. With *C* signal, the picture is reproduced in natural colours. As explained earlier all colours can be produced as a combination of red, green and blue colours, which are known as basic or *primary* colours.

The simplified block diagram of a colour receiver is shown in Fig. 17.23. The (Picture + Sound) signal from the selected channel is processed in the same way by the tuner, I.F. and video detector stages as in case of a monochrome receiver. The sound signal (S) is separately detected, amplified and fed to the loud speaker. The AGC, sync., separator etc., have the same form and function as in monochrome receiver. In addition, the picture tube has *purity magnet, convergence* and *deflection* yokes etc., for colour treatment.

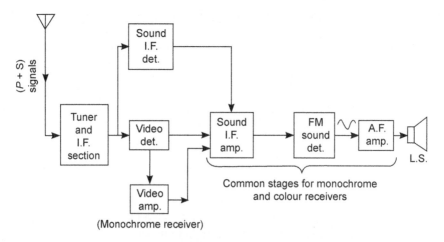

**Fig. 17.23**

At the output of video detector, the composite video and chrominance signals appear in their premodulated form. The *Y*-signal is processed in the same way as in black and white TV receiver, except that the video amplifier has a *delay line* which introduces a delay of about 490 ns (1 second = $10^9$ ns) which is necessary to ensure coincidence of luminance and chroma signals.

*The circuits common to monochrome and colour TV receivers have already been discussed. Here we describe important circuits of colour TV receivers.*

## 17.24 DESCRIPTION OF IMPORTANT CIRCUITS OF COLOUR TV RECEIVERS (FIG. 17.24)

(*i*) *Chrominance Band Pass Amplifier*: This separates the chrominance signal from the composite video signal, amplifies it and passes to the demodulator. The *colour burst* is prevented from appearing at its input by horizontal blanking pulses.

(*ii*) *Colour Demodulators*: The colour demodulators detect the original signal. They are a combination of phase and amplitude detectors, because the output is dependent on phase as well as on amplitude of the chroma signal. Each demodulator has two input signals—the *chroma* which is to the demodulated and a constant output from the local oscillator.

(*iii*) *Burst Separator*: This circuit has to separate colour burst which is transmitted on the back porch of every horizontal sync. pulse. The circuit is tuned to the subcarrier frequency. The burst output is fed to AFC circuit.

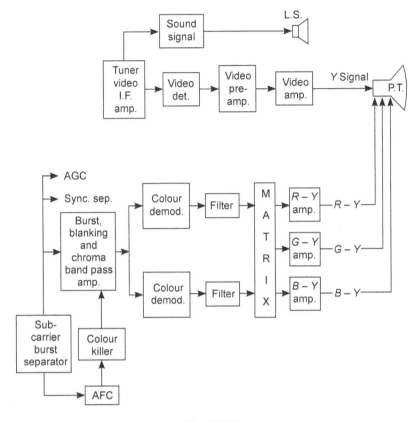

**Fig. 17.24**

(*iv*) *Colour Killer Circuit*: As the name indicates, this circuit becomes ON and disables the band pass amplifier during monochrome reception. In other words, it prevents any spurious signal from getting through the demodulators which may cause any colour interference on the screen. When the colour killer circuit is off, the chroma band pass amplifier is ON for colour information.

(*v*) *Matrix*: This circuit combines signal in specific proportions. At the receiver, the picture tube acts as matrix for inputs of *Y* signals and *R–Y*, *G–Y*, *B–Y* signals to produce *R*, *G* and *B* lights on the screen.

**Table 17.1**  Trouble Shooting of TV Receivers

| Troubles | Faults and Remedy |
|---|---|
| 1. No sound, no picture but normal raster. | Loose antenna or lead connections. Check RF tuner and I.F. Sections. |
| 2. Sound distorted, picture normal. | Defective speaker, defective capacitors in Audio Amplifiers. |
| 3. Sound normal, no picture but normal raster. | Check "picture tube circuit". |

| 4. Insufficient vertical height. | Check vertical sweep circuit. |
|---|---|
| 5. Insufficient width of raster. | Weak output transistor. |
| 6. Fold over in picture, sound normal. | Check horizontal deflection circuit. |
| 7. Picture not centered but sound normal. | Check centering controls of the picture tube. |
| 8. Weak picture and sound but of good quality. | Check tuner and I.F. sections. |
| 9. Hum. | Check loudspeaker and Audio Section. |
| 10. Snow in the picture. | Insufficient signal strength. The lead from antenna to receiver may be open. |
| 11. Bars in the picture. | Horizontal bars are produced when frequency of signal is less than 15625 Hz. Vertical bars are produced when frequency is more than 15625 Hz. |
| 12. Ghosts in the picture. | The antenna receives also the reflected signal from neighbouring buildings, mountains, hills, etc. This produces the duplicate of the picture called *ghost*. A high directional antenna is the solution. |
| 13. White vertical bars. | The reason is the production of oscillations in the horizontal scanning currents. |
| 14. Black vertical bars. | The reason is the radiation of harmonics of the horizontal scanning currents, RF chokes. "Ferrite heads" may be used in amplifier. |
| 15. White spots on the screen after making the set off. | A short circuit between control grid and cathode of the PT. |
| 16. Tilted picture. | Wrong adjustment of Deflection coils of P.T. |
| 17. Black shadow on the border of the screen. | Wrong position of deflection coils on neck of the P.T. |
| 18. Sound OK, outline of the picture looks like gears (*Gear tooth effect*). | AFC circuit is faulty. |
| 19. Sound OK, but a tree shape is seen on the screen (*chrismas tree effect*). | Wrong adjustment of the horizontal oscillator core. |
| 20. Size of the picture not OK (*Trapezoidal effect*). | Short circuit in the vertical deflection coils. |
| 21. The loud Speaker gives a motor boat noise (motor boat effect). | EHT stage is faulty. |

**Problem 17.1.** *Find time for scanning one horizontal line for frames repeated at 60 Hz and 525 lines per frame. If scanning is progressive and without interlacing.*

**Solution.** Time for scanning one line

$$\frac{1}{60 \times 525} = \text{sec.} = 31.75 \ \mu s \quad \textbf{Ans.}$$

**Problem 17.2.** *In a TV picture tube, the cathode is at +230 V and control grid is at 190 V, find grid bias.*

**Solution.**          Grid bias $= 190 - 230 = -40$ V   **Ans.**

**Problem 17.3.** *The period of an equalizing pulse is 31.77 μs. What is the frequency repetitive rate.*

**Solution.** Frequency repetive rate (FRR)

$$\frac{1}{31.77 \times 10^{-6}} = 31476.23 \text{ Hz} \quad \textbf{Ans.}$$

**Problem 17.4.** *The time required for horizontal blanking is 16% of each horizontal time. If horizontal time is 63.5 μs, fmd horizontal blanking time for each line.*

**Solution.** Horizontal blanking time for each line

$$\frac{63.5 \times 16}{100} = 10.2 \text{ μs} \quad \textbf{Ans.}$$

**Problem 17.5.** *If 400 picture elements are scanned in 50 μs. Find time needed to scan 4 picture elements.*

**Solution.** Time to scan 4 elements

$$= \frac{50}{400} \times 4 = 0.5 \text{ μs} \quad \textbf{Ans.}$$

**Problem 17.6.** *A picture has 625 vertical and 50 horizontal elements. Find total number of elements.*

**Solution.** Total number of picture elements $= 625 \times 500 = 31250$   **Ans.**

**Problem 17.7.** *Find % modulation for the following frequency deviation in FM sound signal.*

*(a) 5 kHz*         *(b) 10 kHz*         *(c) 25 kHz*

**Solution.** *(a)*   $\frac{5}{25} \times 100 = 20\%$   **Ans.**

*(b)*   $\frac{10}{25} \times 100 = 40\%$   **Ans.**

*(c)*   $\frac{25}{25} \times 100 = 100\%$   **Ans.**

**Problem 17.8.** *The high tension supply to a colour TV is 24.5 kV and the total beam current from the three guns is 2000 μA. Find power of the picture tube.*

**Solution.**         Power $=$ Current $\times$ Voltage

$$= (2000 \times 10^{-6})(24.5 \times 10^3)$$

$$= 49 \text{ W} \quad \textbf{Ans.}$$

**Problem 17.9.** *A colour TV has a 53 cm screen. The aspect ratio is 4/3. Find height and width of TV screen.*

**Solution.**        $(3x)^2 + (4x)^2 = (53)^2$

$$x = 10.6$$

$$3x = 3 \times 10.6$$

$$= 31.8. \text{ cm } \textbf{Ans.}$$

$$4x = 4 \times 10.6$$

$$= 42.4 \text{ cm } \textbf{Ans.}$$

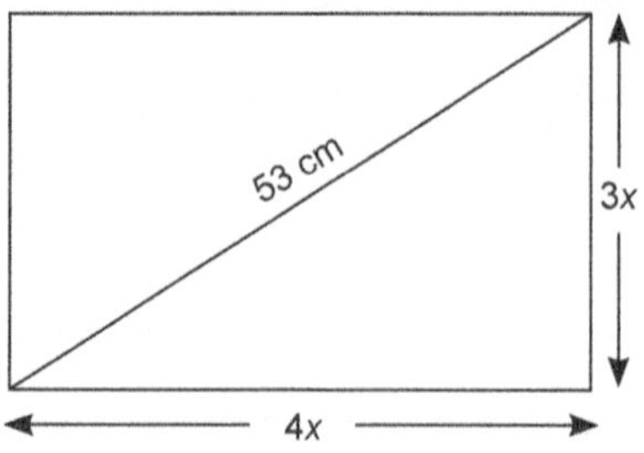

**Fig. 17.25**

**Problem 17.10.** *Find the number of possible colours with a 4 line digital input monitor.*

**Solution.** Number of possible colours

$$= 2^4 = 16 \quad \textbf{Ans.}$$

**Problem 17.11.** *The audio carrier for a TV channel is 529.75 MHz and IF is 33.4 MHz. Find frequency of the local oscillator. (L.O.)*

**Solution.** Frequency of L.O.

$$f_0 = f_c + \text{IF}$$

$$f_0 = 529.75 + 33.4$$

$$= 563.15 \text{ MHz} \quad \textbf{Ans.}$$

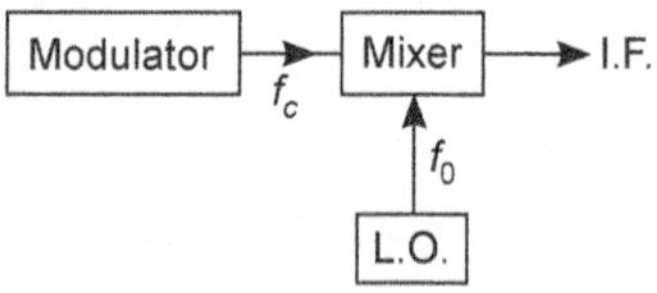

**Fig. 17.26**

**Problem 17.12.** *Find vertical and horizontal resolution for TV. Assume K = 0.7, total no. of lines = 625, 20 lines are lost per field and aspect ratio = 4/3.*

**Solution.**        Total lines $= 625$

Lines lost $2 \times 20 = 40$

Lines $= 625 - 40 = 585.$

Vertical resolution $= 585 \times 0.7 = 409.5$  **Ans.**

Horizontal resolution $= 409.5 \times \dfrac{4}{3} = 546$  **Ans.**

## 17.25  SOLID STATE VIDEO CAMERA

The solid-state camera does not use a camera tube described in the earlier pages. In turn, it uses a fully solid-state device in the integrated circuit (I.C.) package as the device for converting light variations from the object into corresponding electrical signals.

The CCD (Charge Coupled Device) photosensitive array package in the IC form is the device used. The array consists of a matrix of CCD elements arranged in rows and columns. Each CCD element, comprising of a photoconductor whose output is coupled to a capacitor through a FET switch, represents one picture element. One row of CCD elements is represented in Fig. 17.27.

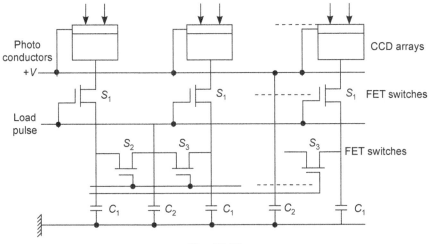

**Fig. 17.27**

The circuit operates as follows: If the FET switches $(S_1)$ for all the elements of a particular row are switched on for a short-time, capacitors $(C_1)$ will charge to a voltage depending upon the light falling on the respective photoconductor. Turning on switches $(S_2)$ transfers these charges to capacitor $(C_2)$. Now if switches represented by $(S_3)$ are switched ON, the charge corresponding to a particular picture element passes on to the following adjacent CCD element. Thus by alternately switching ON $(S_2)$ and $(S_3)$, the charge can be shifted along the row and finally taken out. This is equivalent to a line scan in electron tube based cameras. By taking the output from each row in sequence, a field scan can also be achieved and thereby we can generate a video output signal that represents light intensity falling on the CCD-array. In a colour CCD video camera, a colour striped filter is bonded to the surface of CCD array.

Solid-state video cameras are more rugged and compact than their camera tube counterparts. They donot suffer from the lag problem of vidicon tube based cameras. They do suffer from 'Blooming' problems. Blooming occurs due to saturation of individual photocells when illuminated by very bright light.

## 17.26 SPECIAL TVs

Here, we shall describe the following TVs

(a)  Projection TV              (b)  Closed circuit TV

(c)  Flat Panel TV              (d)  Digital TV

(e)  Three-dimension TV         (f)  HDTV

(g)  LCD/LED TV                 (h)  Plasma TV.

(i)  Satellite TV

These are described briefly:

## 17.27 PROJECTION TV

The projection TV is the outcome of the quest of generating a picture whose size considerably exceeds the dimensions of the TV picture tube target available. Such a large sized picture (1 m × 1 m or even larger) may not be required for home but it certainly has a no. of professional applications. Today, a projection TV is used in a university where the students will follow the lectures on otherwise dull subjects with greater interest. It is used in disco theatres, where the young ones will flock to see their favourites.

It is used in hospitals and medical colleges where the medical researchers will be able to follow more closely the achievements of their colleagues in other advanced countries where the video filming of new techniques is very common. Similarly, there are many more similar applications.

A projection TV set up comprises of a projector and screen [Fig. 17.28]. The projector houses the three projection tubes for R, G and B (red, green and blue) signals (in case of a colour TV projector) and the associated optics. Both floor

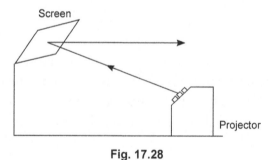

Fig. 17.28

mounted and ceiling mounted projectors are commercially available. These projectors in addition to showing the normal TV programme can also be hooked up to a video cassette recorder (VCR) or a personal computer (P.C.) for a variety of applications.

## 17.28 CLOSED CIRCUIT TV

In a Closed Circuit TV (CCTV) setup, the video signal at the camera output is directly fed to a TV monitor at a remote location by means of a cable. Since the video signal is not modulated and is routed through the cable, the TV monitor is nothing but a TV receiver without RF and IF sections. In a typical CCTV set-up, one video camera may feed more than one monitors [Fig. 17.29]. Projection TVs with large screens can also be used in place of TV monitors.

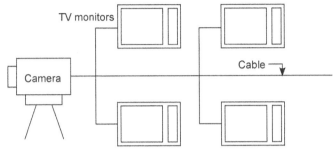

**Fig. 17.29**

The CCTV finds extensive applications in a large number of areas like education, industry, medicine, traffic control and so on. One such application is in the concept of one teacher for many class rooms. A lecture delivered before a camera could be viewed in a number of class rooms on different monitors or projection TVs. The CCTV also finds wide use at airports, railway stations, harbours etc., for making public announcements like arrival and departure schedules and also for silent paging of passengers and personnels. In hospitals, CCTV network ensures remote observation of patients in a ward at a central location. In another application in surgery, all phases of an operation can be transmitted live to a large group of medical students and researchers outside the operation theatre. The CCTV can also be used for observation of blasts from a distance and also for observation and subsequent control of crushers, hoists, etc., in the mining industry. The problems created by today's vehicular traffic can only be tackled through a high speed information system, of which a CCTV forms an integral part.

## 17.29 FLAT PANEL TELEVISION

A flat panel television receiver is the one in which the conventional CRT type picture tube is replaced by a flat panel display. The dimensions of a CRT picture tube put a limit on the degree of compactness achievable in a television receiver. Flat panel TVs are ultracompact due to small sized flat panel displays.

**Fig. 17.30**

The initial initiative and thrust to develop flat panel displays was due to their extensive applications in watches, pocket calculators and so on. Later on, applications of these displays to pocket size TVs has really hastened the pace of development.

There are two distinct approaches being followed in the development of flat panel displays for TV applications. The first and the earlier one is that of using a simple dot matrix with tricolour film dots. The second approach is that of using an active dot matrix. The LCD displays being used in pocket sized television receivers follow the simple dot matrix approach. Fig. 17.30 shows the picture of a small size flat panel display TV receiver.

## 17.30  DIGITAL TELEVISION

When we talk about Digital television, we not only imply use of digital television receivers that have been successfully commercialized, we also mean an analog television studio being replaced by a digital television studio. Though analog video cameras are being replaced by digital video cameras and analog viodeo tape recorders (VTRs) are being replaced by their digital counterparts, a complete changeover can take place only after universal acceptance of a common digital TV standard.

Digital technology has a large number of advantages. Digital equipment produces predictable results and the resulting pictures are stable and immune to noise. The signal if in the digital form can be regenerated as often as desired without any signal degradation so that long transmission lines and multiple tape generation are no longer problems. The digital signals can be stored in a memory for any length of time and can be read from the memory and written into the memory at different speeds, it can be time stretched, delayed, compressed and read in a form different from the one in which it was written. It has made possible the special visual effects that we see on television.

## 17.31  THREE DIMENSIONAL TELEVISION (3 DTV)

In comparison to a two dimensional picture which appears flat and appears to lie in the plane of the screen, in three dimensional television viewing, all the three dimensions-length, breadth and depth are depicted on the screen. The picture appears to have all the qualities of a live scene as seen with natural vision.

We, the human beings are able to see in three dimensions as the left eye and right eye see a slightly different image of the same object while viewing it. These images are transmitted to the brain where they are processed and interpreted as a single composite image in 3-D instead of two overlapping flat images. This in fact is the starting point for achieving a 3-D television broadcast and reception.

The stereo video signals can be obtained from a pair of video cameras suitably located to take a binocular view of the scene to be televised. These video signals are kept separate throughout the transmission and reception processes. Various techniques have been demonstrated for the stereo video display. One such technique is to use a television receiver with a shadow mask picture tube. All the phosphor dots of one colour (say Green) are excited by one of the electron beams intensity modulated by the video signal corresponding to one of the cameras. Similarly, all phosphor dots of another colour (say Red) are excited by another electron beam intensity modulated by the video signal corresponding to the second video camera. The third beam is kept off and the third colour phosphors (Blue in this case) are not excited. A viewer wearing green filter glasses in one eye and red filter glasses in the other eye will view the scene in stereo in yellow. In the improved version of the same principle, the picture tube has only two types of phosphor dots-Blue and Yellow. In such a case, a viewer wearing blue and yellow filters in respective eyes will see a 3-D view of the scene in monochrome as yellow and blue are complementary colours.

The 3D television is not yet commercialised.

## 17.32 HDTV (HIGH DEFINITION TELEVISION)

It is a new generation TV. Displaying images with much greater resolution on screens having an *aspect ratio* of 16 : 9 (instead of 4 : 3 at present) will enable the viewers to experience the visual quality comparable to 350 mm film and a sound quality of a CD system. The HDTV system demands complete transformation from the new system. Today there are 3 HDTV systems. The Japanese MUSE system. European EUREKA system and American DIGITAL System.

The Fig. 17.31 (*a*) shows HDTV system , the Fig. (*b*) shows NTSC TV system for comparison.

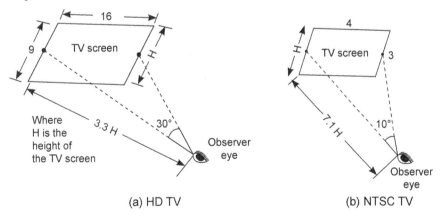

(a) HD TV                                    (b) NTSC TV

The Fig. 17.31 (*c*) shows Japanese MUSC circuit for HDTV which is the most popularly used.

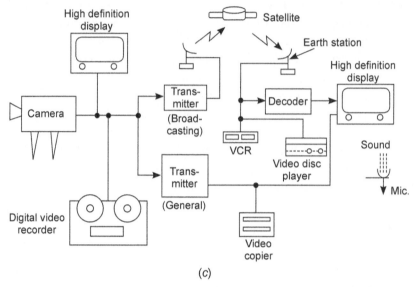

(*c*)

**Fig. 17.31**

## 17.33 LCD/LED TV

We today witness slow death of CRT (Cathode Ray Tube) TVs. Now we have LCD/LED TVs, which replace CRT completely. The LCD TV uses a LCD (Liquid Crystal Display), which gets activated, when an electric current is applied to it.

**Fig. 17.32**

The LED TV works on the same principle with the difference that the CRT has been replaced by an LED (Light Emitting Diode). The LCD TVs use CFL (Compact Fourescent Lamp) as back light. So difference between LCD and LED TVs is only the back light they use. Moreover, the LED TV has less power consumption but costly than the LCD TV. Here we shall describe LCD TV.

## LCD TV

An LCD is in liquid state but has crystalline structure which when a voltage is applied, changes the arrangement of molecules. This characteristic is used to make a shutter which alternately shifts off and passes the light.

An LCD TV produces picture by the above principle, but picture elements equivalent to scanning lines of ordinary TV using CRT are minute liquid crystal plates. The key to clear image is, how fast the liquid crystal molecules can be changed in response to the slight fluctuation of voltage.

The Fig. 17.33 shows LCD-TV circuit, it uses advanced I.Cs.

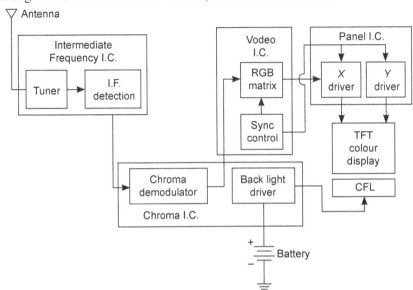

**Fig. 17.33**

The **IF (Intermediate frequency) signal IC** contains circuits for video and sound, synchronising circuit, sync. circuit, AFC and AGC. It has a tuner which can receive VHF (Very High Frequency) as well as UHF (Ultra High Frequency) signals.

*The chroma signal IC* contains colour demodulator, and RGB (Red, Green, Blue) generating circuits. As the liquid crystal is driven by AC supply, the polarities of video signals for red, green and blue colour are reversed by "ac RGB video generating circuit".

The **video IC** contain RGB matrix and sync. control circuits. It is necessary to change the video signals to red, green and blue for each period of horizontal scanning.

The panel IC Board contains *X* and *Y* driver circuits. The panel has two ICs for *X* driver and two ICs for *Y* driver.

The outputs of *X* and *Y* drives goes to T FT (Thin Film Transistor) colour display.

At the backside, a CFL (Compact Fluorescent lamp) is used which gets supply from a battery.

The RGB matrix of video I.C. may be used into two ways:

(*i*) Transparent electrodes are placed over the entire screen. The electric current flows in both the directions to drive LCD molecules to create picture. The picture quality is not so good.

(*ii*) The each tiny picture element has a thin film transistor (TFT), which applies voltage on LCD to create images. In this method the picture quality is good but the cost of the TV is higher.

**Note :** The LED TV is similar to LCD TV, except that the former uses a LED replacing the picture tube.

## 17.34  PLASMA TV

In a CRT television, a gun fires a beam of electrons (negatively-charged particles) inside a large glass tube. The electrons excite phosphor along the screen of P.T. which causes the phosphor to light up. The television image is produced by lighting up different areas of the phosphor coating with different colours at different intensities.

Cathode ray tubes produce crisp, vibrant images, but they do have a serious drawback: They are bulky. In order to increase the screen width in a CRT set, you also have to increase the length of the tube (to give the scanning electron gun room to reach all parts of the screen). Consequently, any big-screen CRT television is going to be heavy and take up a sizable part of a room.

Recently, a new alternative has popped up: the plasma flat panel display. These televisions have wide screens, comparable to the largest CRT sets, but they are only about 6 inches (15 cm) thick. Based on the information in a video signal, the television lights up thousands of tiny dots (called pixels) with a high-energy beam of electrons. In most of the systems, there are three pixel colours– red, green and blue – which are evenly distributed on the screen. By combining these colours in different proportions, the television can produce the entire colour spectrum.

The basic idea of a plasma display is to illuminate tiny coloured fluorescent lights to form an image. Each pixel is made up of three fluorescent lights — a red light, a green light and a blue light. Just like a CRT television, the plasma display varies the intensities of the different lights to produce a full range of colours.

The central element in a fluorescent *light is a plasma, a gas made up of free-flowing ions* (electrically charged atoms) and electrons (negatively charged particles). Under normal conditions, a gas is mainly made up of uncharged particles. That is, the individual gas atoms include equal number of protons (positively charged particles in the atom's nucleus) and electrons. The negatively charged electrons perfectly balance the positively charged protons, so the atom has a net charge of zero.

If many free electrons are introduced into the gas by establishing an electrical voltage across it, the situation changes very quickly. The free electrons collide with the atoms, knocking loose other electrons. With a missing electron, an atom loses its balance. It has a net positive charge, making it an ion.

In a plasma with an electrical current running through it, negatively charged particles are rushing toward the positively charged area of the plasma, and positively charged particles are rushing toward the negatively charged area. In this mad rush, particles are constantly bumping into each other. These collisions excite the gas atoms in the plasma, causing them to release photons of energy.

The Fig. 17.34 explains the plasma technology *i.e.*, how atoms emit light.

The *Xenon and neon atoms*, used in plasma screens, release light photons when they are excited. Mostly, these atoms release ultraviolet light photons, which are invisible to the human eye. But ultraviolet photons *can be used to excite visible light* photons. The xenon and neon gas in a plasma television is contained in hundreds of thousands of tiny cells positioned between two plates of glass.

The main advantage of display plasma technology is that you can produce a very wide screen using extremely thin materials and because each pixel is lit individually, the image is very bright and looks good from almost every angle. The image *quality isn't quite up to the standards* of the best cathode ray tube sets, but it is fairly good.

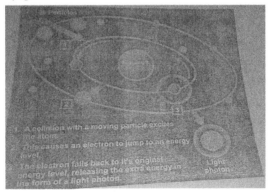

**Fig. 17.34**

The biggest drawback of this technology is its very high price. But as prices fall and technology advances, they may start to replace out the old CRT sets.

## 17.35 LCD Vs LED vs PLASMA TVs

The three TVs are compared as follows:

1. *Contrast ratio*. A contrast ratio is the ability of the TV to show the high-contrast areas on the screen. The high contrast areas are the blacks and the whites in the picture. If a TV has a good contrast ratio, it means that it shows the differences between the darker and the lighter areas of

the screen remarkably well. The acid test of a TV's contrast ratio is in the dark scenes, where it needs to differentiate the shades of black.

Its a well accepted fact that plasma TVs generally have better contrast ratios than LCD and LED TVs. This is because plasma TVs have individual plasma cells, which automatically turn themselves OFF/ON in the darker parts of the screen. LCD TVs on the other hand have a single liquid crystal and when the TV has to show the dark pictures, the CFL (back-light of the LCD) only dims itself and hence the blacks aren't very convincing. On the other hand, LED TVs powered with an LED back-light have a slight edge as the individual LEDs twist themselves into an 'off' position in the dark part of the screen thus giving a more convincing black than the LCD TVs.

On the contrast ratio parameter, the plasma TV clearly has the upper hand. The LED TV comes next in the race and the LCD TV comes last.

2.  *Viewing angle*. A viewing angle is the angle from which the image on the TV can be viewed. All TVs can be viewed when you are right in front of it, but since all viewers cannot sit right in front of the TV and have to sit around it, the viewing angle parameter becomes very important. The winner for this point again is the plasma TV. The image of the plasma TV remains good and solid for viewers at almost any angle. LCD TVs fall back in the race again as there is often a loss of colour and detail, when the TV is viewed from an odd angle. The LED TV minimizes this problem faced by the LCD TV due to its decentralized back-lights.

On the viewing angle parameter, the plasma TV is the winner again, followed by the LED TV in second place and the LCD TV in the last place.

3.  *Colour*. It is pretty clear that good and bright colours of the TV will be an important consideration in the mind of the TV buyer. But there is no clear winner on this parameter as all the TVs seem to show a good quality picture, when it comes to the colour. Differences may exist between two TV models of the same type or two different TV brands – a higher priced model showing better colours – but the picture quality remains largely similar. It is said that if the LED TV has a coloured back-light, it will show the best picture, but this point remains contentious.

All the TVs rank similar on this point, but the discussion may be given a twist by introducing the coloured back-lit LED TV.

4.  *Motion (Speed)*. The LCD TVs have shown great improvement on this parameter in recent years, but it cannot be debated that the technology of the plasma TV itself has an edge here. As the plasma screen has the

individual cells that can refresh at a much faster rate, the motion flow has been a bugbear for LCD TV makers for a while now.

LED TVs largely use the same technology as the LCD TVs and hence indicates better performance of the TV when it comes to fast motion sequences.

5. *Power consumption.* Till now, it seemed that the plasma TV comes out to be the clear winner in this debate, but this is one of its most telling fall-points. The plasma TV consumes a lot more power that the LCD TV, as every sub-pixel on the screen needs to be lit. On the contrary, the LCD TV needs much lesser power to light up the back-light. But the LED TV is the clear winner for this point as its LED back-light is more power efficient and needs lesser electricity to light up.

The LED TV is the winner on this point, followed by the LCD and the plasma.

6. *Life span.* Another fall point of the plasma TV which the plasma-makers are trying tooth and nail to correct is, its life-span. The quality and brightness of the plasma screen takes a beating in a very short-time. But plasma-makers have now come up with more efficient TVs, which have potentials to last for a longer time. The lifespan of the LCD and the LED TV depends largely on the lifespan of its back-light, but on an average the lifespan of these TVs is supposed to be more than that of the plasma TV.

7. *Price.* Plasma TV is number 1 at the moment. Its price is low but faces fierce competition from the LCD TV in the future. LED TV is still very far behind in the pricing race.

The LED TV is the most costly.

## 17.36 SATELLITE TELEVISION

The TV programs from a telecasting station cannot be viewed beyond a certain distance. Using the communication satellites launched in space, the TV programs from a telecasting station can be viewed even for away from the place. The *Geo stationary satellites* placed in space around the earth have enabled us to see national as well as international TV programs. If three satellites are placed in space at a certain distance above the equator at 120° from each other, they can cover almost the entire world (See Fig. 17.35).

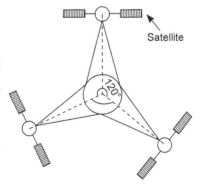

**Fig. 17.35**

## The Process (Fig. 17.36)

The programs are prepared in a TV studio and are fed to the earth station in the form of electronic signals. The earth station converts these signals into assigned uplink microwave frequency (5.9 to 6.4 GHz) and feeds it to the transmitting dish antenna. The dish antenna beams the signals in the direction of the satellite. The satellite is equipped with its own antenna which receives the uplink signal and feeds it to the receiver in the *transponder*. The receiver amplifies the signal, changes it into another frequency known as *downlink frequency* (3.7 to 4.2 GHz). This is to be mentioned that these frequencies allocations are made by International Telecommunication Union (I.T.U.) in Geneva. Here we are to remind the readers that a *transponder* is an equipment on the satellite which contains a receiver, amplifier, frequency converter and the transmitter.

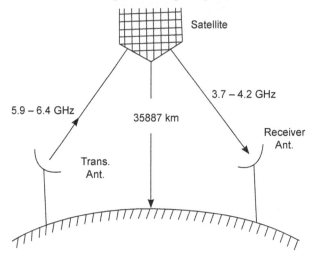

**Fig. 17.36**

A satellite has a number of such transponders for handling different channels' signals. These channels are used for telecommunication, radio and TV transmissions and for many other functions. The commercial broadcasters lease the transponders with assigned specific frequencies for transmitting the signals from the ground station.

The transmitter of the transponder retransmits the signals back to earth through its transmitting antenna. These signals are sent to earth at a down link frequency (between 3.7 to 4.2 GHz).

**Note.** The values of 'up' and 'down' frequencies for satellite TV are different from that required for other satellite communications.

## Collection and Use of Signals

The receiving earth station collects these down link signals using a parabolic dish antenna (Fig. 17.37). The. downlink signals are weak, so a high gain, large diameter antenna is required to collect the sufficient amount of the signal. The signals falling on the antenna are reflected to its "focal point" (*F*) and collected by its " feed horn".

The signals are then fed to a "low noise (high gain) block (LNB) amplifier and a down converter", which amplifies the signal and converts the down link frequency to a lower frequency range (950 – 1450 MHz) and the output

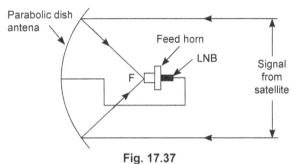

**Fig. 17.37**

is connected to a receiver through a co-axial cable. See Fig. 17.38

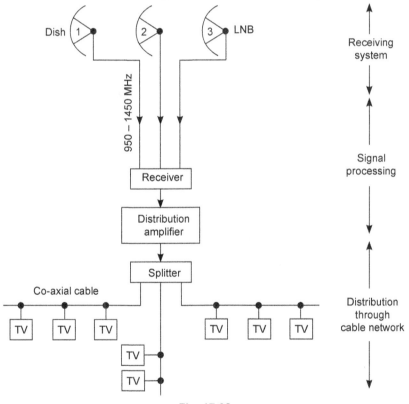

**Fig. 17.38**

The signals from the receiver are further processed and distributed through "cable TV network". For this purpose the signal is fed into a "distribution amplifier", which boosts the signal sufficiently high so that it may reach to individual TV receiver. The output from the amplifier is fed to splitter through coaxial cable and sent to individual TV sets.

## 17.37  HOME THEATRE TV SYSTEM

This is a system which provides effects of a THEATRE in a home. We get in our living room a big sound and big picture giving best audio video experience comparable to a fine theatre.

**Components/Accessories of a Home Theatre TV System along with typical specifications are given below. (Fig. 17.39)**

1.  A large colour TV (CTV) 53 cm (21″)/63 cm (25″)/73 cm (29″) with full and flat square tube, menu driven remote control.

    **NTSC** (National TV System Committee) or **PAL** (Phase Alternation by Line) playback 4.43/3.58 MHz. Compatible system 100 program memory, S band and hyperband tuner (cable ready), "on screen" display and a separate enclosure speaker system, 2 AV inputs +2AV outputs.

2.  *Speaker System.* A *sub woofer* with a powerful "built in" amplifier, alongwith 5 sleek and compact speakers (tweeters, and squawkers (2 front + 1 rear 2 side speakers) together giving a combined output of 1000 W. PMPO (Peak Music Power Output)— Such that if we are watching a helicopter on the screen, we will feel that it is hovering in the room, we are sitting.

**Fig. 17.39**

3.  *ACD/VCD-PIayer.* Audio CD player -Video CD player—3 discs playing continuously—9 scene digest remote control—playback control (Audio CD/Video CD 1.1/2.0 version).

    In addition any audio system (walkman, two in one) can be connected and a PMPO of 1000 W can be obtained.

    The stereo phonic effects are obtained, even if a monophony signal is transmitted by the cable operator.

## 17.38 TV STUDIO

The important TV studio equipment is listed below:

    1. Microphones        2. Loudspeakers

    3. VCRs/VCPs        4. TV cameras

    5. Picture tubes        6. Video Display Units—monitors, CRO etc.

    7. Film Projectors/cinema to graph

    8. Studio lighting.

All above equipment has already been discussed except the last two.

(*i*) *Film Projector*. The important parts of a film projector are shown in the Fig. 17.40.

As the lenses produce an inverted image, the film is placed *up side down* in the projector to get an erect image on the screen.

As the *shutter* of the projector rotates, the gate opens and closes alternatively. The film is placed slightly away from the focal length of the projecting lens, which produces an enlarged and erect image on the screen. The screen is placed at a distance greater than twice of focal length of the lens. During operation of the projector, the film reel unwinds from the upper reel and after passing through the gate it rewinds on lower reel of the projector.

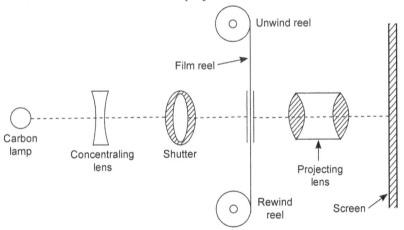

**Fig. 17.40**

The projector projects at the rate of 16 films per second on the screen. Due to *persistence of vision* we cannot visualise this small difference of time and illusion of " image continuity" is maintained.

(*ii*) *Studio lighting*. In a TV studio, lighting is necessary in addition to the illumination required at the scene to be televised. The lighting scheme:

1.  should give equal brightness throughout the whole studio

2.  should not produce shadows and glare.

    About 50 to 100 light fittings are provided in the main studio of wattage between 500 to 5000 W. Generally pendent lights are preferred.

## 17.39  PLAN OF A TV STUDIO

The Fig. 17.41 shows plan of a TV studio. Here $T$ = Tube light, $F$ = Fan, L.S. = Loudspeaker, VDU = Video display unit.

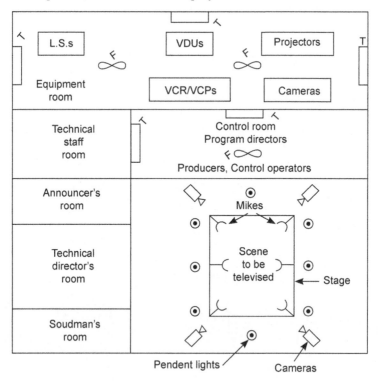

**Fig. 17.41**

TV studio may cover a total of 2 to 3 hectares of land but its "programing hall" measures about $40 \times 20 \times 15$ m. About 3 to 4 stages can be installed at a time. Lights are mostly pendent type but adjustable lights are also there. At a time, 3 to 4 microphones and cameras are at work. One stage is used to transmit the program while the others for rehearsals. All conversations among the staff is through "signals by hand", speaking is avoided or head phones can also be used. Director of the program can visualise all activities going on in the hall from his room.

The control room of the studio has the following officials to control the program.

1. *Control operator*. He controls the camera's output. Facing him, there is a control panel. By various controls he tries to obtain best results from the camera.

2. *Technical Director*. A master control panel is facing the technical director who is fully responsible for the program. All camera control operators send their camera output to the control panel of the director and the director selects the best output for transmission.

3. *Sound man*. The sound man receives output of all the microphones working at the stage. He selects best sound for transmission. The Announcer's sound also first of all reaches to the soundman.

4. *Monitors room*. In the control room or by its side, there are 2 monitors:

   (*i*) *Air monitor*. The camera output selected by the director is monitored on this monitor.

   (*ii*) *Preview monitor*. This monitor shows the final picture going to be transmitted to TV receivers.

## SUMMARY

1. The red, green and blue are called primary colours. Any colour can be prepared by mixing primary colours in right proportions.

2. Colour Triangle gives the percentage of primary colours to be mixed to obtain a particular secondary colour *e.g.*, 30% red + 59% green and 11% yellow gives white colour.

3. Various TV systems used are: PAL., NTSC and SECAM. In India, PAL (Phase Alternate Line) system is popularly Used.

4. A colour camera basically comprises of three Black and White Cameras.

5. A colour picture tube basically comprises of three monochrome picture tubes (guns) of red, green and blue colours in a prescribed configuration.

6. The colour picture tubes are of three types: Delta tube, gun in line and single gun tube.

7. The TV signal is a composite video signal, which contains audio, video and other signals.

8. There are many types of colour TVs such as LCD, LED, Plasma TV etc.

9. Home Theatre TV system gives theatre effect in home.

10. The TV studio equipment are : Microphones, Loudspeakers, TV, Camera etc.

□□□

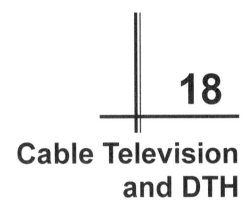

# Cable Television and DTH

Feeding many TV receivers from a single antenna is known as cable TV (CATV). This is very popular system today. A parabolic dish is used as antenna. The TV receivers are joined through *co-axial cables*. The dish antenna receives signals from a satellite.

In DTH (Direct to Home) service, the consumers receive signals directly from satellite through their own dish antenna. The cable operator has no role to play.

## 18.1 CABLE TV (CATV)

The feeding of many TV receivers from a single antenna is called cable TV. A single antenna can feed a complete locality. The antenna is installed at the top of a building and the signal is fed to the houses through cable. The antenna is generally *dish antenna* which receives national as well as international signals through satellite and this signal is sent to houses through cable. The signal should be sufficiently strong so that the picture is clear at the TV sets at each home. Also there should be no interference between the signal received and the signal sent to the viewers. The weak signal is amplified before sending to the subscribers. The signals from VCR etc., may also be sent to the subscribers through cable. Generally *modulation is not required* and the cable used is a *co-axial cable*.

The C ATV system needs miles of cables. They suffer interference in the way moreover, with distance, the signal goes on weaker and when it reaches a TV receiver, it is distorted. To solve this problem, amplifiers are installed at regular distances to keep the signal strong. This also poses problem, if there is a breakdown in any of the amplifiers, The Fig. 18.1 shows block diagram of cable TV network.

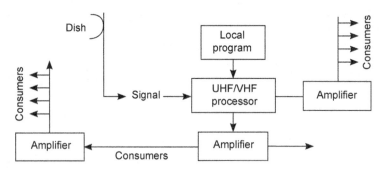

**Fig. 18.1**

The signals suffer an attenuation @ 15 dB per 500 m cable; so use of repeaters/ amplifiers is necessary. A *UHF/VHF signal processor* is used, which is positioned at a central place.

It is necessary to convert UHF channels to VHF channels, because signal loss at VHF is lesser. Moreover, the processor can accept signals from dish as well as the locally made programs. The local made programs include announcements, local advertisements etc.

The connection to consumer may be direct or through a set top box (a converter). The set top box selects channels and controls the operations. But the set up box is usually operated through a remote.

**Note:**

(*i*) The signals can also be sent through cable directly from VCR (Video Cassette Recorder) without any modulation.

(*ii*) If there is a breakdown in any of the amplifiers, there is a total breakdown of the signal ahead so the system needs regular checkup.

## 18.2 CATV THROUGH INTERNET

The CATV through Internet is possible using telephone lines as a medium of transfer and using a dial up **modem** (modulator-demodulator).

For cable internet access on a PC (personal computer), a cable MODEM is required at the user's end. A cable modem is an external device that is connected to the computer to provide high speed data access via CATV network. A cable modem sends and receives data from the internet, using the cable network. The modem translates signals in the same way, a telephone modem does from the telephone line. It translates radio frequency signals to (and from) the cable into internet. The modem is connected to the computer through *N.I.C.* (network interface card). This provides connectivity between cable and computer and translates the signals from the modem so that the computer software displays these signals.

The CATV network has high bandwidth about 500 to 720 MHz. It handles more than 100 channels. The CATV through internet is a two way network.

This is done by upgrading the amplifiers. The internet signals are digital signals and they are to be interfaced with analog CATV network. This interface is known as "Cable modem termination service" (CMTC), which consists of input interface, cable modem and microprocessor.

About 5 to 10 PCs (personal computers) may be connected with the internet through *DSL* line. The feature films and other programs can be down loaded over the internet and can be put on the cable network; so that the consumer can enjoy them on their PC at their home.

The cable connection can also be made through optical fibre cable (OFC) replacing the coaxial cable. This will improve the reception further. The data rate may be as high as 100 megabits per second, so the 3 hour movie can be down loaded in few minutes.

The internet service providers offer telephony, internet access and television services together, resulting in the economics to the consumers.

## 18.3 DTH (DIRECT TO HOME) SERVICE

The DTH service is a satellite transmission technology. It has numerous merits for the sender as well as for the receiver. Due to digital technology, it can handle many channels over a single delivery platform as compared to the analog technology. (Fig. 18.2a)

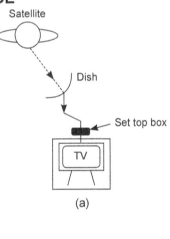

(a)

The DTH service is that in which large number of channels are digitally compressed and are sent through high power satellite. The programs can be directly received at home. What the consumer is to do is the installation of a small dish antenna at the roof of his house. The transmission eliminates cable operators completely as the consumer receives signals directly form the satellite. The consumer also needs a digital receiver. To receive the multiplexed signals as compared to CATV, this has easy control and clear picture.

Transmissions in DTH is done in **Ku (or *C*) band** which is most appropriate. As the signals are digital, this band provides high resolution picture and also better sound. All merits of digital transmission are also applicable.

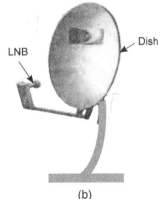

(b)

**Fig. 18.3**

The Fig. 18.2(b) shows the DTH dish set with LNB (low noise high gain block) *i.e.* amplifier.

The set up box is installed near TV, which selects the desired channel and also controls all operations through remote.

## 18.4 MERITS OF DTH COMMUNICATION

As DTH is a digital technology, all merits of a digital communications are applicable. It has numerous merits to the subscribers few are listed below:

1. Digital quality picture

2. Stereophonic sound

3. Uninterrupted viewing

4. Capacity upto 500 channels

5. Interactive TV

## SUMMARY

1. Through cable TV, informations can be sent to many consumers, simultaneously through cable.

2. Usually, coaxial cable is used which runs to all consumers.

3. For this, a dish is used at the roof top. The signal is received from the satellite. This signal is amplified and sent to the consumers.

4. The cable TV system is most popular for sending TV signals and is a means of entertainment.

5. The DTH stands for "Direct to Home". In this a consumer installs a small dish as an antenna at the roof of his house and receives the signals directly from the satellite (without any cable).

# Facsimile (FAX)

Facsimile telegraphy is a technique to send printed messages such as typed or hand written from one station to another. As compared with the ordinary telegraphic system, the following points are to be noted:

(*i*) By this system, the messages written in the language not known can also be sent, whereas it is not possible in the ordinary system.

(*ii*) A bandwidth of 120 Hz is needed in the facsimile system, where as the ordinary telegraph system needs a very high BW of the order of 2200 Hz.

(*iii*) For facsimile, trained operator is required, whereas in ordinary system this is not essential.

(*iv*) The equipment needed in FAX is costly as compared to the conventional telegraphy.

In wider sense, the picture telegraphy and facsimile telegraphy systems are same, Now-a-days these systems are used over *lines* as well as over *radio* (through space). In line facsimile, *AM signals* are used where as in radio facsimile, *FM signals* are invariably used. Our modern transmitters and receivers are capable to handle AM as well as FM signals.

The word "Facsimile" means "exact reproduction" . In facsimile transmission, an exact reproduction of the picture or document is provided at the receiving end.

## 19.1 FACSIMILE (FAX)

Fax, short for facsimile transmission is a way of sending text of pictures over telephone lines.

A device called a facsimile (FAX) machine is used for sending and receiving the messages or pictures. The machine is connected to a telephone. To send a document, the sender inserts it into the machine and dials the telephone number of the receiving fax machine. When the connection is made, an electronic scanner in the transmitting machine converts the images of the document into a set of electric signals. The signals travel over the telephone lines to the receiving machine where the signals are reconverted into a copy of the original document.

The first fax machine began operations in France between Paris and Lyons in 1866. The signals were sent through telegraph wires. It could fax documents and drawings. The first photograph was faxed from Paris to London in 1907. The rate is the same as for an ordinary telephone call.

## 19.2 FAX vs T.V. TRANSMISSION

A fax is a machine that can transmit photographs, written documents, etc. The same is done in television transmission. The following is the difference between the two:

1. In TV, the scene includes "movement" whereas, such is not the case in fax transmission. The TV can send moving pictures. The FAX can transmit only the still pictures.

2. The transmission in TV is much faster than in FAX.

3. Due to the above, TV transmission needs much larger bandwidth than the FAX. This is the reason that FAX messages can be *sent over ordinary telephone lines.*

## 19.3 APPLICATIONS OF FAX

Important applications of FAX are given below:

1. *Business*: The business documents like purchase order, invoice, agreement etc., available at one place can be made available at other place in no time through fax without relying on postal department. The most common example is news paper printing. News, photographs, reports available at one place can be sent to other place to be published in the next morning edition. The photographic plates of a complete news paper can be sent in no time to any corner of the world through fax. It has become very easy to print a news paper from many places simultaneously.

2. *Weather*: The weather reports, maps from one place can be sent to TV stations for broadcasting. Similarly, they can be sent to airports, seaports and to any place through fax for their use.

3. *Legal applications*: The finger prints, photographs, signatures, and documents of a criminal can be sent from one place to another through FAX for quick verification.

4. *Personal applications*: Personal documents such as certificates, photographs and other documents urgently needed by a person at a distance from his residence can be obtained through FAX in no time. No postal service/courier can be too fast.

## 19.4 BASIC FAX SYSTEM

FAX machines provide an easy way to send documents to any phone number equipped with another fax machine. The process of sending the document is called *facsimile*. It is a method by which a page (printed, written or photographic) is converted into electrical signals, sent quickly through the *public switching telephone network* (PSTN) and recorded as a copy in the remote fax machine. In most cases, faxes are a cost-effective-alternative to overnight delivery and a faster option than regular mail.

The Facsimile process involves three basic steps:

1. **Reading** or *converting the documents into electrical signals.*

2. **Sending** or *transmitting the signals through a telephone system to another fax unit.*

3. **Converting** *a received signals into a "facsimile" of the transmitted documents."*

A fax machine works by scanning each outgoing page, converting the image into a series of light and dark dots. This pattern is then translated into audio tones, and sent over regular phone lines. The receiving fax hears the tones and prints the total compilation of dots. The resulting document is black and white like the original page.

## 19.5 OPERATION OF FAX

Fax involves the following steps of operation which can be very well understood from the block diagram shown in Fig. 19.1.

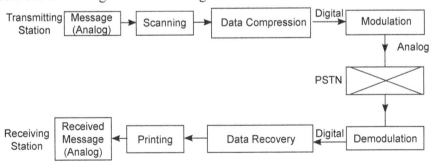

**Fig. 19.1**

1. *Scanning*: **(Reading and converting the document into electrical signals)**. The transmitted document originally consists of two-dimensional structure. We have to make it one-dimensional structure in order to consider document as the concept of line and then, we convert the line into the dot structure, we regard it as "SCAN".

   The scanner is the part of a fax unit that converts the marks on the page being sent into electrical signals. A photosensor looks at a very small "picture element (Pixel), determines whether it is black or white, and generates a strong or weak electrical pulse. To read a page electronically, all of the spots in a thin strip 1/ 100th inch high across the top of the page are read, one at a time, starting with the upper left-hand corner. A stepper motor moves the page down and the photosensor reads the next strip below. Successive strips are read until the whole page is converted into electrical pulses.

2. *Modem* (*Modulation and Demodulation*): A modem is a modulation-demodulation device that accepts the digital information and modulates it into an analog signal to be sent over a voice telephone line. At the receiver, the analog signal is converted back into the digital.

   We have to modulate the scanned picture element to convert digital signal into analog signal at transmitting station in order to transfer through the PSTN (*Public Switching Telephone Network*).

3. *Printing i.e., converting the received signals into a facsimile of the transmitted documents. Normally we use The Print Head Operation.*
   The Print head has a row of very small resistor-element spots across the recording paper with very-very small width. The thermal recording paper touches this row of resistors. A pulse of current through a resistor causes it to become hot enough to mark the paper in a spot about 0.005 inch in diameter. The spot temperature must .be changed from non marking temperature to marking temperature and return to non marking before the paper steps to the next recording line. Marking temperature is about 200° F. The *thermal paper* is a special paper that turns black on heating. Either thin-film or thick-film technology is used for making the *print heads*.

The signal obtained from scanning cannot be directly transmitted, as it is difficult to amplify low frequencies. Hence process of modulation is carried out. The Fig. 19.2 shows the transmission and reception process. The signal is "amplitude modulated" with a carrier. After this, we use AM/FM converter, as frequency

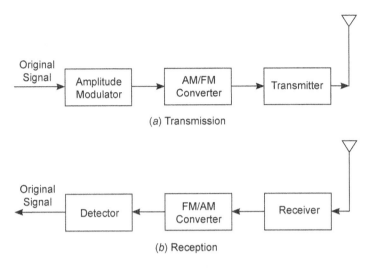

*(a)* Transmission

*(b)* Reception

**Fig. 19.2**

modulation is superior to that of the amplitude modulation and the signal is transmitted. At reception side we again use an FM/AM converter and the detecter, which detects the original signal and the carrier is separated. The carrier frequencies used for both the modulations are between 1200 Hz to 1800 Hz.

## 19.6 TYPES OF FAX MACHINES

Mainly two types of faxes are used (Fig. 19.3):

1. Thermal Paper Fax
2. Plain Paper Fax.

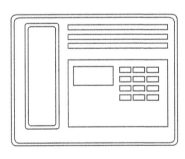

**Fig. 19.3**

1. *Thermal Paper Fax*: Thermal printing works by firing minute dots of heat at a special heat sensitive paper. This triggers a chemical reaction so that where the heat hits the paper, it turns black forming the required image.

2. *Plain Paper Fax*.

This can be classified as:

(*a*)  *Thermal Transfer*: Thermal transfer adopts a similar process where minute dots of heat are fired at a heat sensitive ribbon, which in turn transfers carbon like material on to plain paper in the form of an image.

(*b*)  *Laser Fax*: Laser faxes use fundamentally the same technology as laser printer which is, similar to that used in *photocopier.* The only difference is that instead of receiving digital pulses from the computer the laser fax receives them over a telephone line.

(*c*)  *LED Fax*: Light Emitting Diode (LED) printing is very similar to laser printing. Instead of firing a laser beam at a drum, LED uses a row of infrared lights to produce a charge which then goes on to produce the image.

(*d*)  *Inkjet Faxes*: Inkjet printing creates an image by firing jets of ink from a matrix of nossels onto a page.

## 19.7  CONVERSION OF OPTICAL SIGNAL INTO ELECTRICAL SIGNAL

This is the basic process which occurs in FAX. Following devices are used for this purpose:

1.  *Photodiode* (Fig. 19.4a): This is the simplest device which can be used to convert light into electrical signal. When light from picture or printed page falls on a photodiode, an electric current flows through the diode. This current is made to flow through a resistive circuit to produce a voltage which is later on amplified. The output of the amplifier is proportional to the light intensity falling on the diode.

2.  *Photo Multiplier tube* (Fig. 19.4b): The photo multiplier tube has so many "photo electrodes" . The light is made to fall on the first electrode which emits electrons. These emitted electrons strike the next electrode and more no. of electrons are emitted from the later. The process continued and thus multiplied output of electrons *i.e.*, a large photo current reaches at the collecting electrode (anode). The amplification in this way takes place within the device and very little amplification is further needed. The device rather needs much higher voltage to operate and thus it is costly.

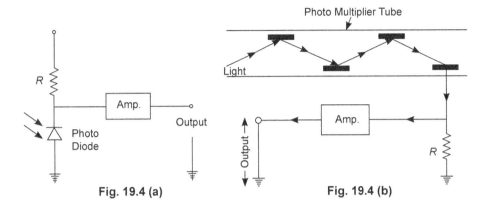

Fig. 19.4 (a)          Fig. 19.4 (b)

## 19.8 SPOT LIGHTING

Whatever the type of photodetector used to convert light into the electrical signal, the light should be restricted on the particular part of the picture or document to be transmitted. There may be following methods.

(a) In the first approach, the light is focussed through a lens on the part which is to be transmitted. The reflected light from the part reaches the photo diode, which converts the lightness (or darkness) of the part into voltages. A pure white part generates a high voltage and a dark part generates a low voltage. (See Fig. 19.5a)

(b) In another approach, though the whole surface of the document is illuminated, but reflected light from a restricted area (which is to be transmitted) is allowed to reach the photo diode. The light out of this area is not detected by the photodetector. (See Fig. 19.5b)

The first method is superior as it can be achieved by using a LASER beam which gives a better control on the light intensity.

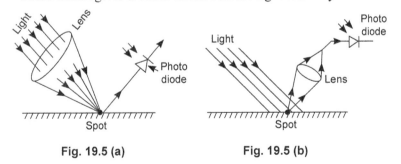

Fig. 19.5 (a)          Fig. 19.5 (b)

## 19.9 FAX TRANSMITTER

The message is "scanned", usually by *optical scanning* in the transmitter. Two methods are adopted for optical scanning:

1. Cylindrical Scanning.

2. Tape Scanning.

1. *Cylindrical Scanning* (Fig. 19.6): The message is wrapped around a cylindrical drum to allow complete scanning. The cylinder (drum) is rotated and the scanning spot is kept fixed. The light reflected from the drum is fed to a photo diode, which converts the picture signal (light) into electrical signal. The electrical output from the photo diode is amplified and transmitted. The drum is rotated at about 60 RPM and scanning rate is kept at 5 lines/minute.

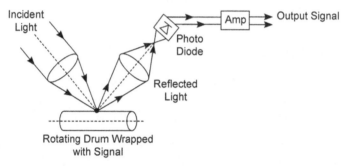

**Fig. 19.6**

2. *Tape Scanning*: This is also an optical scanning method. In this method message is taken directly from the printed tape. Due to width limitations, the method has lesser in use. The Fig. 19.7 shows the method.

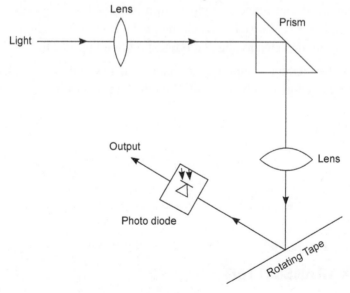

**Fig. 19.7**

## 19.10 SCANNING SPOT

The shape of scanning spot in optical scanning may be rectangular [Fig. 19.8 (*a*) or a trapezoidal (*b*)]. The shape of scanning spot decides the size of the wave shape of the output signal.

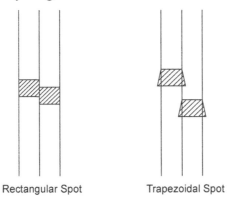

Rectangular Spot         Trapezoidal Spot

**Fig. 19.8**

Usually a rectangular spot is used. The size of the spot is very small and covers an area of constant illumination on the message. The change in the average illumination gives the signal output.

## 19.11 FAX RECEIVER

In the receiver, "reverse" is done, of what happened in the transmitter. While transmission, the picture message was converted into an electrical signal, in the receiver, the electrical signal is converted back into picture signal. The equipment (drum etc.) in the receiver is however, almost same as used in the transmitter. We may get an output free of distortion, for this it is necessary that the signal is properly " synchronised".

## 19.12 SYNCHRONISATION OF THE SIGNAL

(*a*) When the signal is a written message (*e.g.*, a page of a book), synchronous motors which rotate at constant speeds are used both at transmission and receiving ends. These motors are operated under controlled frequency supply.

If $f$ is the frequency of the supply fed to a synchronous motor with no. of poles $P$, the synchronous speed of the motor will be

$$N_s = \frac{120 \times f}{P}$$

The synchronous speed $N_s$ will remain constant so long frequency of the supply $f$ is kept constant by some means.

(*b*) If it is a picture (photograph) signal, a synchronising signal of 1 kHz is sent for synchronization.

When the signal modulates a carrier e.g., in the case of F.D.M. (Frequency Division Multiplexing), the carrier is sent along with upper side band (U.S.B.). The carrier helps in the total recovery of the synchronising signal later on.

## SUMMARY

1. The fax telegraphy is a technique to send printed message from one station to another.

2. In comparison to TV transmission, the fax needs a much shorter bandwidth, this is the reason that fax messages can be sent over ordinary telephone lines.

3. The fax finds application in business, weather calculations and personal use.

4. The fax process involves three basic steps: converting the document into electrical signal, transmitting the electrical signal through telephone wires to another FAX machine and there converting the received signal into the original document.

5. The picture is converted into electrical pulses by a photodiode or photo multiplier.

6. In transmission, AM-FM converter is used. In receiver, FM-AM converter is used.

**20**

# Radar Systems

Radar is basically a means of gathering information about distant objects or targets (aeroplane etc.). This is done by sending electro-magnetic waves at them and analyzing the *echo* received.

## 20.1 THE RADAR

The Fig. 20.1 shows and Indian radar "Guidance", of 20 km range. The term "RADAR" means "Radio Detection and Ranging". At first it was used as a method to detect an approaching aircraft but now it is used for many other purposes.

The radar has multifarious jobs' to perform. It can be used as a navigational aid and in many ways in military. In military, radar is used for aiming guns at ships and aircrafts, directing missiles, searching submarines, attacking enemy crafts etc.

**Fig. 20.1**

In an aircraft, a radar can provide information of the path in case of darkness or poor visibility; similarly in a ship it can give information of other ships, land masses etc.

Radars can also be used for safe landing of aircrafts. On approaching the airfield, the pilot is guided by the radar. Radar helps the *meteorologists* in forecasting weather. As the rain drops are able to reflect radar signals, height of the clouds can be measured. Radar can be used for discovering ores, oil fields and buried metals.

Pilots sometimes fly at very low levels to avoid radar. When they do so, radar reflection given off by the plane become indistinguishable from those given off by buildings and other tall objects, and the plane can go undetected. Some planes like the American B-2 are made of materials that do not reflect radar waves but instead, absorb them. As no radar waves are reflected back to the radar ground station, the plane becomes invisible to radar.

## 20.2 BASIC PRINCIPLES OF RADAR

Basically a radar consists of a *transmitter* and a *receiver*, each connected to a *directional antenna*. The transmitter sends VHF (*i.e.*, microwave) power through the antenna. The receiver collects as much energy as possible from the *echo* reflected from the target and analyses the information received and later on displays it. The receiving antenna is same as the transmitting antenna. All this is accomplished through TDM (Time Division Multiplexing).

The Fig. 20.2 shows the fundamental block diagram of a radar to explain its functioning. For each transmitted pulse, the cycle of events is as follows: In response to an internally generated trigger signal, the transmitter generates a short *rectangular* pulse. As soon as a small fraction of the pulse power is fed to the *duplexer*, the receiver is disconnected and the transmitter is connected to the antenna. The antenna moves in a predetermined pattern called *scanning pattern*. Either way, the antenna is highly directional and sends out the generated pulses towards the target. The *scanning speed* is high but it is small compared to the time taken by the pulses to return from the normal range of the targets. Thus when echoes are received from the target the antenna still points in the right directions to *receive* them. Now the duplexer disconnects the transmitter from the antenna and reconnects the receiver with the antenna. The received echo is *processed* in the receiver *i.e.*, amplified and demodulated. The receiver is a *super-hetrodyne type*. The pulses (and noise of course) are then fed to the displaying device. Now one cycle is complete and the set is once again ready for transmission of the next cycle, while the antenna goes on scanning along its path.

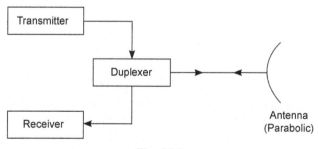

**Fig. 20.2**

The radar is able to show position of a target because the information about the horizontal and vertical direction of the antenna is available. In addition, distance of the target can also be calculated from the total time taken by the pulse on its forward and return journey. We know that speed of electro magnetic pulses is $3 \times 10^8$ m/s, and if a pulse returns in 1 $\mu S$, *i.e.*, it covers *go and return* distance in 1 $\mu S$, the distance of the target can be calculated from the following formulae:

$$\text{Speed} = \frac{\text{Distance}}{\text{Time}}$$

$$3 \times 10^8 \text{ m/s} = \frac{\text{Distance}}{1 \text{ } \mu S}$$

$$\text{Distance} = 3 \times 10^8 \times 1 \times 10^{-6} (1 \text{ } \mu S = 1 \times 10^{-6} S)$$

$$= \textbf{300 meters}$$

As the pulse has covered go + return distances, the distance of the target is

$$= \frac{300}{2} = 150 \text{ meters.}$$

However, a limit exists on the smallest number of pulses that can be sent per second. Practically the system sends out several hundred pulses per second, with a duration of 1 $\mu S$.

## 20.3 FREQUENCIES AND POWERS USED IN RADARS

The radars mostly operate in frequency at a range of 100 MHz to 250 GHz. Long range radars operate between 1 to 4 GHz. The radars designed for high *resolution* operate above 20 GHz. This permits to limit the size of antenna within practical length.

Table 20.1: Frequencies and Powers used in Radars

| Band | Frequency Range (GHz) 1 GHz = $10^9$ Hz | Max. power available in the transmitter (in MW) (1 MW = $10^6$ W) |
|---|---|---|
| UHF | 0.3 – 1.0 | 5.0 |
| L | 1.0 – 1.5 | 30.0 |
| S | 1.5 – 3.9 | 25.0 |
| C | 3.9 – 8.0 | 15.0 |
| X | 8.0 – 12.5 | 10.0 |
| Ku | 12.5 – 18.0 | 2.0 |
| K | 18.0 – 26.5 | 0.6 |
| Ka | 26.5 – 40.0 | 0.25 |
| The output power of a radar transmitter is given in terms of maximum power. Generally radar power lies within the range of 100 to 500 kW. | | |

## 20.4  MEASUREMENT OF AZIMUTH (BEARING OR POSITION) AND RANGE

Since a radar works on the principle of wave reflection (*i.e.*, echo) it uses *quasi optical microwaves* in the range from 1000 to 10,000 MHz. These can be shaped by dish (parabolic) reflector into sharply focussed beams like those of search lights. This beam is made to scan the sky through rotation of the radar antenna and the reflector. When the beam is reflected from a target, the direction of the antenna at that moment gives azimuth or bearing (*i.e.*, position) of the target at that moment. The azimuth or bearing in degrees is usually shown on the indicator.

The distance of the target can be measured directly from the CRT. In Fig. 20.3(a), the pulse is seen leaving the antenna simultaneously with indication on CRT. The pulse is known as "*main bang*". After reflection from the target, the '*echo*' is recorded on the CRT. The distance of the target can be calculated from the gap between the *main bang* and *echo* obtained on the CRT, for which CRT can be calibrated in a suitable scale.

The Fig. 20.3 (*b*) shows that range of several targets may be indicated on the face of a CRT simultaneously.

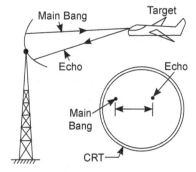

(*a*) Measurement of distance of the target

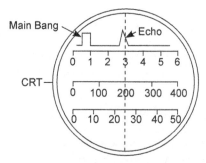

(*b*) Simultaneous measurement of range of several targets on a CRT

**Fig. 20.3**

## 20.5  FUNCTIONS (APPLICATIONS) OF A RADAR SYSTEM

A radar system has to perform two basic functions: (*i*) Searching  (*ii*) Tracking.

A radar system *searches* a target and once target is found, it follows or *tracks* the target to ascertain its position and range (*i.e.*, distance) etc. Sometimes the same radar performs both the functions or two radar systems may be employed for the above two tasks. Thus there are two radar systems.

(*i*)  *Search radar systems*: For searching a target, the radar system should be capable of *scanning* its region rapidly. Narrow beam antenna pattern takes long time for scanning therefore *broad beam scanning* is a better solution. Once the approximate position of the target is obtained the information may be passed on to the tracking radar which will quickly follow it. Another solution to this problem is to use two "fan shaped" beams, so that one is in the direction of *azimuth* and other in the direction of *elevation*. The two antennas rotate together: one *searches* and other acts as a *height finder*. *Air traffic control* (ATC) radars are its example.

(*ii*) *Tracking radar systems*: Once a target is found, it may be then *followed* or *tracked* to find its actual position in the space. For this, *conical scanning* may be employed. Tracking may be of two types:

    (*a*) *Tracking in angle* means to find the angular position of the target.

    (*b*) *Tracking in range* means to find the range (as well as the angle) of the target.

*Airborn* radar is a system, in which a single radar can perform both the functions. First the system works in *search mode* once the target searched, the system may be switched on to the *track mode*. The difficulty with this radar is that when it is "tracking" it cannot "search" and the radar is *blind* in all directions expect the one.

The *Track While* Scan (TWS) is a radar which can perform both functions simultaneously but this radar can be used only for small areas.

Radars are also available which can track satellites, spacecrafts, and space rockets. They use special way of tracking. In this system only *searching* is needed as the approximate *position* of the target (*e.g.* moon) is known. For this purpose, a very high transmitting power, sensitive receivers and antennas are required.

It should be mentioned here that the radars for deep space tracking are *monostatic* (transmitting and receiving antenna located at the same point). In exceptional cases, they are *bistatic i.e.,* transmitter and receiver are separate.

## 20.6 ANTENNA SCANNING PATTERNS (FIG. 20.4)

As already mentioned radar antennas are often made to scan a given area in the space. For different applications, different scanning patterns are popular. Few are briefly described below:

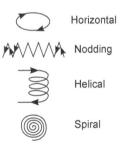

**Fig. 20.4**

    (*i*) *Horizontal scanning:* This is simplest, in which antenna is made to scan in the horizontal plane only. This pattern is used in *ship to ship* radars.

    (*ii*) *Nodding scanning:* In this, antenna scans in both *i.e.,* horizontal as well as in vertical planes. This method can be used to scan a limited area of the space.

(*iii*) *Helical scanning:* In this system also, antenna scans in both the planes and through a large area in the space. The antenna returns to its starting point at the end of each cycle. The typical speed in this pattern is 6 r.p.m.

(*iv*) *Spiral scanning*: This method is used when a smaller area in circular shape is to be scanned. In radars "scanning antenna" is used. The purpose of using scanning antenna is to find direction of the target w.r.t. to the target. The direction of antenna, at the instant when echo is received gives direction of the target.

## 20.7  SIMPLE RADAR SYSTEM

The process of sending out wave energy in short pulses is known as *pulsing* and the same is used in radars. The essential parts of a simple radar system is shown in Fig. 20.5. The *indicator* unit consists of a cathode ray tube (CRT) similar to that used in a TV set. The CRT displays the original transmitted pulse as well as the return echo pulse along a scale calibrated in terms of distance. There may be other indicators for drawing a map of the area searched by the radar antenna, but the system shown here is only to give information about the *range and direction of the target.*

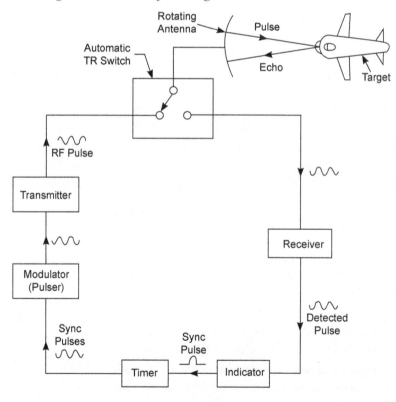

**Fig. 20.5**

The *Timer* in the block diagram assures that a single pulse from the *pulser* turns on the transmitter exactly at the same time as the indicator begins to draw the pulses along its scale. The *transmitter* (high power magnetron) then sends out the short pulses of very high frequencies through a *Transmit-Receiver* (TR) switch. At this moment, switch is connecting transmitter and terminating the receiver off the circuit. These pulses are sent to the rotating antenna, which radiates them into space. At the same time the transmitted pulse is displayed on the indicator scale. If the transmitted pulse strikes a target some where in the space, an echo will be received by the antenna. The echo will be maximum when the target is exactly in the centre of the transmitted pulses. In practice, this is achieved by turning the antenna till a maximum echo is obtained.

In the mean time, the fast acting automatic TR switch has already turned towards the *receiver* where echo pulses are amplified and demodulated by a *superheterodyne* circuit. The output of the receiver is given to a display device (say a CRT). The distance between the *original* pulse and the *echo* can be read out directly (say in kilometers) thus giving *range* of the target. The direction of the antenna at this time gives direction of the target.

## 20.8 CLASSIFICATION OF RADARS

The various radar systems can be classified as below:

1. Pulse radars
2. Moving target indication (MTI) radar
3. Beacons radar
4. CW radars
5. Tracking radar
6. Laser radar

Here we discuss them briefly:

## 20.9 PULSE RADAR (FIG. 20.6)

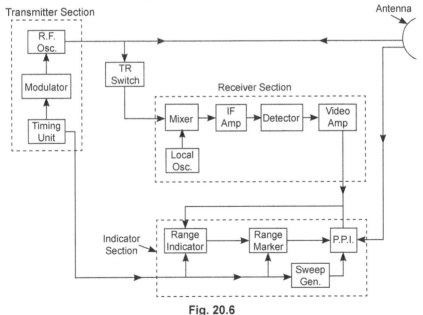

**Fig. 20.6**

## (*a*) Description of Main Blocks

1. *Transmitter*: In this section, a high power *oscillator* (usually a magnetron oscillator) generates RF oscillations. The voltage pulses are formed in a modulator which are applied in its plate cathode circuit. The pulse operation is controlled by a *timer* unit, which is basically a multivibrator. The trigger pulses obtained from the timer unit are also applied to the indicator section for synchronisation.

   The RF energy of the oscillator is transmitted through an *antenna* which is a specially shaped and designed parabolic reflector.

2. *Transmits Receive (TR) switch*:

   The TR switch which periodically makes contact with transmitter and receiver employs a gas discharge tube. The high power pulses from oscillator (magnetron) break down the gap in the tube and short circuits the line to the receiver.

3. *Receiver*: The radar receiver is a superheterodyne receiver. The intermediate frequency (IF) is kept usually 30 to 60 MHz. A diode is used as a *detector*. The video amplifier raises the signal amplitude to the required level to be applied to the indicators.

4. *Indicators*: The indicator panel of the radar is provided with the various measuring instruments and indicators *e.g.* Pulse position Indicator (PP1). It is provided also with a sweep generator for range measurements.

## (*b*) Description of Main Components

(*i*) *T.R. Switch (Duplexer) (Fig. 20.7)*: ATR switch or duplexer is a circuit which enables the same antenna to be used for both *transmission* as well as *reception* without any interference. In radars, generally *branch type* duplexer is used. It has two switches (TR and anti TR) which alternately connect transmitter and receiver with the antenna. *Cold cathode vacuum tubes* are generally used for these switches.

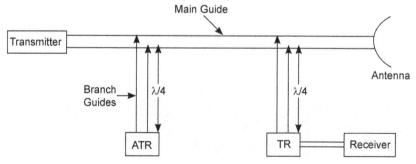

**Fig. 20.7**

*Operation*: A *main guide* connects the transmitter to the antenna. With this main guide, TR and ATR (Anti TR) switches are connected through the *branch guides* of quarter wave lengths.

When the transmitter produces RF pulses, both switches get short circuited. The short circuiting of TR switch prevents RF power from entering the receiver. At the termination of the transmission, both switches are open circuited. The ATR switch throws a *short circuit* thus preventing any signal from entering into transmitter. The guide leading through TR switch to the receiver becomes continuous, thus the echo reflected back from the antenna can go to the receiver.

(*ii*) *Modulator*: Modulator is a component of radar transmitter. The function of modulator is to switch *on* and *off* the output tube as and when required. There are two types of modulators in use:

(*a*)  Line pulsing modulators        (*b*) Active switch modulators.

(*a*)  **Line pulsing modulator (Fig. 20.8):** In this, anode of the output tube is modulated directly by a system that generates and provides large pulses of supply voltage. This is achieved by slowly charging and discharging a transmission line. But practically a *Pulse Forming Net work* (PFNW) replaces the line. As shown in the Fig. 20.8 the PFNW behaves as a transmission line for frequencies below $f = 1/2\pi\sqrt{LC}$ where $L$ and $C$ are inductance and capacitance respectively. For switch(s), a *thyratron or a SCR may be employed.*

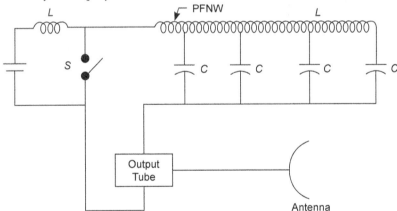

**Fig. 20.8**

(*b*)  **Active switch modulator:** The line pulsing modulator cannot give variable pulse length, therefore active switch modulator may be used. In this, pulses are generated at low power level and later are amplified by an amplifier, but this modulator is less efficient, more complex and bulky, so rarely used.

(*iii*) *Antenna*: The majority of radars use *dipole* or *horn type* parabolic *reflector* as antenna. The beam width in the vertical direction is worst than in the horizontal direction, but this is immaterial in *ground to ground* or even *air to ground* radars. It has the advantage of having small size and weight, reduced wind load, moreover they need small motors.

The various shapes used for radar antenna are shown in the Fig. 20.9.

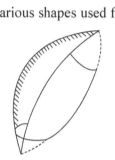 

**Fig. 20.9**

## 20.10  MTI RADAR

In Moving target Indicating (MTI) radar system, only moving targets are displayed on the screen, and the echo obtained from the stationary targets is not displayed at all. The MTI radar employs Doppler's effect explained below.

### Doppler Effect

The scientist *Doppler* in 1842 gave a postulate known as "Doppler Effect", which says:

"The apparent frequency of an electromagnetic (*i.e.*, light, sound) wave depends on the relative motion of *source* and the *observer*."

If source and the observer are moving away from each other, the apparent frequency will decrease and if they are moving towards each other, the apparent frequency will increase. (This is due to the doppler effect that causes a change in the *pitch* of a whistle from a passing train.)

In case of radar, involving a moving target, the signal undergoes a change due to doppler effect. The target acts as a *source* of the reflected waves. Now we have a moving source (target) and a stationary observer (radar). If the target is moving towards the radar, the overall effect is doubled. The doppler effect is observed only in *radial* motions and not in the *tangential* motions. If target is rotating, the radar can distinguish its *leading* edge from its *trailing* edge. The rotation of the planet *Venus* has been measured by employing this effect as due to dense cloud cover over this planet, this cannot be done by a telescope.

Basically, a MTI radar system compares a set of received echoes with those received during previous sweep. The echoes whose phase remains constant are cancelled out as they are obtained from stationary targets, but the echoes obtained from moving targets show a *phase difference* due to *doppler* effect and hence not cancelled. It also helps in the detection of moving targets whose echoes are very small due to long distance than those of nearby stationary targets.

The transmitted frequency is the sum of the outputs of two oscillators $(f_c + f_s)$. The first oscillator is called *Stalo* (stable) and other a *coho* (coherent) (See Fig. 2010). The stalo is a local oscillator and coho operates at intermediate frequency (I.F.) and provides *coherent signal*. The mixers I and II are identical and both use the *stalo* thus phase relations of their inputs are preserved in their outputs also. The output of the I.F. amplifier and signal of coho oscillator is fed to a detector (phase discriminator) circuit.

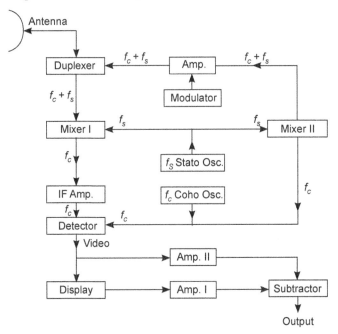

**Fig. 20.10**

The coho oscillator is used to generate RF signal as well as the signal for the detector. Since the output of the detector is phase sensitive, an output will be obtained for all fixed and moving targets. The phase difference between *transmitted* and *received* signals will be constant for fixed targets and it will be changing for moving targets as per the doppler effect.

## Blind Speeds in MTI Radars

In an MTI radar, the phase difference between transmitted pulse and the echo remains constant from pulse to pulse in case of stationary targets. The phase difference, however, varies in case of moving targets. That is, why a MTI radar can distinguish between stationary and moving targets.

If the target happens to be moving with a velocity such that its velocity component along the radar axis results in a phase difference of $2\pi$ radians or its integral multiple between successive sets of pulses to the radar, the targets appear as stationary. These target velocities are said to be **Blind speeds**.

If the target velocity is such that it moves a distance equal to $n \cdot \lambda/2$, where $n$ is an integer and $\lambda$ the transmitted wave length along the radar axis, the phase difference will be $n \cdot 2\pi$ radians, the blind speed is expressed by

$$V_b = \frac{n\lambda/2}{1/fr} = \frac{n\lambda fr}{2}$$

when $n = 1$ the lowest blind speed $= \lambda\, fr/2$.

Where, $n$ is an integer, $\lambda$ = wave length transmitted and $fr$ = pulse repetition frequency (PRF).

## 20.11  BEACON RADAR (RESPONDERS)

The *beacon* is a small radar set, consisting of a receiver, transmitter and an *all directional* antenna. When another radar transmits a coded signal to a beacon *i.e.*, when a beacon is *interrogated,* the beacon *responds* by sending back its own coded pulses. A beacon is called a *responder* and it may respond in the same frequency as interrogated by the *interrogating* radar or in a special beacon frequency in which case a separate receiver is required by the interrogating radar. It is to be noted that a *beacon* does not transmit pulses continuously as a *pulse radar* but it only *responds*, when *interrogated* by a radar.

One of the functions of a beacon is to *identify*. For example, a beacon may be installed on target such as an aircraft and it will transmit a coded pulse when interrogated. These pulses appear on the plan position indicator (P.P.I.) of the *Interrogating radar* and inform it about identity of the target. When used as such it is called *Identification Friend or Foe* (IFF). An IFF operates satisfactorily with naval *targets*. It however, fails if large number of fast planes: some friends and some foes (enemies) are involved.

Another application of a beacon is that it can work as a "light house". A beacon light house can operate over much larger distances. A lost ship can calculate its position by interrogating a beacon lighthouse.

A Beacon is generally used to identify whether an approaching target (*e.g.* plane) is a friend or a foe.

## 20.12 CW RADARS

A CW radar sends *continuous waves* (sine waves) rather than pulses. It also uses the *Doppler effect* to detect the frequency change caused by moving target and displays it as a *velocity* of the target in kilometer per hour.

Two types of continous wave (CW) radars are described below:

### (a) Unmodulated Continuous Wave (CW) Radars

An un-modulated CW radar has been shown in Fig. 20.11. Since the transmission is continuous, a *duplexer* is not needed. Separate antennas for transmission and reception have been shown. The arrangement increases isolation between transmitter and the receiver.

As shown, a small portion of the transmitter output is mixed with output of a local oscillator and the sum is fed to the mixer. This also receives the *shifted signal* from the receiving antenna and produces an output difference frequency (plus or minus) called *doppler frequency*. The output of this mixer is amplified, and signal from the detector gives the doppler frequency, after amplification the output is given to the indicator.

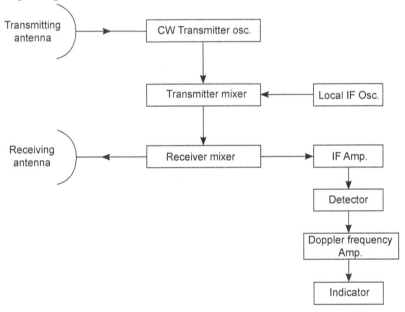

**Fig. 20.11**

### *Advantages*

(*i*) Un-modulated CW radar gives accurate measurement of relative velocities, using low transmitter power and low power consumption.

(*ii*) This radar is comparatively of lesser size than the widely used pulse radars.

(*iii*) It is uneffected by stationary targets.

(*iv*) The receiver of the radar is ON all the time, it can be operated even at zero range.

(*v*) It can find the direction of the target in addition to its speed.

### Disadvantages

(*i*) It has limitations of its maximum range, which is very smaller than of pulse radar.

(*ii*) It is incapable to indicate the range of the target, it can show only its velocity.

### Applications

(*i*) It is used in aircraft navigation.

(*ii*) As it is unaffected by stationary objects, it is able to detect presence of aircrafts.

(*iii*) It can measure speed of missiles, automobiles etc.

## (b) Frequency Modulated CW (FMCW) Radar

The greatest limitation of unmodulated CW radars is that they can not measure the range. If the transmitted signal is frequency modulated, the same radar will be able to measure range, because then it will be able to distinguish one cycle from the other. Using FM will require an increase in BW of the system.

In this way, a FMCW (Frequency modulated continuous wave) radar is an improved form of a CW unmodulated radar. As the main application of a FMCW radar is to measure height of an aircraft or plane, it is generally known as an *Altimeter*. The Fig. 20.12 shows the block diagram of a FMCW radar being used as an altimeter.

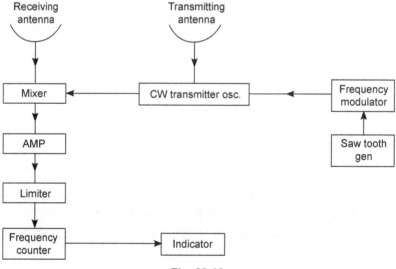

**Fig. 20.12**

As an Altimeter, if the target (in this case, earth) is stationary w.r.t. to a plane whose height is to be measured, a frequency difference proportional to the height of the plane will exist between transmitted and the received signals, because the signal now being received was sent at a time, when the frequency was different. If the *rate of change of frequency* w.r.t. time due to FM process is known, the difference in the time between *sent* and the received signals may be found and thus height of the plane above earth can be easily calculated. Thus output of the mixer which produces the frequency difference can be *amplified* and fed to the *frequency* counter and then to the indicator, whose output is calibrated in kilometers.

As an altimeter, FMCW radar is preferred to the pulse radar, as the FMCW radar has no limit on the minimum range, as short heights are involved in this measurement. Further simple, low power and smaller equipment may be used with small antenna. A typical FMCW radar uses a 1-2 W transmitter power easily obtained from a diode and has a range of about 10 km or more.

## 20.13 TRACKING RADAR

The tracking radar does the job of continuously tracking a moving target. It is usually a ground base system used to track the air born targets. The tracking radar antenna sends out a very narrow beam (called pencil beam), whose beamwidth could be any where between fraction of a degree to one degree both in azimuth and elevation to get the requisite resolution for tracking purpose. One can visualize that it is imperative to use a search radar and acquire the target with comparatively much larger bandwidth before a track action is intiated.

## 20.14 LASER RADAR

A laser radar uses an optical beam instead of microwaves, in other words, the e.m. energy transmitted lies in the optical spectrum in laser radars where as in microwave radar, it lies in the microwave region. In laser radar, the frequencies involved are very high. It is possible to generate laser pulses as narrow as a fraction of a picosecond.

## 20.15 RADAR DISPLAYS

The output of a radar may be displayed in any of the following devices

  (*i*)  A scope

 (*ii*)  Plan position indicator

(*iii*)  Monitor

(*i*) *A scope* (*See Fig.* 20.13): The operation of this system is more or less similar to that of a CRO. A sweep wave form is supplied to the horizontal deflection plates of the CRT which moves the beam slowly left to right across the face of the tube. The flyback period is rapid just as in a TV and the beam rapidly returns to its starting point. As with a CRO, in the absence of a signal, the display is a horizontal straight line.

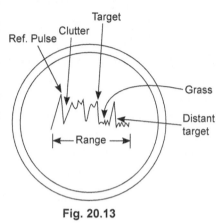

**Fig. 20.13**

(*ii*) *Plan Position Indicator* (*PPI*): [Fig. 20.14 (*a*), (*b*)] The PPI displays a map of the targetted area. The signal obtained from the receiver after demodulation is applied to the grid of the CRT. Long persistence phosphors are used at the face of PPI screen, so that it does not flicker. The PPI display is used with search radars.

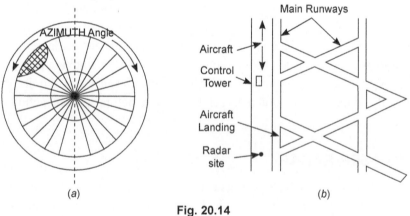

(*a*)                                                    (*b*)

**Fig. 20.14**

(*iii*) *Monitors*: They are just like TV receivers with a screen to display and a picture tube and all other components are missing. The output is given to the picture tube. The panel has all control of a TV receiver like brightness, contrast, focus, etc.

## 20.16  TERMS RELATED TO RADAR SYSTEMS

The important terms related to radar systems are given below:

1. *Pulse Repetition Rate* (*PRR*): The PRR stands for no. of pulses transmitted per second. The PRR is determined by the time difference

between two pulses which should be large enough, so that the echo of a particular pulse lies within the time interval between the transmitted and succeeding pulses. But if the time difference is much larger, there is every possibility that the radar fails to detect the target, *i.e.*, the radar is not transmitting any pulse, when the antenna points in the direction of the target.

2. *Range*: The range of a radar is the time lapse between the transmitted pulse and received echo. The maximum possible range of radar is given by

$$[R_{max}]^4 = \frac{P.G.A.\sigma}{(4\pi)^2.S \min}$$

Where      $P$ = Transmitted power in Watt.

             $G$ = Gain of the transmitting antenna

             $A$ = Effective aperature area of receiving antenna in m$^2$

             $\sigma$ = Target cross-section area in m$^2$

             $\pi$ = 3.14

             $S_{min}$ = Minimum detectable signal

3. *Resolution:* This is the ability of a radar to distinguish between two closely spaced targets.

4. *Range ambiguity:* The inability of a radar to distinguish between two targets at different ranges is called the range ambiguity (See Figure 20.15). The radar can not distinguish between targets represented by the positions *A* and *B*.

**Fig. 20.15**

5. *Clutter*: All those objects, which reflect the radar signal, and which are other than the target are termed as clutter.

The echo corresponding to the stationary target is also called clutter.

6. *Blind range*: When a target lies at such a distance, that the radar receives the echo while it is transmitting, it fails to detect the target. In such a case the target is called to be at the blind range.

7. *Precision*: The standard deviation in the estimate of certain parameters such as dopplers shift, range etc amidst noise is termed as precision. The precision is a function of S-N ratio and the wave form.

8. *Monostatic and Bistatic radar systems*: In monostatic radar system (Fig. 20.16) single antenna is used for both transmission and reception. A duplexer is used to separate the two functions, both in time and power amplitude domains.

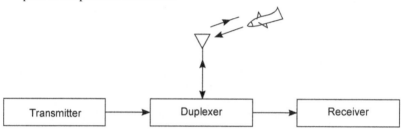

**Fig. 20.16**

In Bi stable radar system, separate antennas are used for transmission and reception. These antennas may be sometimes miles apart, see Fig. 20.17.

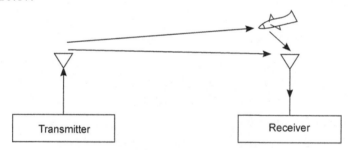

**Fig. 20.17.** Bi-stable radar system

9. *Duplexer*: A device which disconnects the receiver from the antenna and connects it with the transmitter and vice versa.

10. *Nautical miles (Nmi)*: a unit of distance and one nautical mile (Nmi) is equal to 1852 m, this unit is used in radar systems.

11. *TWT (Travelling Wave Tube)*: a microwave amplifier.

12. *CFA (Cross Field Amplifier)*: a micro wave power amplifier similar to a magnetron.

13. *Grass*: In radar system, the noise which is not constant in amplitude and position is termed as grass.

## SUMMARY

1. Radar stands for "Radio detection and ranging". It detects and gives range (distance) of a plane. It can also follow the enemy plane.

2. American radar B-2 cannot, be detected by enemy radar. If a plane flies too low, even then it cannot be detected.

3. The radar has a transmitting as well as receiving antenna. The transmitting antenna sends VHF signal which is reflected by the enemy plane. The radar receiver antenna receives the reflected signal and displays the same on the monitor. The time between transmission and reception of signal helps in the calculation of distance (range) of the enemy plane.

4. The various radars are: pulse radar, MTI radar, Beacon radar, CW radar etc.

5. The MTI radar is based on the doppler effect which says that the apparent frequency of an e.m. wave depends on the relative motion of source and the observer.

6. The most popular video display unit used for radars is "plane position indicator".

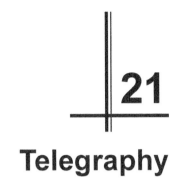

# Telegraphy

**21**

Telegraphy is a communication system of sending an information/signal or message from one place to another. The messages are expressed with the help of words and words are made of letters or *characters*. Further the characters are made up of combination of "signal elements". This representation of characters by elements is known as Telegraphic *code*. At the transmitter, the message is "coded" and sent through lines. At the receiver the message is "decoded" to reobtain the original message. In *manual* telegraphy, a telegraphic "key" is operated to code the message and in "speed" telegraphy, codes are formed on a perforated paper.

## 21.1 TELEGRAPHY

The "telegraphy" is a technique of sending massages from one place to other. At sending end, the massage is "coded" and sent through wires. At receiving end, the same is "decoded" and the original massage is obtained.

## 21.2 CODE

This may be defined as a system of signs used for transmission of a message. The important codes are:

1. Morse code
2. 5/7.5 unit code
3. ASCII (Askee) code

These codes are discussed below:

## 21.3 MORSE CODE

In Morse code, the letters are represented by "dot and dash". The number of "dot and dash" is different for different letters. The maximum and minimum number of signal elements present in a letter is 4 and 1 respectively and the "duration" of dot or dash is *inverse* of the speed.

The important points in using the code are:

   (*i*)  The duration of a dash (–) is *three* times the duration of a dot (.) signal.

   (*ii*) The duration of space between signal elements is equal to the time period of a *dot* element. The duration of space between two letters is equal to a *dash* and the duration of the space between two words is four times of a *dot*. The morse code is given in the Fig. 21.1.

**(A) Letters**

| | | | |
|---|---|---|---|
| *a* | · — | *n* | — · |
| *b* | — · · · | *o* | — — — — |
| *c* | — · — · | *p* | · — — · |
| *d* | — · · | *q* | — — · — |
| *e* | · | *r* | · — · |
| *f* | · · | *s* | · · · |
| *g* | — — · | *t* | — |
| *h* | · · · · | *u* | · · — |
| *i* | · · | *v* | · · · — |
| *j* | · — — — | *w* | · — — |
| *k* | — · — | *x* | — · · — |
| *l* | · — · · | *y* | — · — — |
| *m* | — — | *z* | — — · · |

**(B) Digits**

| | | | |
|---|---|---|---|
| 1. | · — — — — | 6. | — · · · · |
| 2. | · · — — — | 7. | — — · · · |
| 3. | · · · — — | 8. | — — — · · |
| 4. | · · · · — | 9. | — — — — · |
| 5. | · · · · · | 10. | — — — — — |

**(C) Punctuation**

Full stop         · — · — · —

Comma             — — · · — —

**Fig. 21.1**

The transmission may be by

    (*i*)  Single current method

    (*ii*)  Double current method.

In *single current method*, the current flows through the line in same direction during the signal period only and no current flows during the space periods.

In *double current method* the current always flows through the line, and during space periods it flows in opposite direction.

These methods are also called "Two conditions method" as only two conditions are possible on the line either *"current or on current"* or *"positive or negative current"*.

**Illustration:** If we have to send the word "dc" and we know that

*d* is formed by dash-dot-dot (— · ·)

and *c* is formed by dash-dot-dash-dot (— ·— ·)

In single current method, duration of pulse for dash is about 3 times of the dot and no current is to flow between *d* and *c*.

In double current method, the reverse current is to flow between the period of *d* and *c*.

The two pulse trains have been shown to transmit the word dc. [See Fig. 21.2 (*a*) and (*b*)].

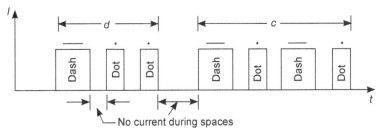

(*a*) Single Current Method

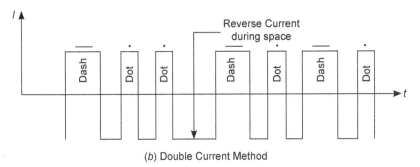

(*b*) Double Current Method

**Fig. 21.2**

## 21.4 FIVE (5/7.5) UNIT CODE (TELEPRINTER CODE)

The morse code is uneven and of non uniform length, because the no. of elements in all the characters is not same. The five unit code is of uniform length and each character consists of 5 elements. The Five unit code is a *binary code* and hence can represent $2^5 = 32$ combinations. But 32 combinations are not sufficient to represent 25 alphabets, 10 numerals and various punctuation marks. This problem is removed by using two combinations as *shift signals* (as used in typewriters for capital letters), one of them is called as *letter shift* and the other is called as *Figure shift.* All other combinations represent two characters-one *letter* and *one* Figure. When the *letter shift* is transmitted, the machine prints only *letters* till it receives *Figure shift,* to print Figures. In this way, the capacity of the code becomes double.

The one set of 5 unit code is *Murray code* which is also known as CCITT code (consultative committee for International Telegraph and Telephone) Code. When the 5 units of this code is preceded by start element (*i.e.,* space) and followed by a stop element (*i.e.,* mark), the code is known as "start stop code" or more popularly as a *Teleprinter code.* The duration of start element is always same as of the 5 code element where as duration of stop element may be 1.5 times, the total length of the code may thus be 7.5 units.

A part of this code is given in Table 21.1.

**Table 21.1** Five Unit Code

| Letter Shift | Figure shift | Start | Code Element | | | | | Stop |
|---|---|---|---|---|---|---|---|---|
| | | | 1 | 2 | 3 | 4 | 5 | |
| A | – | 0 | 1 | 1 | 0 | 0 | 0 | 1 |
| B | ? | 0 | 1 | 0 | 0 | 1 | 1 | 1 |
| C | : | 0 | 0 | 1 | 1 | 1 | 0 | 1 |
| Q | I | 0 | 1 | 1 | 1 | 0 | 1 | 1 |
| W | 2 | 0 | 1 | 1 | 0 | 0 | 1 | 1 |

## 21.5 ASKEE OR ASCII (AMERICAN STANDARD CODE FOR INFORMATION INTERCHANGE) CODE

The ASCII (pronounced as askee) code is a 7 bit code. It is a alphanumeric (alphabet, numbers etc.) code very much used by teletypewriters (TTY). The TTY is a widely used input/output device for computers. The TTY has a alphanumeric key board for entering program and data, and a printer to get answers out of the computer. The ASCII code allows standardisation of interface hardware such as key board, printers, video displays etc.

The table 21.2 shows the ASCII code. As told already it is a 7 bit code with the format

$$X_6X_5X_4 - X_3X_2X_1X_0$$

The first 3 bits are taken from the column in which a symbol appears and the last 4 bits are corresponding to the row. *e.g.*, the letter *B* appears as $100–0010 B = 100–0010$.

Table 21.2  ASCII Code

|       | 000 | 001 | 010 | on | 100 | 101 | 110 | 110 |
|-------|-----|-----|-----|-----|-----|-----|-----|-----|
| 0000 | NULL | $DC_0$ |    | 0 | @ | P | | |
| 0001 | SOM | $DC_1$ | ! | 1 | A | Q | | |
| 0010 | EOA | $DC_2$ | " | 2 | B | R | | |
| 0011 | EOM | $DC_3$ | # | 3 | C | S | | |
| 0100 | EOT | $DC_4$ | $ | 4 | D | T | | |
| 0101 | WRU | (STOP) | % | 5 | E | U | | |
| 0110 | RU | SYNC | & | 6 | F | V | | |
| 0111 | BELL | LEM |   | 7 | G | W | | Unassigned |
| 1000 | $FE_0$ | $S_0$ | ( | 8 | H | X | | |
| 1001 | HT/SK | $S_1$ | ) | 9 | I | Y | | |
| 1010 | LF | $S_2$ | * |   | J | Z | | |
| 1011 | $V_{TAB}$ | $S_3$ | + |   | K | [ | | |
| 1100 | FF | $S_4$ |   | < | L | / | | |
| 1101 | CR | $S_5$ |   | = | M | ] | | |
| 1110 | SO | $S_6$ | * | > | N | ↑ | | |
| 1111 | SI | $S_7$ | / | ? | O | ← | | |

| Abbreviations | | | |
|---|---|---|---|
| NULL | Null idle | CR | Carriage return |
| SOM | Start of message | SO | Shift out |
| EOA | End of address | SI | Shift in |
| EOM | End of message | $DC_0$ | Device control Reserved for data Link for escape |
| EOT | End of transmission | $DC_1 – DC_2$ | Device control |
| WRU | "Who are you?" | ERR | Error |
| RU | "Are you .... ?" | SYNC | Synchronous idle |
| BELL | Audible signal | LEM | Logical end of media |
| FE | Format effector | $SO_0 – SO_7$ | Separator (information) |
| HT | Horizontal Tabulation | | Word separator (blank, normally non-printing) |
| SK | Skin (punched card) | | |
| LF | Line feed | ACK | Acknowledge |
| $V_{TAB}$ | Vertical tabulation | | Unassigned control |
| FF | Form feed | ESC | Escape |
| | | DEL | Delete idle |

## 21.6 PUNCHED PAPER TRANSMISSION

For high speed transmission, the signals are obtained in the form of "holes" or "punches" on a paper. This punched paper is rapidly passed in the transmitter and sent through the lines. For this also, the Morse code is used. (See Fig. 21.3)

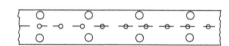

| Dot | O  o  O |
| Dash | O o   o O |
| Space | o |

(a) Morse code on paper          (b) Punched paper with morse code

**Fig. 21.3**

(*i*)  *For dot.* two bigger holes on the sides of a small hole.

(*ii*)  *For dash.* Two smaller holes on the sides of two small holes.

(*iii*)  *For space.* A small hole in the centre.

## 21.7 ORDINARY TELEGRAPHIC SYSTEM (WIRE TELEGRAPHY)

In a telegraphic communication, following important equipment is used:

1.  *Transmitter.* Here signals are formed from the message. In other words, message or information is converted into signal for transmission.

   In ordinary telegraphic system, a simple *telegraphic key* is used as a transmitter. By pressing or releasing the key according to the morse code, any message may be converted into the electric signal.

   The key may be:

   (*a*)  *Single current key (SC key).* The Fig. 21.4 (*a*) shows its pictorial view and Fig. 21.4 (*b*) shows it symbolic representation.

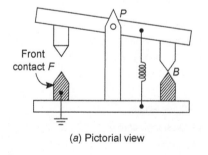

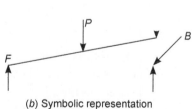

(a) Pictorial view                    (b) Symbolic representation

**Fig. 21.4**

(b) *Double Current Key (DC Key)*. This key is obtained by mechanical coupling of two single current keys. This key is not popular now-a-days and double current operation is obtained from a single current key with two set of batteries as shown in Fig. 21.5.

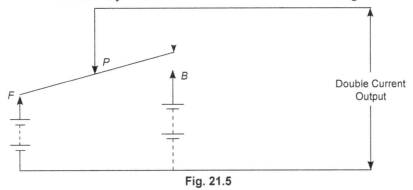

**Fig. 21.5**

2. *Receiver* (*Fig.* 21.6). A morse sounder is used as receiver. It is an electro magnetic device. In converts the incoming electrical signal, corresponding to dots and dashes of the morse code into audible sound. This is further translated into language by a trained operator.

When current flows through the electromagnet coil, it pulls both the armature and the lever down. The armature is prevented from touching the pole piece. The armature in the attracted position is an unstable condition of the sounder. For going back of the armature, the attractive force should be less than the spring tension.

The attractive force of the magnet is proportional to the square of the current [ $F \propto I^2$ ] and independent of the direction of the current. The sounder therefore is not affected by the reversal of current. In other words the sounder is non *polarized.*

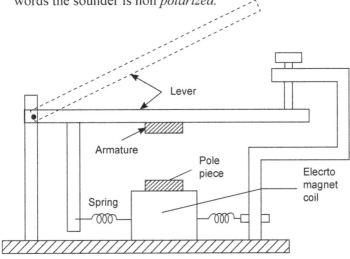

**Fig. 21.6**

3. *Relays.*        Generally the distance between the transmitting and receiving stations is much more and the current reaching the receiver is not sufficient to operate the sounder. For this purpose the telegraph

**Fig. 21.7**

relays are used which take very small current to actuate. The Relays in turn operate the sounder. The relays may be polarized or non polarized type, usually polarized relays (Fig. 21.7) are used which can distinguish between the direction of current through its coil. The relay has an armature and two contacts. The armature rests at one contact for one direction of current and at other contact for the reverse direction of current.

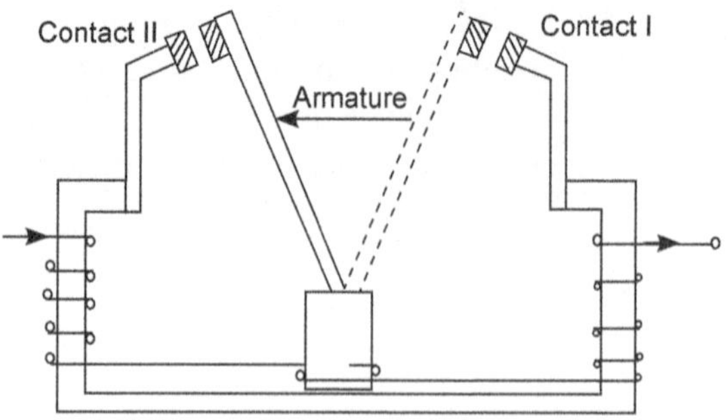

## 21.8 TELECOMMUNICATION LINES (FIG. 21.8)

The telegraphic messages are sent by overhead wires on poles. It has the following components:

(*i*) *Conductors or wires.* Generally galvanised iron or steel wires are used to carry the telegraph signals. The conductivity of iron is poor but as currents are smaller, can be easily carried by it. Also the mechanical strength of iron is sufficient. For ordinary frequencies, one conductor is required as the *earth* acts as the return conductor. At high frequencies, two conductors are required, the second conductor acts as return conductor.

(*ii*) *Poles.* In rural areas, wooden poles may be used. These poles are made of well seasoned teak wood. Further, they are treated against dampness and attack of insects, Otherwise galvanised iron pole may be used. They are of approximate 8 meters in length and out of which about 2 meter is kept under ground. Between poles, a span of about 100 m is used.

(*iii*) *Cross arms.* The poles are fitted with cross arms which are to carry insulators. The cross arms used are of wood or a channel section for wooden and iron pole respectively.

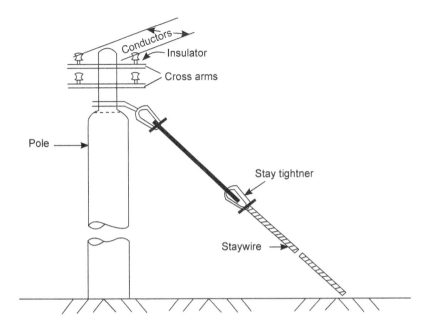

**Fig. 21.8**

(*iv*) *Stay.* Due to weight of cross arms, and wires etc. the pole may turn down. As a balance force, a stay wire is tightened with the pole in the opposite direction to keep it "erect".

(*v*) *Insulators (See Fig.* 21.9). The insulators are fitted on the cross arms to carry conductors. The insulators are made of some insulating material like porcelain and prevent the direct contact of conductor with the cross arms on the pole. If wire or conductor comes in direct contact with the

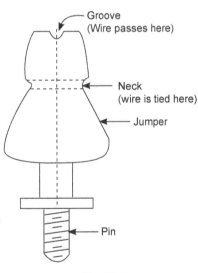

**Fig. 21.9**

pole, the signal currents will also flow through the pole and it will create problems. They are generally pin type insulators, with the help of the pin they are fitted with the cross arms. They are *galzed* to prevent accumulation of water on its surface.

## 21.9 CARRIER TELEGRAPHIC SYSTEM (WIRELESS TELEGRAPHY)

Now-a-days, the old *Duplex* and *Quadruplex* telegraph systems are being replaced by carrier telegraphy system. It enables us to send more number of *channels* simultaneously through the same line. The system is popular as "Multichannel Voice Frequency" (MCVF) system. Generally 6, 12, 18 or 24 channels are simultaneously used. This is worked with *teleprinter* on both sides. The system is more or less similar to radio communication.

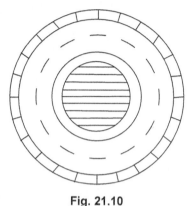

**Fig. 21.10**

For MCVF transmission, two wires are used for sending and two wires for return of the signal. Coaxial cable (Fig. 21.10) consisting of concentric conductors are used in "wide band carrier telephone system". At very high frequency, signals can be sent by this system.

### MODULATION SYSTEMS

Two types of modulation systems are used in *carrier* telegraphy.

   (*i*)  Amplitude Modulation (AM).       (*ii*) Frequency Modulation (FM).

As described earlier, in AM, the amplitude of the carrier is varied according to telegraph signal and in FM, the frequency of the carrier is varied according to the telegraph signal to be sent. In modulated wave, we get the carrier and sidebands.

If different carriers having different frequencies are used for different telegraph signals; different side bands of the different frequencies would be obtained and thus number of signals can be simultaneously sent over the same telegraph lines. Similarly at the receiver these different signals can be demodulated and detected. This principle is more or less same as of sending *radio signals*.

Here it is important that in telegraphy the signal is a *square* wave which besides the fundamental frequencies contains *harmonics also*; (Fig. 21.11) therefore a modulated wave should be filtered out of the higher frequency of sidebands and this is what is actually done in telegraphy. However, this is not done in radio communications.

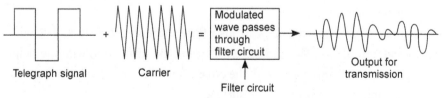

**Fig. 21.11**

As all the harmonics are filtered out except fundamental frequency, as a result a bandwidth of 60 Hz is required for transmission of the signal and of 120 Hz for side bands. Generally 180 Hz bandwidth is used per channel transmission.

The multichannel system uses carrier frequency between 400 Hz to 3 kHz. For higher channel, the carrier frequency also increases, *e.g.* For channel No. 1, this value is 400 Hz, for channel 12 this is 1720 Hz and for channel No. 24, it is 3 kHz.

## 21.10 PICTURE TELEGRAPHY

This is a technique to send pictures from one place to another. For this purpose "photo telegraph machines" are used. The photo electric cell is an indispensible component of all these machines.

The principle is more or less same as used in *Television*. The picture is divided into small elements or lines and they are scanned. The scanning (Fig. 21.12) is same as in sending TV signals. A picture of normal size is divided into $2 \times 10^5$ elements. A photo electric cell generates a "video signal" according to the intensity of the light.

The picture to be scanned is rolled over a drum which is rotated at a determined speed by a motor. The speed of the scanning motor should be same at the transmitting and receiving ends. In other words, the two scanning should be properly synchronised. (See Fig. 21.13)

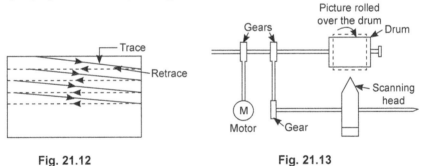

**Fig. 21.12**                          **Fig. 21.13**

## Transmission (Fig. 21.14)

The picture is rolled over the drum and the scanning head carries the photo electric cell. As a result of scanning, the output of the cell is a *video signal* proportional to the intensity of the light.

The video output of the photo cell *modulates* a carrier of suitable frequency and the obtained modulated signal is then amplified.

The normal telephone circuits deal with a frequency band of 300–2200 Hz and the carrier frequency used is of 1200 Hz. This carrier is modulated by *picture frequency* produced as a result of the scanning.

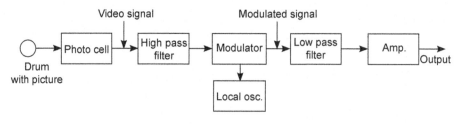

**Fig. 21.14**

To keep off the noise, the transmission is done by "*Sub Carrier Frequency Modulation*" For modulation, 7 kHz frequencies are utilised and the output obtained is of 1400 Hz (for white light) and 2400 Hz (for coloured light).

## Reception

The AM signal at the receiving end is first passed through a high pass filter circuit. Its output is amplified and then demodulated. The output of the demodulator is fed to a low pass filter circuit. The output is used to move the oscillograph, which records the message on a photographic film rolled on a drum. (Fig. 21.15).

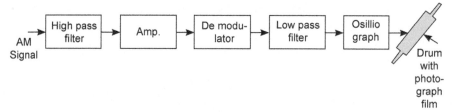

**Fig. 21.15**

## 21.11 TELEGRAPHY SYSTEMS

For short distances, telegraph messages are transmitted over the line as direct current (dc) pulses. For this purpose, a battery may be used. For long distances however, the dc signals are converted into ac signals at transmitting ends and converted back into dc signals at receiving ends. For sending and receiving telegraphic messages, the following systems are used:

(*i*) *Simplex Working*. This is the system of sending one message at a time through a line *i.e.*, one station "sends" and other station "receives" the message.

(*ii*) *Duplex Working*. This is the system of sending two messages at a time through the same line but in the opposite directions, *i.e.*, both the stations send one message and receive the other at a time. It is further of two types:

(a) **Full duplex** is a simultaneous two way communication. [See Figure 21.16 (a)].

(a)

(b) **Half duplex** is a two way communication but in only one direction at a time. [See Fig. 21.16 (b)].

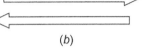

(b)

**Fig. 21.16**

(iii) *Diplex Working.* This is the process of sending two messages in the same direction through a line.

(iv) *Multiplex Working.* This is the process of sending many messages through the same line. There are however, two important multiplex systems:

(a) *Time Division Multiplexing (TDM).* In this system, same circuit is used on time distribution basis. "Boudot multiplexing" is the example. Several operators share same transmission on a time distribution basis.

(b) *Frequency Division Multiplexing (FDM).* In this system, same circuit is used on frequency distribution basis. The "Voice Frequency Telegraphy" is example with 24 channels. Each signal modulates a particular carrier frequency allotted to the channel. The various operators share same transmission on a frequency distribution basis.

The DUPLEX system used commercially is discussed below:

## 21.12 DUPLEX SYSTEM

As mentioned already, in duplex system two messages may be sent simultaneously over the same line. In this system, both the stations transmit one message and receive another simultaneously. According to operation, the duplex system can be classified as:

1. Differential duplex      2. Bridge duplex.

1. *Differential duplex system.* This may be sub classified as:

(a) Oppositional differential duplex

(b) Combinational differential duplex.

(a) *Oppositional differential duplex.* This method uses a differential relay which has two differentially wound coils which operate only, when current in one coil exceeds in the other coil equal to the operating current of the relay.

The Fig. 21.17 shows circuit for its working. The relay is non polarised with two identical and differential wound coils. The single current key (SCK) has two contacts I and 2. The "Duplex Balance" (DB) is an impedance network. The relay coils which are connected in series with the line are called *Line Coils* (LC) and the coils which are connected in series with the duplex balance are called *balance coils* (*BC*).

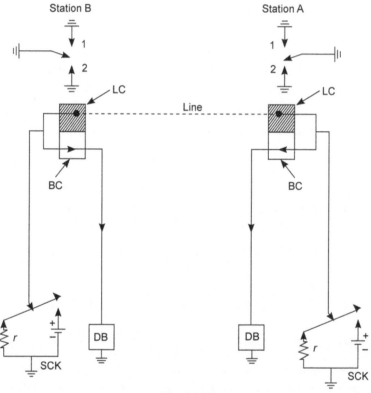

**Fig. 21.17**

(b) *Combinational differential duplex.* In this method the batteries at both the stations are connected in such a way that they combine in the line when keys of both the stations are working, hence the method has got the name.

The circuit is same as above with the difference that there is no resistance and on both sides of both the SCKs, a battery is connected. The working of the circuit is also more or less same with the difference that in combinational method, the battery voltage at the two stations should be equal, which is a bit difficult task as the battery voltages are obtained from different sources. If equal voltage is not maintained, this may give rise to faulty operation. Moreover, in combinational method, the line circuit carries double the current and power sent is more than in the oppositional method.

2. *Bridge duplex system.* The Bridge duplex system is based on the principle of well known *wheat stone bridge*, that at the balanced condition of the bridge, the product of the impendances of the opposite arms of the bridge is same.

The Fig. 21.18 shows the circuit for the bridge duplex system. The resistances $P$ and $Q$ on one station and $R$ and $S$ on the other station form a wheatstone bridge. The receiving relays on both sides are connected as shown. The capacitors, $C_1$ and $C_2$ enable the line to be charged rapidly and capacitor $C_3$ is used as distortion correcting device.

The bridge circuit is preferred over the differential duplex circuit on long submarine cables due to possibility of working over higher speeds.

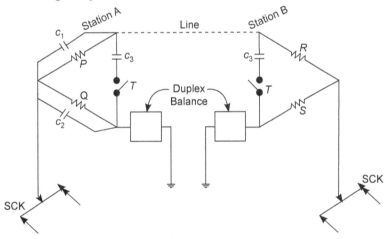

**Fig. 21.18**

## 21.13 TELEPRINTER

So long, we have studied manual telegraphy systems. The *teleprinters* (tele means distance) are automatic machines for sending and receiving messages. The "morse key" and other manual equipment have been replaced by automatic transmitters and receivers. In this way, we can handle the increasing telegraph "traffic". The teleprinters use "start/stop" telegraphy.

In this telegraphy whenever the code for a character is to be transmitted, the transmitter is *started and stopped* after each one revolution. During the revolution, it sends the required code impulse.

## Teleprinter Code

For teleprinter, a 5 (5/7.5), unit code also known as CCITT (Consultative Committee for International' Telephone and Telegraph) code is used. In this, each "character" consists of 5 elements.

Each code group is preceded by a start pulse which is a "spacing pulse", having same duration as the individual code element. The code group is followed by a stop pulse which is "marking pulse". Generally length of the start stop code is 7.5 units.

The teleprinter has automatic functioning replacing the morse code, keys, sounders etc. and have high speed operation.

## 21.14  COMPONENTS OF A TELEPRINTER

The important components of a teleprinter are the following:

1. Key board
2. Transmitter
3. Receiver
4. Printer.

1. *Key Board*. The teleprinter key board is similar to the keyboard of a typewriter, with some difference like:

   (*i*) The teleprinter mechanism is operated by the electrical pulses which it receives from the transmitter,

   (*ii*) The page remains stationary, but the carriage containing the keybars move.

   (*iii*) In teleprinters, the printing machine is beyond the reach of the operator and all the mechanical (non printing) functions are controlled by signals from the transmitter. For this, there are 5 functional keys (signals):

   (*i*) **Letter shift.** On receiving this signal, the machine prints only the letters.

   (*ii*) **Figure Shift.** On receiving this signal, the machine prints only figures.

   (*iii*) **Space.** This signal provides usual space between the words.

   (*iv*) **Carriage return.** This signal moves the carriage to left.

   (*v*) **Line Feed.** The signal provides space to start the next line.

   The teleprinter cannot print capital and small letters, as the no. of  keys are limited. There are 26 characters, 5 functional keys (described above) and two more keys ('*Run out*' and '*Here is*') on its key board. Under the key board, there are 7 "key bars". When an operator precesses a key on the keyboard, these keybars are pressed.

2. *Transmitter*. The function of the transmitter is to send the 5 code elements, preceded by "start pulse" and followed by a "stop pulse" as soon as a key bar is pressed.

   If an operator presses a key bar for a very short time, which is less than the time required by the transmitter to send all the 7.5 units, it is necessary that operator is prevented to set up any new combination, this is achieved by locking of the key board. Contrary, if the operator keeps a key bar pressed for a time longer than required, to ensure the transmission of a character only once, the locking takes place within the transmitter.

3. *Receiver*. The function of the receiver is just reverse of that of the transmitter. As explained already the transmitter receives an information from the key board, converts it into corresponding electrical pulses and sends into the line. The receiver senses and stores this electrical signal. The amount of distortion a receiver can tolerate is 30% as recommended by Consultative Committee for International telegraph and Telephone (CCITT). This is known as *margin* of the receiver.

4. *Printer*. The printer prints the received message on a page or on a tape.

## 21.15 BLOCK DIAGRAM OF A TELEPRINTER

The Fig. 21.19 shows various functional blocks involved in a teleprinter circuit connecting two *key board send receiver* (KSR) units which are used for direct communication between two operators. The key board and the printer both are mechanically driven from a common motor through a set of shafts.

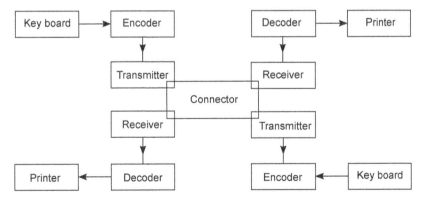

**Fig. 21.19**

## 21.16 TELEGRAPHY TERMS

The important terms related to telegraphy are:

1. *Speed.* This is equal to the reciprocal of the time in seconds used up in the transmission of unit element. It is measured in "BAUDS".

   *e.g.*, A teleprinter operates at 600 letters per minute for a 7.5 unit code.

   $$\text{time for one letter} = \frac{1 \text{ minute}}{600} = \frac{60 \text{ seconds}}{600} = 0.1 \text{ sec.}$$

   $$\text{time for unit} = \frac{0.1}{7.5} = 0.013 \text{ sec.}$$

   Speed is the reciprocal of $0.013 = 1/0.013 = 77$ Bauds

   The speed in bauds may also be defined as the " inverse" of the period of the shortest information unit (which corresponds to 1 bit) used in the code transmission. The BAUD is thus the maximum information rate of the code transmission *i.e.*,

   $$\text{speed} = \frac{1}{T} \text{ Baud.}$$

   where $T$ is the duration of shortest information unit in minutes.

   The digit 0 or 1 is called a "bit". The digital communication is carried out through bits.

2. *Distortion.* Ideally the received signal should be the "exact replica" of the sent signal. When there is a difference between the two, distortion is said to have occured.

   Distortion is defined as the maximum difference between any two instants of modulation. Distortion is always expressed in *percentage*.

   The line has four constants: Resistance ($R$), Inductance ($L$), capacitance ($C$) and conductance ($G$). The condition for zero distortion is $\dfrac{L}{R} = \dfrac{G}{C}$ or $LC = RG$. The condition is rarely maintained and hence the telegraph signal is distorted (Fig. 21.20).

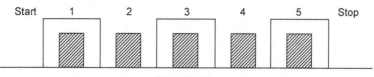

Fig. 21.20

   This distortion may occur in the transmitter itself or during transmission. The distortion however may be of the following types.

   (*a*) *Characteristic distortion.* This distortion is due to the value of constants of the line being different at different frequencies and also due to the signal consisting of fundamental and harmonics. This distortion can be easily predicted.

(b) *Fortuitous distortion.* This distortion is due to some defect in the circuit or instrument, and also due to external interference. This distortion can not be easily predicted.

(c) *Bias distortion.* This distortion occurs when the transmitter and receiver are not biased. This distortion occurs when mark-space (or space mark) transition does not take place at proper instant, which results in lengthening (or shortening) of the mark/space elements. This distortion can be predicted for a good teleprinter. This distortion should not be more than 6 percent.

3. *Margin.* The margin of a teleprinter is defined as the percentage of distortion which is tolerable. This should not be more than 30% as recommended by CCITT.

$$\text{Mathematically, margin} = \frac{d_c - d_s}{2d_c} \times 100 = \frac{d_c - d_s}{d_c} \times 50\%$$

where $d_c$ = duration of the code element

$d_s$ = duration of selection

From expression, it is obvious that margin can be increased by reducing the time of selection.

Ideally if $d_s = 0$, the value of margin = 50%, which is the theoretical maximum value.

**Problem 21.1.** *If duration of the smallest element in a telegraphic code is 10 ms, calculate the speed in Bauds in Morse code, 5 unit code and 7.5 units teleprinter code.*

**Solution.** The duration of smallest element

$$= 10 \text{ ms} = 10/1000 = 1/100 \text{ s.}$$

Speed in Bauds = 100 Answer (in all the codes)

**Problem 21.2.** *The duration of smallest element is 20 ms. Find speed in characters/sec and in words/minute in the following codes.*

1. Morse code

2. Five unit code.

**Solution.**

1. *Morse Code.* It requires an average 8.6 units to transmit one character.

(a) Time required to transmit one character

$$= 8.6 \times 20 \text{ ms} = 8.6 \times \frac{20}{1000} \text{ s} = 8.6/50 \text{ s}$$

The speed = 50/8.6 = 5.8 character/s. **Ans.**

(b)  Taking each word consisting of 5 characters per word and one space between successive words

$$\text{Length of word} = (8.6 \times 5) + [5 \times 2] = 53$$

$$\text{Time to transmit one word} = \frac{53}{100} \times \frac{20}{60} = 1/60 \text{ min}$$

$$\text{Speed} = 60 \text{ words/min} \qquad \qquad \textbf{Ans.}$$

## 2.  Five unit code

(a)  This code does not require any spacing

$$= 5 \times \frac{20}{1000} \text{ ms} = 0.1 \text{ s.}$$

$$\text{Speed} = 10 \text{ characters/s.} \qquad \qquad \textbf{Ans.}$$

(b)  Time taken to transmit one word

$$= \frac{100 \times 6}{1000} \text{ s} = 0.01 \text{ min}$$

$$\text{Speed} = 100 \text{ words/min} \qquad \qquad \textbf{Ans.}$$

**Problem 21.3.** *A printer using 7.5 unit code is transmitting 400 letters/min. Each word contains 5 letters and the adjacent words are separated by a space signal. Calculate the rate of transmission in Bauds and in words/min.*

**Solution.**

(a)  Duration of one letter = 60/400 = 0 .15 s.

Duration of one unit = 0.15/7.5 = 0.02 sec.

The transmission rate = 0.15/0.02 = 50. Bauds          **Ans.**

(b)  Each word consists of 5 letters and adjacent word separated by a space, time taken to transmit each word

(5 + 1) × 0.15 Sec. = 3/200 min.

The rate of transmission/word/min = 200/3 = 66.6 words/min.

                                                                                    **Ans.**

**Problem 21.4.** *A teleprinter has each character represented by one start element, seven character elements, one parity element and one stop element, all of equal length.*

*If teleprinter takes 100 millisec to transmit one character, calculate the speed in Bauds.*

**Solution.** The teleprinter requires 10 units (1 + 7 + 1 + 1) to transmit one character.

Time taken for one character = 100 ms =100/1000 = 0.1 s

Duration of each unit = 0.1/10 = 0.01 s

Hence, speed in Bauds = 100 Bauds.          **Ans.**

## SUMMARY

1. Morse code is generally used for sending telegraphic signals.

2. Other codes are "Five Unit Code" and ASCII code.

3. Punch paper transmission is also used at some places.

4. Ordinary telegraph system uses a transmitter (Morse key) and a receiver.

5. The wireless telegraphy employs A.M. and F.M. techniques.

6. The picture telegraphy (and telephony) has the same principle as used in TV transmission.

7. Telegraphy system may be simplex or duplex.

8. Teleprinters are automatic machines for sending and receiving messages.

9. Telegraphic speed is measured in Bauds.

   Speed $= \dfrac{1}{T}$ Bauds, where $T$ is the time of shortest information.

10. Margin is the percentage of tolerable distortion.

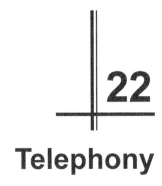

# 22

# Telephony

The word "telephony" is derived from the greek word "tele" meaning distant and "phone" meaning sound.

The telephony is a branch of line communication, which deals with transmission of spoken messages. The speech from one person is converted into electrical signal in the microphone of his telephone instrument, the electrical signal then travels through wires to the telephone exchange and conveyed to loud speaker of the instrument of the "called person", in which the electrical signal is converted back into sound.

## 22.1 TELEPHONY

Telephony is a technique of sending spoken massage from one place to other. The speech of the person is converted into electrical signal by the "microphone" and sent through wires. The listener's "loudspeaker" converts back the electrical signal into original speech.

## 22.2 A TELEPHONE HAND SET

A telephone hand set consists of two major parts – *Transmitter* and the *receiver.* Both are contained in a handy bakelite piece, which can be taken in hand for use.

Its transmitter is nothing but a *microphone* in which the person speaks. The microphone converts his speech into electrical signal. The receiver is a *loud speaker* and converts the electrical signal back into original speech, therefore, it is kept near the ear. The human voice has a frequency of about 3.5 kHz, therefore, the mike and speaker of the telephone should be able to handle these frequencies.

## The Handset has Following Components

(*i*)  *Transmitter or microphone* (Fig. 22.1). The transmitter is an improved

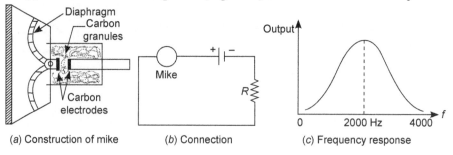

(*a*) Construction of mike        (*b*) Connection        (*c*) Frequency response

**Fig. 22.1**

form of the carbon microphone used in P.A. (Public Address)  system. It is *immersed* type as the carbon electrodes are totally immersed in carbon granules from all sides. As different persons handle it in different way while using, this ensures uniform pressure on the electrodes in all positions thus variation of dc resistance is also negligible. The electrodes are also made of carbon and the diaphragm is of light aluminium alloy called *duralumin* (Al + Cu + Mg), thus weight of the moving parts is very less and it increases sensitivity of the instrument.

The frequency response curve of this microphone is also better than P.A. microphone [Fig. 22.1 (*c*)].

(*ii*)  *Telephone receiver*: The telephone receiver used in a hand telephone is shown in the Fig. 22.2. It consists of a permanent magnet with two pole pieces on which coils are put with large number of turns. The electrical signal currents received from the transmitter flow through these coils. A diaphragm is put in front of these coils. The diaphragm moves to and fro according to the signal strength and thus original sound can be

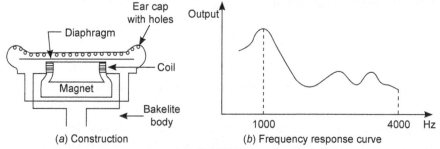

(*a*) Construction        (*b*) Frequency response curve

**Fig. 22.2**

heard. The whole arrangement is contained in a bakelite body which is covered with an ear cap with holes so that the sound may come out. The frequency response curve for the receiver is shown in Fig. 22.2(*b*). It has a peak value for the output at about 1 kHz.

(*iii*) *Dial* (*Fig.* 22.3). A dial on the telephone's face is an important device as this produces the *train of impulses.* There is a *Finger plate* with 10 holes, fixed on the main spindle. At the backside is a *number ring* with numbers, 1 to 10 (1, 2, 3, 4, 5, 6, 7, 8, 9 and 0). When a number is to be dialled, the finger is inserted into a particular hole of that number and rotated upto the *finger stop* and released. When the dial returns, it sends required number *of pulses.* Pulses of any number between 0 to 10 can be produced, the frequency being 10 pulse per second *e.g.,* when we dial "7", on its return journey a "train of 7 pulses", is produced by the ring.

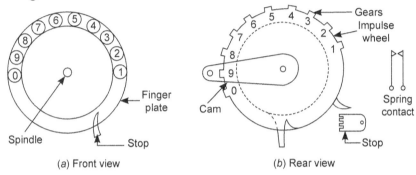

(a) Front view        (b) Rear view

**Fig. 22.3**

At the backside of the dial, there is also an *impulse wheel* with teeth. If we dial '5', It means 5 slots will pass beyond the *spring* contact and during return of the dial, these

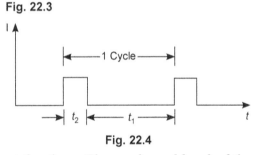

**Fig. 22.4**

slots will operate the contact *five* times. Thus *make and break* of the contacts will produce a train of 5 pulses per second.

The nature of impulse currents produced by the dial is also shown (See Fig. 22.4). If $t_1$ is the time for which no current flows and $t_2$ is the time for which the flow of current takes place, the frequency will be equal to

$$f = \frac{1}{t_1 + t_2}$$

Generally, $t_1 = 90$ mS and $t_2 = 10$ mS.

*Note:* All modern telephone sets have press buttons for digits in place of number ring.

(*iv*) *Connection with lines* (Fig. 22.5). The telephone transmitter and receiver are connected with the lines through transformers.

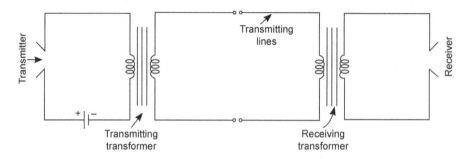

**Fig. 22.5**

*Impedence matching* is to be kept in mind for good results. Thus *one* impedence matching is needed between transmitter and the lines and the *second* in between line and the receiver.

## 22.3 ELECTRONIC TELEPHONE

The hand telephone set described above makes use of electro magnetic devices (such as relays, contactors, electromagnets etc.) and some electro-mechanical devices. They suffer from the following drawbacks:

1. They are delayed devices and take time to actuate and operation.

2. They are heavy and equipment become even heavier. They take more floor space, whereas the world is advancing towards *minitiarisation.*

3. Electromagnetic and electromechanical devices need periodic maintenance.

4. Due to their wear and tear, their life is short.

Therefore, we are nowadays using electronic telephone set employing *electronic switches i.e.,*diodes, transistors etc. It is important to note that:

 (*i*) Electronic switches can make ON and OFF a circuit very fast. Their frequency of operation may be $10^9$ per second.

 (*ii*) They are light in weight, free of maintenance and very accurate in operation.

 (*iii*) They have long life with no wear and tear and are more efficient.

## 22.4 TELEPHONE EXCHANGE

A telephone exchange is the centre from where calls are made with the desired subscriber. Thus, it interconnects the different subscribers. The telephone exchange may be.

1. Manual exchange

2. Automatic exchange.

1. *Manual Exchange*: Where the telephone connections are established by human operators, it is called a manual exchange.

   Each subscriber is provided with an individual set consisting of transmitter, receiver, bell etc. When a person calls, the bell rings at the instrument of the person who has been called. The transmitter needs direct current for energizing, accordingly the manual exchange may be:

   (*i*)  *Local battery exchange*: (*L.B. exchange*). When a local battery is provided in the set it self.

   (*ii*) *Central battery exchange*: (*C.B. exchange*). When it is energized by battery kept in the exchange. Manual exchanges are cheaper but have maintenance problem. At branch exchanges, these are still in use.

2. *Automatic Exchange*.

   Where the telephone connections are established automatically, they are called automatic telephone exchange. These are very costly but need little maintenance and, therefore, replacing the manual exchanges. The automatic exchanges are of two types.

   (*i*)  *Cross Bar Exchange*: In this exchange, connections are made through relays which are automatic electromechanical switches. A cross bar switch helps for interconnecting different lines through relays Fig.22.6 shows the schematic arrangement of such an exchange how the relay $T$ connects line $A$ with line $P$.

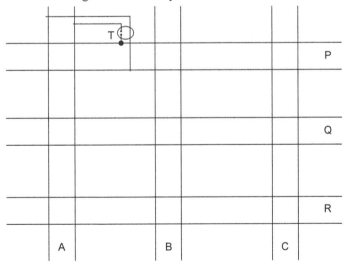

**Fig. 22.6**

(*ii*)  *Electronic Exchange*: Their basic operation is same as cross bar exchange with the difference that mechanical relays are replaced by electronic switches *e.g.*, transistors.

Modern exchanges have all *computerised* control. The merits of electronic exchanges are listed below:

(*i*) They ensure trouble free performance.

(*ii*) The use of mechanical parts is more or less eliminated and hence increased reliability.

(*iii*) The operating time is very less. The speed of electronic switches may be $10^9$ operations per sec.

(*iv*) They need less space and consume very little power.

## 22.5  SYSTEM OF ELECTRONIC EXCHANGES

The electronic telephone exchanges generally use the following systems:

(*a*)  *Time Division Multiplexing* (*TDM*) (Fig. 22.7). If a speech wave is *sampled* at a frequency greater than the highest frequency present in the speech, we get a sample wave as shown in which the speech signal is present and same may be obtained at the receiver.

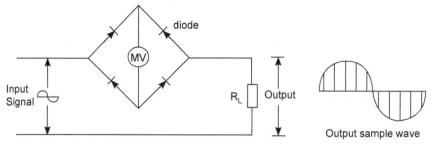

**Fig. 22.7**

For generating such a wave at the transmitter a square wave generator (Multi-vibrator MV) producing 10 kHz carrier wave with a bridge rectifier circuit is used. Across the load $R_L$, the sampling wave is obtained. At the receiver the same circuit detects the sample wave across its load, provided both the circuits (at transmitter as well as at receiver) work in *synchronism*. If two persons are provided with such a circuit, they can hear each other. Moreover, a number of signals can pass through the same transmission line.

(*b*)  *Frequency Division Multiplexing* (*FDM*): This is also known as **carrier** telephone system. In this system, a carrier frequency is modulated by the speech signal and as a modulated wave, we get original carrier along with sidebands. Only the side bands which carry the signal are sent. At the receiver the original speech signal is detected or **demodulated**. By using different carrier frequencies, we get different side bands of different frequencies and, therefore, any number of signals can be transmitted

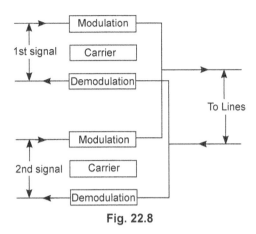

**Fig. 22.8**

simultaneously through the same lines. Similarly, the signals can be received at the receivers by using different filter circuits.

The Fig. 22.8. shows simplified diagram of an electronic telephone exchange using FDM, only two signals have been shown but any number of signals can be sent through a pair of lines.

## 22.6 TERMS RELATED TO TELEPHONY

Below we are giving important terms related to telephony:

(*i*) *Traffic and Trunk*: The traffic is the aggregate of calls passing over a group of circuits. Note that the group of circuits is called trunk.

(*ii*) *Side tone:* This is the mild tone heard by a person in the receiver of his telephone while speaking in his mouth piece.

Too much side tone is not desirable, but if the tone is eliminated, the telephone appears to be dead. The telephone sets are equipped with an "anti sidetone coil" to keep the dial tone to the prescribed level.

(*iii*) *Busy hour:* The busy hour of an exchange is the hour (60 minutes), during which traffic is heaviest. The telephone traffic is not same but varies with time according to the need of the subscriber and also to the location of the exchange.

(*iv*) *Busy hour calling rate:* This is defined as the average no. of calls originated per subscriber during the busy hour.

If no. of subscribers in an exchange are $S$, the calls in the busy hour are $C$, the calling rate, $R = C/S$.

(*v*) *Holding time:* The Holding time $T_H$ of a call is the sum of the time taken by the exchange to set a call ($T_E$) and the time of conversation by the subscriber ($T_S$) or

$$T_H = T_E + T_S$$

(vi) *Traffic intensity (Erlang)*: This is defined as the rate of flow of traffic during busy hour.

If $C$ are the no. of average calls during busy hour at an exchange and the average hold time per call is $T_H$ then the traffic intensity is

$$I = C \cdot T_H$$

It is expressed in traffic units (*TU*) or *Erlang*

If $C = 1$, $T_H = 1$, we have $I = 1$.

Thus the traffic intensity may be defined as the rate of flow of traffic, when during busy hour only one call lasting for an hour passed through the exchange.

(vii) *Lost calls* $(C_L)$: The no. of calls which are allowed to fail during busy hour due to any reason are called *lost calls*.

(viii) *Offered calls* $(C_O)$: The total no. of calls passing through the exchange in the busy hour are called offered calls.

(ix) *Grade of service:* The ratio of the *calls lost to the calls offered* is called *grade of service*, mathematically.

$$G = C_L/C_O$$

This is given in percentage *e.g.* The grade expressed as 0.005 means 5 calls lost out of 1000 *i.e*, $G = 0.05\%$.

**Problem 22.1:** *If 1000 subscribers originate 60 Erlangs of traffic in the busy hours at an exchange with an average holding time of 2 min 42 sec.*

(*i*) What is the calling rate

(*ii*) If 35 calls are lost due to some reason, what is the grade of service.

**Solution.**      Holding time = 2 min 42 sec.

$$= 2.7 \text{ min} = 0.05 \text{ hr.}$$

$$\text{Total calls offered} = \frac{60}{0.05} = 1200$$

$$\text{Calling rate} = \frac{1200}{1000} = 1.2/\text{Subscriber. \textbf{Ans.}}$$

$$\text{Grade of service} = \frac{\text{Calls lost}}{\text{Calls offered}}$$

$$= \frac{35}{1200} = 0.03 \text{ or } 3/100$$

*i.e.,* three calls lost in 100 calls **Ans.**

**Problem 22.2.** *The traffic intensity of an exchange is 3.5. If no. of calls are 170, calculate average duration of each call.*

**Solution.** Duration of a call $= \dfrac{3.5}{170}$ hour

$$= \dfrac{3.5}{170} \times 60 = 1.24 \text{ min } \textbf{Ans.}$$

**Problem 22.3.** *An exchange with 10,000 lines has a total no. of calls as 15000 in the busy hour. The average holding time is 2 min 30 sec. Calculate*

   (*i*)  *Calling rate*

   (*ii*)  *Traffic flow rate*

**Solution.**

   (*i*)  Calling rate $= \dfrac{15000}{10000} = 1.5$ calls per subscriber **Ans.**

   (*ii*)  The holding time = 2 min 30 sec .

$$= 2.5 \text{ min} = 2.5/60 = 0.042 \text{ hr.}$$

     ∴ Traffic flow rate

        = No. of calls in busy hour × holding time

        = 15000 × 0.04 = 600 Traffic units (TU) **Ans.**

**Problem 22.4.** *There are 10000 calls in an exchange in the busy hour and average holding time is 2 min 51 sec. Find*

   (*i*)  *Rate of traffic flow*

   (*ii*)  *No. of simultaneous calls in the busy hour.*

**Solution.**

   (*i*)  2 min 50 sec = 2.85 min

     Rate of traffic flow $= 10{,}000 \times \dfrac{2.85}{60} = 475$ *TU.* **Ans.**

   (*ii*)  The simultaneous calls in the busy hour = 475 **Ans.**

## 22.7 TELEPHONE MODULATION SYSTEM

Speech produces much more complex modulation pattern as compared to other signals but the principle of *amplitude modulation* in telephony remains the same.

By having a number of carrier frequencies, telephonic conversations can be "stacked" one above the other in terms of frequencies and, therefore, takes place without interfering with one another. Upto 300 simultaneous conversations may be carried out in this way over a single pair of cable.

An illustration of such a telephonic modulation system is given in Fig. 22.9.

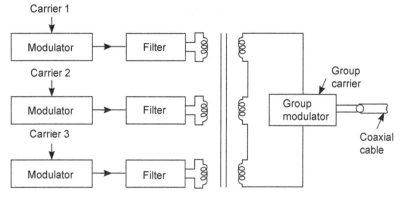

**Fig. 22.9**

## 22.8 TELEPHONE LINES AND CABLES

(*a*) The simplest type of telephone line (apart from the single overhead wire with ground as return conductor) is the twin wire over-head line. This wire is usually made of galvanised steel, which provides better mechanical strength. The bare (naked) wires are used but if insulation or protection is required, a poly vinyl chloride (PVC) may be used as a covering over the wires. The power lines (50 Hz) going near by, can create an interference in telephone wires but this can be eliminated by *transposing* the telephone wires. The Fig. 22.10 (*a*) shows how interference can be picked up by telephone lines from power lines through inductive (*L*) and capacitive (*C*) couplings. The Fig. 22.10 (*b*) shows *transposition* of the telephone wires. By transposing the telephone wires, the constants (resistance, inductance and capacitance) of the wires become uniformly distributed over the entire length of the line and therefore, interference is reduced.

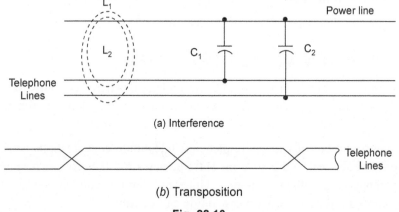

(*a*) Interference

(*b*) Transposition

**Fig. 22.10**

(*b*)  Where open lines are not feasible, wires are formed into multicore cable assembly which is usually carried *underground* in ducts. These wires are made of annealed (soft) copper, which is flexible, and each strand is insulated with high grade paper tape. A lead sheath is then given as an outermost covering.

(*c*)  For high frequency signals (*e.g.*, carrier telephony and television), co-axial cables are used.

**Note:**

(*i*)   The telegraph and telephone lines basically are the medium frequency lines. The voice messages (conversation) are transmitted between 250 Hz to 2800 Hz and radio messages over telephone lines are transmitted at between 50 to 7200 Hz frequencies.

(*ii*)  The telegraph and telephone lines may be "open wires" running on poles with insulators or cables run overhead, underground or laid in the ocean (ocean cables, for overseas communications). The cables have low inductance but high capacitance. The cables can handle frequencies in MHz.

(*iii*) Some telegraph systems use earth as one side of the circuit, while other use a complete metallic path (pair of wires) for the purpose.

(*iv*)  By carrier communication, many telephone messages can be sent simultaneously through telephone lines. Usually *amplitude modulation* is used and the carrier frequency used for this purpose is between 50 – 500 kHz.

(*v*)   As mentioned above underground cables can handle frequencies in MHz. Now co-axial cables are developed which enable telecommunication in MHz and that too can handle no. of channels over the same line. However, "repeaters" (amplifiers) are to be used at short intervals along the route.

## 22.9  PLCC (POWER LINE CARRIER COMMUNICATION)

Whenever some fault or a breakdown occurs in a power line, the message to rectify the fault is to be sent to the substation. The matter being of utmost importance needs quickness in sending the message. The ordinary telephone system remains almost engaged, therefore, that cannot be used. The power lines are used for this purpose. The message is modulated by IF carriers and sent as radio waves through the power lines. This is the reason that the system is known as "power line carrier communication" (PLCC) system.

The carriers to modulate the message are of 50 to 500 kHz. This modulated signal is amplified and transmitted over the power lines of above 66 kV. At the receiving end, the signal is separated. It is to be noted that the signal received has exactly the same frequency as the original signal transmitted.

### Advantages of PLCC

1. It is cheaper, as the same power lines are used for telephone purposes and separate telephone lines are not to be laid.

2. It is more reliable, as the power lines are mechanically stronger than the telephone lines. The power lines are made of high grade copper, where as telephone lines are made of iron, which may be corroded in unfavourable weather conditions.

3. It has short length, as the power lines provide shortest route between the power stations, as they run on high towers which can cross fields, rivers, railway lines etc.

4. It causes less attenuation of the signal, as power lines have very low resistivity.

5. It causes less leakage of power between conductors and ground.

6. The large spacing between the power lines reduce cross talk.

### Disadvantages of PLCC

1. The power lines carry high voltages and currents, proper protection to the PLCC equipment is must.

2. The power communication suffers from noise, as problems of corona etc., exist on the lines. Note that the phenomenon of production of arc/ flash along with the hissing sound between high voltage power lines is called *Corona*.

## 22.10 COMPONENTS OF A PLCC SYSTEM

The important components of a PLCC systems are under:

1. Coupling capacitor

2. Wave trap

3. Coupling filter circuit/LTU

4. Carrier set.

1. *Coupling Capacitor*. (Fig. 22.11). It is of very large size and, therefore, mounted on a pedestal outdoor foundation. It is filled with oil which

offers low impedance to carrier frequencies. It is protected against high voltage surges and also against grounding. The value of capacitor is in between 2000 pF to 10,000 pF and voltage rating from 33 kV to 400 kV.

The protective devices like spark gap (g), ground switch (s) etc., are provided in the base of the coupling capacitor.

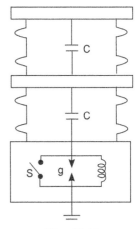

2. *Wave Trap*. (Fig. 22.12). They have coils of inductance of 2 mH to 2H which carry full line currents. The wave traps are connected in series with the power line and have high impedance to carrier frequencies. In this way, they "trap" the carrier but allow power frequencies (50 Hz) to flow through them. They have inductance and one or more capacitors in

**Fig. 18.11**

parallel and hence can be "resonated" to a particular frequency. The tuning (resonant) circuit makes it suitable to be used on more than one channel. They also contain lightening arrester as a protection against surges. The wave traps are generally hanging from the line. The Fig. 22.12 (*a*) shows single frequency tuned and (*b*) double frequency tuned wave trap.

Their standard ratings are 500, 800 and 1200 Amp.

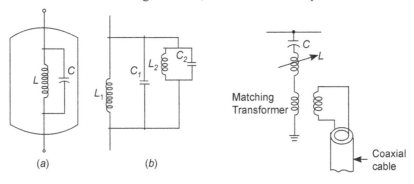

(a)                    (b)

**Fig. 22.12**                    **Fig. 22.13**

3. *Coupling Filter*. This facilitates feeding of modulated output to the coupling capacitor. It is also called as "line tuning unit" (LTU). The Fig. 22.13 shows one channel being fed to the power line. The variable inductance $L$ in series with coupling capacitor $C$ forms a resonant circuit which passes carrier frequency. The transformer is used for impedance matching between co-axial cable and the coupling filter circuit.

4. *The Carrier Set (Terminal)*. This is the "terminal", where speech signal (message) is modulated with the carrier. The demodulation is also carried out in this terminal. In other words, it contains transmitter and receiver of the system.

The Fig. 22.14(a) shows line diagram for a typical PLCC system and Fig. (b) shows outlook of the system.

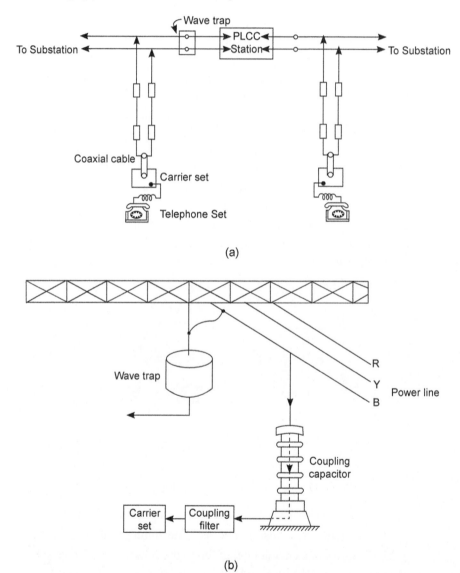

(a)

(b)

**Fig. 22.15**

# SUMMARY

1. By telephone we can talk from a distance.

2. A telephone set has a transmitter and a receiver. When we press 5 on the dial, a train of 5 pulses is generated.

3. Electronic exchanges use electronic switches like transistors. The speed of electronic switches is $10^8$ operations per second or even more.

4. The electronic exchanges use the techniques of TDM and FDM.

5. The telephone lines pick up interference to avoid this, they are transposed.

6. Hard drawn bare copper (HDBC) and Co-axial cables are used for making telephone wires.

7. The power lines above 66 kV are also used as telephone wires, this is called PLCC.

8. Components of PLCC are coupling capacitor, wave trap etc.

9. The voice messages are sent at 250–2800 Hz. Radio messages on telephone are sent at 7200 Hz.

❑❑❑

# 23

# Satellite Communication

A satellite is a station in space which serves many purposes. The "communication" satellites, receive signals from earth, amplify and return back the same to earth. The communication satellite is stationed at a distance of 35800 km from earth and completes one revolution in 24 hrs (as the earth does), such a satellite appears "Geo stationary" or "stationary'.

Many countries have stationed their satellites in space. India has also set up a series of satellites and we are using them for transmission of telephone, TV and fax signals.

## 23.1 SATELLITE COMMUNICATION

A satellite is basically a repeater (amplifier) station. The long distance signals cannot reach to the users, unless they are amplified at regular intervals. This is, what actually a satellite does *i.e.*, it receives a signal, amplifies and after processing, retransmits it. The satellites are used for transmission of telephone, TV and other signals.

## 23.2 CLASSIFICATION OF SATELLITES

A satellite as already pointed out is an "off surface" earth repeater station. The function to be performed by a satellite varies from type to type, but almost all satellites have the capability to receive a signal, process and transmit it again. However, from operational point of view, satellites can be of the following categories.

1. *Observatory Satellites*. These satellites are launched into a nearby elliptical earth orbit. They are meant for gathering information about the earth's crust for hidden minerals, melting of glaciers, onset of rains, cyclones, earthquakes etc. The Indian remote sensing (IRS) satellite is an example of such a satellite.

2. *Communication Satellites*. A communications satellite does not have
   the capacity of generating its own signal. It receives a signal from earth,
   processes it and transmits again to the earth. It may be mentioned that
   the signal in "up link" (upward) journey carries the same information
   as the one in "down link" (downward) journey. These are generally
   launched into an orbit at a distance of 35800 km from the surface of
   earth and the satellite is called as "Geostationary" because, in this orbit
   it completes one revolution around the earth in the same time as the
   earth itself *i.e.*, in 24 hrs. Because of the "speed synchronism", the
   satellite appears to be stationary at all times to an observer on earth.
   This gives it the name as "Geostationary" (stationary w.r.t. an observer
   on earth).

Such satellites work on solar energy and are controlled through computers at
the controlling centres on the earth.

As we know that moon is a natural satellite of earth. This paved the way for
the man made satellites. This is based on Sir Issac Newton's theory about
gravitational forces towards the earth and motion of moon around earth. The
Geostationary satellite rotates in *unison* with the earth, due to which it always
remains at the same place above earth, where it was placed. Based on all
these principles described above, in the year 1957 the first man made satellite
SPUTNIK-I was launched.

A communication satellite carries microwave receiving and transmitting
equipment in order to relay signals from one point to the other point on the earth.
Satellites are placed above *Ionosphere*, therefore, microwave frequencies must
be used to penetrate the ionosphere. Similarly, microwave frequencies are also
required to handle the wideband signals used in present day communication.

**Note:**   Satellites are used by some advance countries for their defence as well.

## 23.3 COMMUNICATION SATELLITE ORBITS

A communication satellite orbit may be:

    1. Equatorial             2. Elliptical

    3. Polar                 4. Inclined

    5. Circular.

The Fig. 23.1 (*a*) shows a "equatorial" (*b*) "polar" and (*c*) an "inclined" orbit.

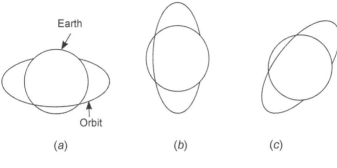

**Fig. 23.1**

The 'polar' orbits are not used because majority of the world population would then remain uncovered, similarly " Inclined" orbits are also not popular for the same reason.

'Equatorial' and 'elliptical' orbits are most commonly used. All *INTELSATS* (International telecommunication satellite) use these orbits due to following advantages:

(*i*)  they give uninterrupted communication.

(*ii*)  Only 3 satellites (displaced at 120°) are needed for full global coverage.

(*iii*)  Very limited earth station tracking is required.

However, they have following disadvantages:

(*i*)  they need greater launching capability.

(*ii*)  they give no coverage to polar regions.

(*iii*)  they have a one way delay of about 300 milliseconds.

**Note :** We may have six possible combinations:

1. Equatorial-circular        2.  Equatorial-elliptical

3. Polar-circular             4.  Polar-elliptical

5. Inclined-circular          6.  Inclined-elliptical

## 23.4 TERMS RELATED TO SATELLITE COMMUNICATION

The important terms related to the satellite communication are following:

1. *Apogee.* This is the *highest* point attained by a satellite moving round the earth. (Fig. 23.2)

2. *Perigee.* This is the *lowest* point attained by a satellite moving round the earth. (Fi.g 23.2)

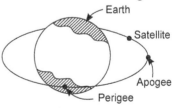

**Fig. 23.2**

The APOGEE and PERIGEE have relevance only, if the satellite has an elliptical path. In case of circular orbits, both are equal.

3. *Geosynchronous*. By this term, we mean that the satellite is moving round the *earth* in synchronism *i.e.*, both complete one rotation in 24 hrs. A geosynchronous satellite is always *geostationary*.

4. *Apogee Boost Motor (ABM)*. A communication satellite when shot into space moves round the earth in an elliptical orbit even if planned to make it geostationary. This is called "transfer orbit". The "apogee boost motor" (ABM) then puts the satellite into a near geosynchronous orbit from transfer orbit. The Fig. 23.3 shows the phenomenon clearly, which shows the satellite in transfer orbit initially and then in the geostationary orbit after successful firing of ABM.

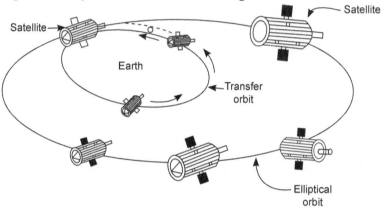

**Fig. 23.3**

5. *Orbital Velocity (v)*. It is the velocity with which the satellite moves in its assigned orbit. It is given by

   where                     $v = R\sqrt{g/R + H}$

   $R$ = radius of earth = 6400 km

   $H$ = height of satellite = 35800 km

   $g$ = acceleration due to gravity = 9.81 m/s.

6. *Escape Velocity*. If we throw a stone vertically upwards, it falls back to the surface of the earth after attaining a height depending upon the velocity of projection. If the stone is given a sufficiently large velocity it may escape the gravitational pull of the earth and may never return to the earth.

   The minimum velocity with which a body must be projected in the atmosphere so as to enable it to just overcome the gravitational pull, is known as escape velocity.

   The escape velocity is given by

   $$v = \sqrt{2gR}$$

   where                     $R$ = radius of earth,

   $g$ = 9.81 m/s.

Like orbital velocity, escape velocity is also independent of the mass of the body to be projected.

(*a*) Velocity of escape at earth is 11.2 km/s (taking radius of earth = 6400 km).

(*b*) Velocity of escape at the surface of moon is 2.35 km/s (taking radius of moon, $1.7 \times 10^6$ m).

7. *Payload:* This is another name for the satellite. In fact, the complete system comprises of a rocket (SLV) and a satellite. The satellite which is later separated from the SLV is called "Payload". The SLV stands for "Satellite Launch Vehicle" *i.e.,* a rocket.

8. *Satellite Launch Vehicle* (SLV). This is the rocket, that carries the satellites. After it is fired, it goes up and puts the satellite into orbit.

9. *Figure of Merit* (FOM). The "Figure of merit" of on earth station is expressed by 10 log (G/7), where G is the gain of the receiving antenna of the earth station and T is the noise temperature of the receiving system.

The FOM is the direct indicator of the overall performance of a satellite system. To work satisfactorily with the present day INTELSATs, an FOM of 40.7 is required. In 6 GHz and 4 GHz Up and Down link frequencies respectively, an antenna with diameter of about 26 Metres and gain of 58 dB is needed. This requires a receiver noise temperature of 53 K to achieve an FOM of 40.7.

10. *EIRP:* The EIRP stands for *effective isotropic* radiated power. This is the actual power radiated by the antenna of a satellite multiplied by its " Isotropic gain".

Note that an *"Isotropic"* antenna radiates energy uniformly in all directions. Alternately, gain of an isotropic antenna may be termed as *"Isotropic gain"*.

11. *C/N Ratio:* This is meant by carrier/noise ratio. It is equal to the ratio of the carrier signal power of the receiving antenna to the noise power.

12. *Gain/Noise Temperature Ratio:* The gain/noise temperature ratio is the ratio of the antenna gain at 6 GHz to the receiving system noise temperature.

13. *Noise Figure or* (*Noise Factor*)*:* The "noise figure or factor"(NF) is defined as the ratio of the signal/noise power supplied to the input terminals to the signal/noise power at the output terminals thus

$$NF = \frac{\text{S/N input}}{\text{S/N output}}$$

The typical value of noise figure for an amplifier is around 15 dB.

14. *Noise Temperature:* The noise power produced by a resistor is directly proportional to the temperature and is independent of the resistance value. The power is specified in terms of noise temperature and denoted by Kelvin, the typical value being 175 K.

The noise temperature is another convenient method of expressing noise power, particularly while dealing with VHF and microwave low noise devices, such as: antennas and receivers. The main advantage of using noise temperature is that it can be added like noise power, the another advantages of using noise temperature over that of *noise figure* is that for the same noise level, the change (variation) in noise temperature is more significant as compared to that in the noise figure, so it becomes easy to visualize the change when using the noise temperature.

## 23.5  ANTENNA BEAM WIDTH AND SIZE

The gain of satellite antenna has to be compatible with the earth coverage required by satellite transmission.

The Table 23.1 shows beamwidth and size of the antenna for the various earth coverages.

**Table 23.1**   Beamwidth and Size (dia) of Antenna

| Coverage | Beamwidth | Size (Diameter) of Antenna dish in cm (at 6 GHz) |
|---|---|---|
| 1. Global coverage | 18° | 23 |
| 2. Continental coverage | 4° | 100 |
| 3. National coverage | 1° | 415 |

## 23.6  DOMSAT

The DOMSAT is an acronym for "Domestic Satellites (system). In this system the geostationary satellites are used for domestic communications. Many countries like America, Russia, Brazil and Indonesia have *domsat* system. In India, **INSAT** series satellites are a part of domsat. It has helped in achieving nation wide TV broadcast in the country.

The Fig. 23.4 shows the operation of DOMSAT. Like other satellite communication setups, in DOMSAT also, the signal from the transmitting station is beamed up at the satellite which in turn retransmits it to a number

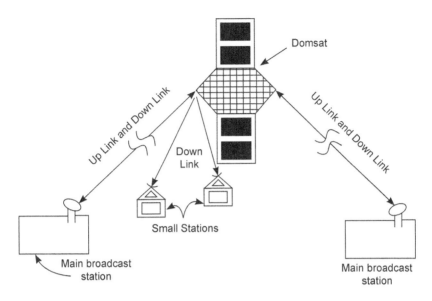

**Fig. 23.4**

of small stations lying within its coverage. In a set up like this, there may be as many as 100 small stations and only a few main broadcasting stations. It may also be mentioned that the main broadcast stations have both transmitting as well as receiving antenna, whereas the small stations have got only the receiving facilities.

A domsat has a much smaller coverage, while "Intelsat" must be accessible roughly to 1/3rd of the world with its antenna having a coverage of about 170 million square kilometers. On the other hand, a circular antenna having a radius of 1490 km would cover entire India.

## 23.7 INTELSAT

The *"Intelsat"* stands for "International tele-communication Satellite". These satellites are meant for global communication. The Fig. 23.5 shows a typical "set up" for demonstrating communication from one country to the another. The signal from the main broadcast station, where the program is generated reaches the earth station nearer to the broadcast station and also the one lying within the satellite coverage *via* a chain of microwave links. It may be mentioned that the broadcast station may or may not lie within the satellite coverage. The signal from the earth station is beamed upto the satellite. The satellite retransmits the signal down towards the earth. It is then picked up by the interested earth station on the other side. The signal then finds its way through a similar chain of microwave links to the main broadcast station from where it is retransmitted either through cables or on a "wireless" media for the interested subscribers.

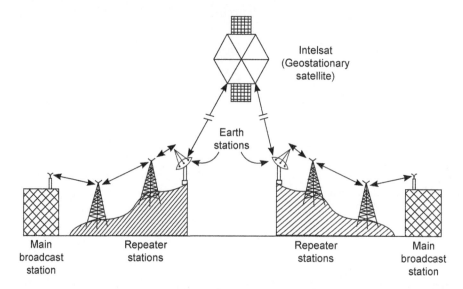

**Fig. 23.5**

It may be pointed out here, that both the earth stations should be within the coverage of the satellite. There may, however, be as many as 30 to 40 earth stations within the coverage of an INTELSAT.

"The Communication Satellite Corporation" (COMSAT) of USA and other world communication agencies signed in 1964 and formed The International Telecommunication Satellite (INTELSAT) Consortium. Till now many International satellites have been launched.

A typical intesat is 15.9 m long and its overall height is 6.4 m. When it is in the orbit, all its antennas point towards earth. It uses solar panels. Like all, this satellite also uses 5.925 to 6.425 GHz up link frequency and 3.7 to 4.2 GHz as down link frequency. It has 11 low noise 6.6 kHz receivers consisting of 8 transistor amplifiers and a low noise mixer. Five of these amplifiers operate at a time and others remain "stand by" .

## 23.8 THE SATELLITE COMMUNICATION SYSTEM

The basic components of a satellite communication system are an "earth stations" and a "satellite station" with accessories. (Fig. 23.6)

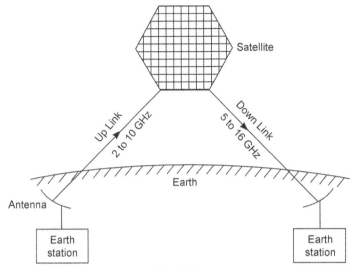

**Fig. 23.6**

The earth station sends signals to the satellite and receives back the signals from the satellite. The frequency range is 2 GHz to 10 GHz for "uplink" and 5 GHz to 16 GHz for "down link". The two frequencies are kept different to avoid any interference between the coming and going signals.

## 23.9 EARTH STATION

There are many systems/equipment/instruments but few important systems on Earth station are briefly described below (See Fig. 23.7):

  (*i*) *Power Supply*: An uninterrupted power supply (**UPS**) is needed to be supplied to the earth station. In emergency, batteries provide the necessary supply which are kept on charging during normal supply hours.

  (*ii*) *Transmitter*: The transmission system has modulators, amplifiers (repeaters which may be Double Conversion type or Single Conversion type) and other accessories. The system also has frequency converters which convert 65 MHz (IF) to 5 GHz for transmission.

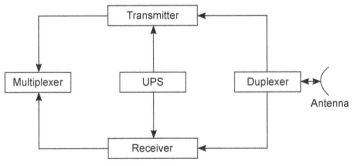

**Fig. 23.7**

(*iii*) *Receiver System*: The receiving system has demodulators amplifiers (which amplify in stages) and other accessories. The system also has frequency converters which converts the incoming frequency again to standard intermediate frequency (65 MHz).

(*iv*) *Antenna*: The earth station has "transmitting" antenna as well as a "receiving" antenna. The transmitting antenna should throw all power towards the satellite and the receiving antenna should receive all the power sent by the satellite to the earth station. They should perform their job without any or little power loss, and also they should not pick up any noise, in other words both the antennas should be *"directional"*. Generally parabolic or horn antenna are used for these purposes. The earth station antennas are 25–35 m in diameter and may be of hundred tons of weight.

(*v*) *Duplexer and Multiplexer System:* These are the systems to send/ receive more than one message simultaneously through the same line.

The Fig. 23.8 shows a complete block diagram of a typical satellite earth station.

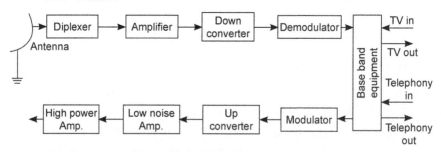

**Fig. 23.8**

## 23.10  SATELLITE STATION

The few important systems at satellite station are described below (See Fig. 23.9):

(*i*) *Power Supply:* The power supply for the satellite is obtained by using a panel of *solar cells* which convert sunlight into electrical energy for working of the satellite. The solar cells are connected in series/parallel combination to get the required

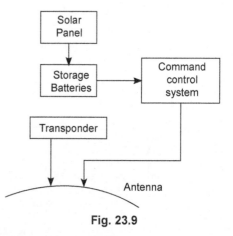

**Fig. 23.9**

voltage and current for the satellite. It is estimated that to generate 100 W of electrical energy, we need about 2200 solar cells.

(*ii*) *Transponder:* A *transponder* is an equipment on the satellite which receives a signal, amplifies it, changes its frequency and retransmits it to the earth stations. The satellite may have more than one transponder. They function at a bandwidth of 35 MHz.

(*iii*) *Command and Control System:* This system is for receiving commands from the earth station and to adjust the position of the satellite accordingly.

(*iv*) *Antenna:* The antenna should be highly directional so that it can receive and transmit the signals.

## 23.11 SUBMARINE CABLE COMMUNICATION

For inter continental communication, satellite and submarine cables both are used. Both systems are in fact complementary.

The extremely low frequency (ELF) communication is used to send information between submarines under ocean. This is done through submarine (or ocean) cables.

The ELF band ranges from 30 Hz to 300 Hz and the corresponding wave length is from 1,000 km to 10,000 km. It means that only 1 Hz bandwidth will be available for sending information. This is clear that ELF does not offer more capacity for sending messages. But at present, this is the only method known for submarine communications. This problem is solved by sending transmission at the rate of 1 bit per second. The 10 bits allow 1024 $(= 2^{10})$ messages to be stored and time required to send ten bits will be 10 seconds. Naturally the system is very costly. This is the reason that submarine communication is employed only in defence.

The principle of submarine cables is very much same as that of *co-axial* cables. The submarine cables are also "co-axial" having repeaters (amplifiers) and equalizers and dc power is fed to them with opposite polarities from opposite ends to reduce insulation problems. However, submarine cables use a single co-axial tube for both directions of transmissions with frequencies like those of microwave links to separate the two directions.

The laying of submarine cables is a critical process requiring special care. Some times, *sea parachutes* are employed for the purpose. Usually light weight cables are used at deeper sea portions. *Armoured* cables are used to protect them from ship anchors, travellers and tidal movements. For laying cable, a submarine cuts a 50 cm *deep trench* for the cable, the trench is later on covered. All the included joints are checked later on by X-rays.

**Table 23.2**    Satellite vs Submarine Communication

| S.No. | Satellite Communication | Submarine Cable Communication |
|-------|-------------------------|-------------------------------|
| 1. | Satellite can be approached by any earth station that lies in its coverage. | Cables can be used in the areas which have been connected through the cable. |
| 2. | Satellites are highly reliable except failure of terrestrial links or failure of earth station. | Cables are equally reliable, except failure of the cable. |
| 3. | In satellite the time delay is 600 ms. The time delay is more due to its large distance from earth (35800 km). The TV transmission is not affected much but it creates problem in telephone messages. | In case of submarine cables, the time delay is only 100 ms. |

## 23.12  HYBRID COMMUNICATION

In fact, today entire TV broadcast goes through satellites. Even telephone messages also go through satellites. The cables find their use where the point of transmission and reception do not face a common satellite. One way to overcome this problem is the *hybrid communication system* which uses a satellite as well as the submarine cables.

Say the signal is to be transmitted from station $A$ to station $C$. (See Fig. 23.10). The $A$ and $C$ do not face a common satellite but $A$ & $B$ ($B$ is an intermediate station) and  $B$ & $C$ separately face a common satellite. Thus signal from $A$ could be made to reach $C$ through $B$. Obviously the whole setup involves a delay of 1200 MS which is definitely undesirable for telephone communication. It will be comparatively better to route the signal from $A$ to $B$ or $B$ to $C$ via a submarine cable.

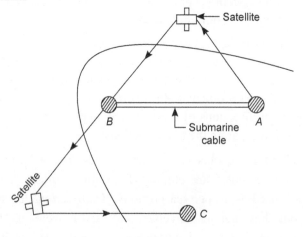

**Fig. 23.10**

## SUMMARY

1. A satellite is a station in space and serves many purposes. The communication satellite receives signal, amplifies and after processing, retransmits it.

2. The satellite orbits may be polar, circular or elliptical.

3. The highest point in the orbit is called "Apogee" and the lowest point is called "Perigee".

4. The minimum velocity, with which a body must be projected in the atmosphere as to enable it to just overcome the gravitational pull is called as escape velocity. Its value is 11.2 km/s at earth.

5. The basic components of a satellite system are "earth station" and a "satellite station". The uplink frequency range is 2 to 10 GHz and downlink frequency range is 5 to 16 GHz.

6. The important systems of an earth station are power supply, transmitting, receiving stations and duplexer system.

7. The important systems of a satellite station are power supply, transponder and antenna.

8. The extremely low frequency communication is used to send information between submarines under ocean. This is done through submarine cables.

9. The satellite and submarine cables are complementary techniques.

10. Satellites are also used to send and receive television and telephone signals.

# Transmission Lines, Cables and Waveguides

Energy can be transmitted either through electromagnetic waves (such as in radio, TV) or it can be transmitted through conductors called **transmission lines**, cables and waveguides.

## 24.1 TRANSMISSION LINE, CABLE AND WAVEGUIDE

A transmission line may be defined as a conductive method of sending electrical energy from one place to another. In electronic communication, these lines are used as a link between antenna and a transmitter (or receiver).

An insulated wire is called a cable, may be laid underground.

The waveguides are hollow pipes used for transmission at UHF.

These are described below briefly:

## 24.2 TRANSMISSION LINES

Transmission lines are of following types:

(*i*) *Parallel (open) wire lines*: This is a very common form of transmission line. The ordinary telephone lines running on overhead poles are its examples. Fig. 24.1 (*a*) shows cross section of parallel wire lines.

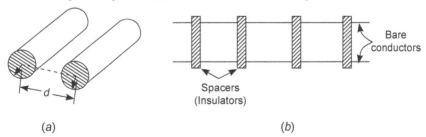

(*a*)

(*b*)

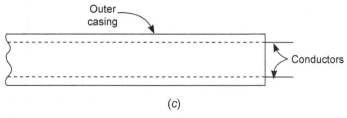

(c)

**Fig. 24.1**

These lines are constructed by keeping insulator spacers between the parallel bare conductors with a spacing of about 15 cm [See Fig. 24.1 (b)]. Impedence of this line is about 400 ohm. The Fig. 24.1 (c) shows a parallel line in its simple construction.

(ii) *Flat twin lead* [Fig. 24.2 (a) and (b)]: This is also an open wire line. The flat parallel lead is one of the most popular transmission lines in use for very high frequency (VHF). The wires are enclosed in a flat plastic cover and, therefore, are unaffected by weather conditions. Its impedence ranges from 75 ohms to 300 ohms. This line is balanced. It is also unshielded and used for noisy locations.

It should not run close to power lines so that it does not pick up hum. As most of the receivers have an input impedance of 300 ohms, a twin lead of 300 ohms is ideal for impedence matching.

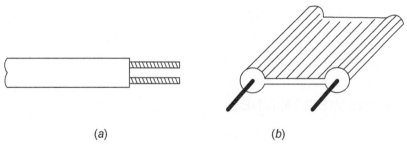

(a)                                                    (b)

**Fig. 24.2**

(iii) *Tubular twin lead* (*Fig.* 24.3): This is also an open wire line, it is similar to the flat twin lead except that in this, the twin conductors are enclosed in a tubular plastic tubing with air as dielectric. It has low losses and suited for VHF bands. The plastic tubing is strong, flexible and protects the lead against bad weather conditions.

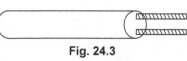

**Fig. 24.3**

## 24.3 SPECIAL TRANSMISSION LINES (RF LINES)

So-long we have studied open wire lines and co-axial cables. For microwave transmission, we have modern transmission lines to replace the conventional lines. These are high power lines used to feed radio frequency power to antenna in kW range. The solid dielectric is omitted to keep losses at minimum. Few special lines are described below:

(*i*) *Microstrip line*. This line consists of a conductor mounted over a metallic ground and separated by a dielectric. However, above the conductor, the dielectric is air. See Fig. 24.4 (*i*) the Fig. 24.4 (*ii*) shows its cross-section and (*iii*) its LC circuit.

Microstrip is similar to open wires, consisting of the top strip and its image below the ground. The dielectric used are teflon ($\varepsilon_r = 4$), alumina ($\varepsilon_r = 8$) and silicon ($\varepsilon_r = 10$), where $\varepsilon_r$ is the relative permittivity.

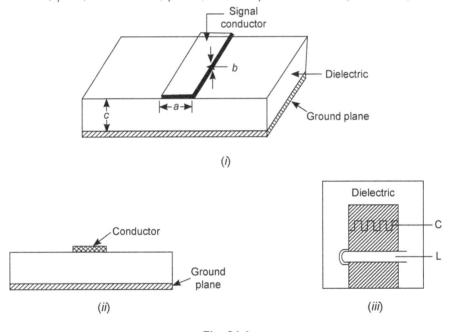

Fig. 24.4

The microstrip lines are constructed by the photographic process used for I.C.s.

Owing to the open type structure of microstrip lines, the electromagnetic field is not confined to the dielectric but partly in the surrounding air as shown in Fig. 24.5, with electrical and magnetic fields around the dielectric.

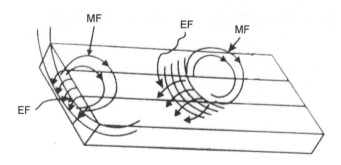

**Fig. 24.5**

The microstrip line suffers from radiation losses from its irregular surfaces and sharp corners. The power handling capacity of this line is also low. The lines are used upto 10 GHz.

The microstrip lines belong to the group of parallel plate lines. These lines are widely used in electronics. Apart from being used as transmission lines for microwave I.C.s., the microstrip are also used for circuit components such as filters, couples, resonators and antennas. As compared to co-axial cables, microstrip offers more flexibility and compactness.

(*ii*) *Strip Line*. This line consists of a conductor placed between two ground planes as shown in Fig. 24.6 (*i*), the Fig. 24.6 (*ii*) shows its cross section. The strip line is a balanced line whereas the microstrip line discussed above is an unbalanced line.

Note that, a balanced line is that in which electric field distribution between the two is symmetrical, whereas in an unbalanced line it is unsymmetrical.

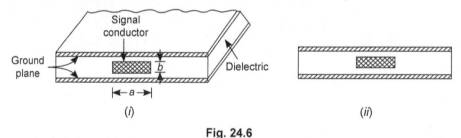

**Fig. 24.6**

The strip lines are used upto 12 GHz (1 GHz = $10^9$ Hz).

The strip line circuits are made using printed circuit board (PCB) methods. The special board materials are used which have low losses at microwave frequencies, examples are "woven fibre glass" and "polyoteflin". The boards may be copper cladded on one or both sides. Two PCBs are used each having a ground plane. The Fig. 24.7 (*a*) shows method of strip line assembly and (*b*) shows its assembled view.

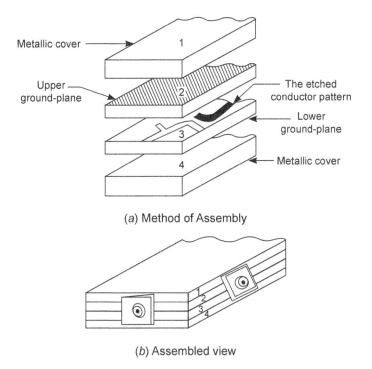

(a) Method of Assembly

(b) Assembled view

**Fig. 24.7**

## 24.4 CHARACTERISTIC IMPEDENCE OF A TRANSMISSION LINE

The input impedence of an infinite length of a transmission line is known as its "Characteristic Impedence". Its unit is **ohm** and it is also sometimes referred as **Surge impedence**.

The characteristic impedance (or surge impedance), $Z_0$ of a line may be defined as the ratio of square root of series impedance and shunt admittance per unit length. It is equal to:

$$Z_0 = \sqrt{\frac{R + j\omega L}{G + j\omega C}}$$

where $R$ = resistance, $G$ = Conductance, $\omega L$ = Inductive reactance and $\omega C$ = capacitive reactance of the transmission line.

It should be kept in mind that the characteristic impedance $Z_0$ does not involve length of the line, but is determined by characteristic of the line for unit length. It is not the impedance, that a line itself possesses.

(i) For an open wire line

$$Z_0 = 276 \log_{10} \frac{d}{r}$$

where $d$ is the spacing between the wires and $r$ is the radius of each wire (See Fig. 24.8).

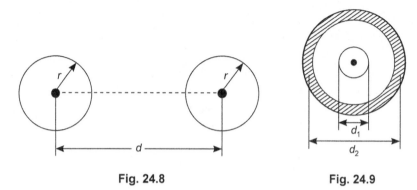

**Fig. 24.8**                              **Fig. 24.9**

(*ii*)  For co-axial cables (Fig. 24.9)

$$Z_0 = 138 \log_{10} \frac{d_2}{d_1}$$

where $d_2$ is diameter of the outer conductor and $d_1$ is the diameter of the inner conductor.

## 24.5 THE REQUIREMENTS OF TRANSMISSION LINES

The desirable requirements of transmission lines are:

(*i*)  The losses along the line should be minimum.

(*ii*)  There should be no reflection of the signal on the lines.

(*iii*)  The line itself should not pick up any stray signals, to prevent this, the line should be *shielded*.

(*iv*)  There should be a balanced matching between antenna and the receiver.

The Fig. 24.10(*a*) shows a balanced match between antenna and the receiver and (*b*) shows an unbalanced match between the two.

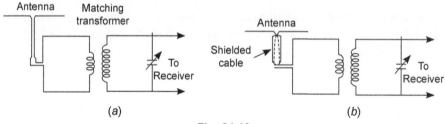

**Fig. 24.10**

## 24.6 STANDING WAVES

When a transmission line is not correctly (properly) terminated, the travelling wave at the receiving end is reflected completely or partially at the termination. There will be forward as well as reflected travelling waves. The combination of incident and reflected waves give rise to sinusoidal (or non-sinusoidal distribution) of power, these are called *standing* or non progressive waves of current and voltage. These have definite maxima and minima of current and voltage along the line (See Fig. 24.11).

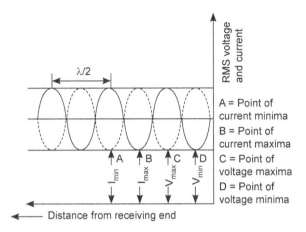

**Fig. 24.11**

## 24.7 STANDING WAVE RATIO (SWR)

The ratio of maximum and minimum magnitude of currents or voltage on a line, having standing waves is called *standing wave ratio* (SWR).

$$SWR = \frac{V_{max}}{V_{min}} = \frac{I_{max}}{I_{min}}$$

So there will be two SWRs.

(*a*) *Voltage SWR (VSWR)*

The ratio of maximum and minimum magnitudes of voltage on a line having standing waves is called voltage SWR.

$$VSWR = \frac{|V_{max}|}{|V_{min}|}$$

The value of VSWR is always greater than 1. When it is equal to 1, the line is correctly terminated and there is no reflection.

(*b*) *Current SWR (CSWR)*

The ratio of maximum and minimum magnitude of current on a line having standing waves is called CSWR.

$$CSWR = \frac{I_{max}}{I_{min}}$$

## 24.8 IMPEDANCE MACHING OF LINES

The transmission lines are used for transmission of power and information. For radio frequency power transmission, it is desirable that max power is transmitted to the load with minimum losses. This requires that the load must be "matched" to the "characteristic impedance" of the line, so that the standing wave ratio (SWR) on the line is close to unity as possible. The mis-matching produces echoes and *distortions* in the signal.

    (*c*) **Methods of impedance matching of lines:**

        The following methods are used for impedance matching of the lines: (*i*) By Quarter wave transformer and (*ii*) By Stub matching.

        (*i*)  The **disadvantage** of the quarter wave transformer is that it is a "narrow band device".

      (*ii*)  Though the single lumped inductors or capacitors can match the line. But it is more common to use the suceptive properties of short circuited pieces of lines. A *stub* is a piece of transmission line. It is possible to connect sections of open or short circuited lines called **stub (Tuner)** with the main line, at some points to obtain impedance matching.

The stubs have the following advantages:

- The length and characteristic impedance of the line remains unchanged.
- •• Adjustable susceptance can be added in shunt with the line.

## 24.9  CO-AXIAL CABLES

Open wire lines create following problems at higher frequencies:

    (*a*)  Radiation losses from the line become excessive high.

    (*b*)  An external shield becomes necessary to protect the lines from external interference.

These problems have been removed in co-axial cables. At high frequencies, co-axial pair of wires has half the inductance and twice the capacitance of the open parallel wires of the same diameter, hence radiation losses are reduced. Moreover, the outer cover protects the line against external interference.

It consists of two conductors arranged co-axially (both have same axis), one conductor is hollow and the other being located inside it, co-axially. The dielectric between the two may be a solid, air or a gaseous medium. A plastic cover over the line provides it protection against bad weather. As the outer metallic shield is grounded, the cable does not pick up any stray signals or noise. With outside diameter of about 1 cm, the impedance of the cable is about 75 ohm. It is costly but suitable for noisy areas and placed where no. of cables are to be laid close to each other. Sometimes, *foam* is also used as dielectric,

The Fig. 24.12 shows dimensions and Fig. 24.13 shows cross section of a typical co-axial cable. The **Interstices Quads** (I.Q.) are provided to give the cable a perfect circular shape.

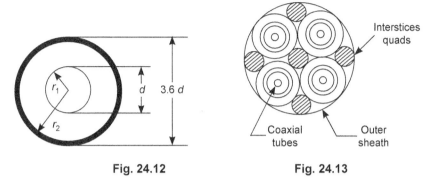

**Fig. 24.12**                              **Fig. 24.13**

The Fig. 24.14 shows simple construction of a co-axial cable.

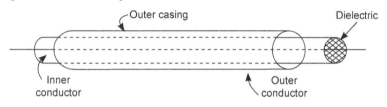

**Fig. 24.14**

A co-axial cable is a two conductor transmission line with a centre conductor surrounded by a braided shield. The inner conductor is supported by some form of dielectric insulation.

The most significant feature of a co-axial cable is its shielded structure. In a shielded structure like this, the electromagnetic fields associated with the two conductors remain confined within the space between the inner conductor and the outer shield. As the frequency of the signal carried by the cable increases, the current concentrates itself inside of the other conductor. As a result, a co-axial structure is a self-shielding with the shielding characteristics improving with increase in frequency. The major application of co-axial cable is in transmission of high frequency broadband signals. These are rarely used at lower frequencies, as their shielding properties become poor at low frequencies and they tend to be more expensive than the open wire lines for the same transmission loss. Fig. 24.15 shows conductor arrangement in different co-axial cables.

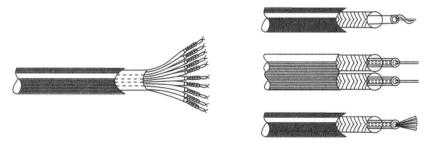

**Fig. 24.15**

## 24.10 DISTORTION LESS CONDITION IN CABLES

To approach distortion less condition $\dfrac{R}{L} = \dfrac{G}{C}$, the methods are:

   (*i*)  Increase L (inductance)    (*ii*)  Decrease C (Capacitance)

  (*iii*)  Reduce R (Resistance)    (*iv*)  Increase G (Conductance)

## 24.11 WAVEGUIDES

Waveguides are basically hollow metallic (rectangular, circular or elliptical) tubes or pipes. They transmit electromagnetic waves by repeatedly reflecting the signal from their inner walls. The waveguides are used as transmission lines at UHF. Important features of waveguides are given below:

   (*i*)  The waveguides are most suitable, at **UHF** (Ultra High Frequencies), where these are even superior than the co-axial cables. At low frequencies the dimensions of the waveguides will be very large and thus will be impractical.

  (*ii*)  The input Impedance (or *characteristic impedance*) of a waveguide is directly proportional to the frequency, where as in conventional transmission lines, it is dependent upon the geometry of the line.

 (*iii*)  Waveguides are usually air filled and can also be *coupled* with co-axial cables.

  (*iv*)  A waveguide acts like a high pass filter. Hence it has a *cut off frequency* and a corresponding *cut off wave* length. Any wave length greater than this does not propagate through the guide and is reflected back. It is determined by the dimensions of the guide. This makes guides suitable only for UHF.

## 24.12 SHAPE OF WAVEGUIDES

The waveguides can be of any shape as shown in Fig. 24.16.

   (*i*)  The rectangular guides are used for general purpose.

  (*ii*)  The circular guides are preferred for rotating antennas (*e.g.*, radar).

 (*iii*)  Elliptical shapes are popular for flexible waveguides. The a and b are called its dimensions.

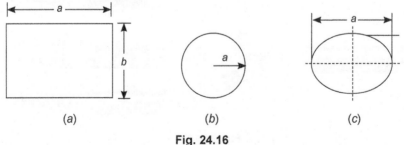

            (a)                        (b)                       (c)

**Fig. 24.16**

## 24.13 TYPES OF WAVEGUIDES

### There are two important types

(a) *Co-planer Waveguide*. The shape of this line is similar to the microstrip line. In this line also, there is a ground plane and a dielectric. On the dielectric, a conductor is deposited in the centre. In addition, there are two more conductors deposited on both sides of the "centre conductor" but with some separation as shown in Fig. 24.17.

This line is not much used.

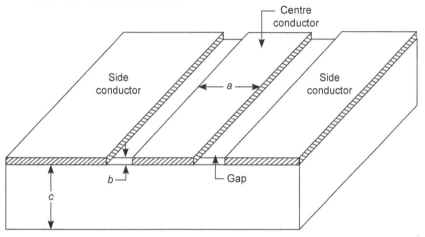

**Fig. 24.17**

(b) *Optical Waveguide (Optical Fiber Cable)*. The optical waveguides are the advanced technique waveguides operated optically. They have a wider scope for communication.

In construction, it is a solid tube of glass or silica of maximum purity. The diameter of the tube is about 0.02 mm, and any length can be obtained by joining the pieces through highly advanced technique. There is a cladding over the glass or silica core. The refractive index of cladding is lower than that of the core but it is also of the same material as the core. See Fig. 24.18. An optical waveguide differs from the ordinary waveguides in the following ways:

(i) In optical waveguides, refraction is obtained through total internal refraction, whereas in case of ordinary waveguide, it was only from the walls.

(ii) The loss of light energy due to dispersion is very less in optical waveguides.

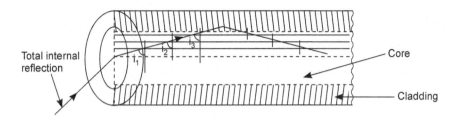

**Fig. 24.18**

$n_1$ refraction index of core

$n_2$ refraction index of cladding ($n_1 > n_2$)

(*iii*)  The optical waveguide proves to be cheaper as it does not need *repeaters* (amplifiers) upto 50 km line, because the *attenuation* is negligible.

(*iv*)  It has low *cross talk* losses.

(*v*)  The cable is free from *electromagnetic interference* (EMI).

(*vi*)  The light wave used is only of wavelength of 0.75 μm and therefore, losses are very low. Moreover, the frequency handling capacity of the cable at this wavelength comes to be very high to the tune of 30 GHz (1 GHz = $10^9$ Hz).

## 24.14  WAVEGUIDE MODES

A waveguide does the same job at microwave frequencies which the transmission lines usually do at relatively lower frequencies. At microwaves frequencies, it is more convenient to talk in terms of electric and magnetic fields propagating in the transmission medium rather than voltages and currents which is in case of transmission lines. A waveguide is nothing but a conducting tube through which energy is transmitted in the form of electromagnetic waves. The waveguide can be considered to be a boundary which confines the waves in the space enclosed by boundary walls. The waveguide can assume any shape theoretically but the analysis of irregularly shaped guides becomes very difficult. Two popular types are the rectangular waveguides and circular waveguides and again out of the two, former is more extensively used.

A rectangular waveguide is characterized by its wide dimension (*a*) and narrow dimension (*b*).

There will be infinite number of possible electric and magnetic field configurations inside the waveguide if there was no upper limit for the frequency of the signal to be transmitted. Each of these field configurations in which a wave may propagate in a wave-guide is called a **MODE**. There are two types of modes; **TM** (Transverse Magnetic) and TE (Transverse Electric). In TM mode, magnetic lines are entirely transverse to the direction of propagation of e.m. wave whereas, the electric field has a component in that direction.

In TE mode (Transverse Electric), the electric field lines are entirely transverse to the direction of propagation whereas magnetic field has a component along the direction of propagation. The propagation modes, both TM and TE are designated by two subscripts. The first subscript indicates the number of half wave variations of the electric field in the wide dimension ($a$) of the waveguide whereas the second subscript indicates the number of half-wave variations of electric field along the narrow dimension ($b$) of the waveguide. For instance, in $TE_{10}$ mode, which is the simplest mode, there is only one half-wave variation of electric field along the wide dimension and there is no electric field variation along the narrow dimension. The Fig. 24.19 shows rectangular wave guide in $TE_{10}$ mode. It may be mentioned that this subscript notation is only for rectangular waveguides. In circular waveguides, the subscripts are there but they do not carry the same meaning as they do in case of rectangular waveguides.

Accordingly there are $TE_{10}$, $TE_{11}$, $TM_{10}$, $TM_{11}$ modes.

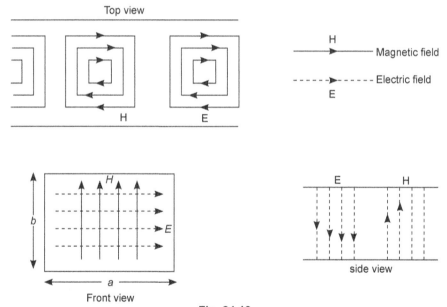

**Fig. 24.19**

## 24.15 PROPAGATION MODES IN WAVE GUIDES

Following are the wave propagation modes in wave guides (Fig. 24.20):

(i) **TE mode:** In *transverse electric* (TE) wave propagation mode, there exists a component of $\overline{H}$ (Magnetic field) and no component of $\overline{E}$ (electric field) exists in the direction of propagation ($z$), *i.e.*

$$E_z = 0 H_z \neq 0$$

(*ii*) **TM mode:** In *transverse magnetic* (TM) wave propagation, there exists a component of $\overline{E}$, but no component of $\overline{H}$ exists in the direction of propagation (*z*), *i.e.*

$E_z \neq 0 H_z \neq 0$

(*iii*) **TEM mode:** In *transverse electromagnetic* (TEM) wave propagation, both $\overline{E}$ and $\overline{H}$ donot exist in the direction of propagation (*z*), *i.e.*

$E_z \neq H_z = 0$

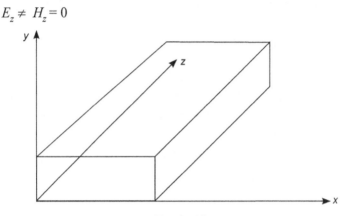

**Fig. 24.20**

In case of TEM Wave, the components of electric field $E$ and magnetic field $H$ lie in the plane transverse to the direction of propagation *i.e.* there are no components of $E$ and $H$ in the direction of propagation.

The TEM wave is a special case, in which all the field components are zero *i.e.* $E_z = 0$, $H_y = 0$, $E_x = 0$, $H_x = 0$ and $E_y = 0$.

This shows that a TEM wave cannot propagate in rectangular (or hollow) waveguides.

The TEM wave only exits in two conductor transmission line or in free space.

## 24.16 COUPLING A WAVEGUIDE WITH CO-AXIAL CABLE

Sometimes a direct coupling is needed between a waveguide and a co-axial cable. Two methods are shown in Fig. 24.21.

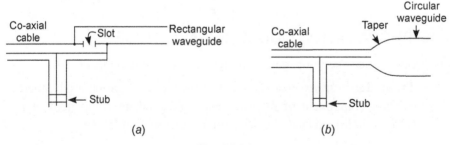

**Fig. 24.21**

The Fig. 24.21 (*a*) shows a slot in a common wall where energy from co-axial cable is coupled into the waveguide. In Fig. 24.2(*b*), coupling is by means of a taper section, in which TEM mode is converted into dominant mode in the guide. The mismatch, if any exists is turned out by using adjustable stub.

## 24.17 CAVITY RESONATORS

A simple cavity resonator consists of a waveguide closed off at both ends with metallic planes. If the propagation is in longitudinal direction *standing waves* will be set up in the closed structure and the closed unit will start resonating.

The cavity resonator is analogous to *LC* turned circuits. They have frequency coverage as that of waveguides. They can be used as tuned circuits of amplifiers, oscillators etc. Their most important application is as cavity wave meter used for measuring microwave frequencies. The Fig. 24.22 shows a simple cavity resonator.

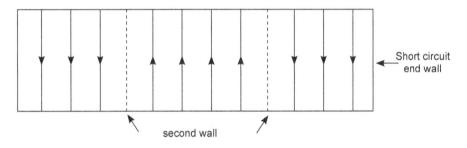

**Fig. 24.22**

## SUMMARY

1. The transmission lines may be open parallel  wires or co-axial cables.

2. The lines used at radio frequencies are microstrip and strip lines. These can be used at 10–12 GHz.

3. The impedence of an open line = 276 log *d*/*r* and of a co-axial cable = 138 log $d_2/d_1$. There should be a balanced impedence match between antenna and the receiver.

4. Waveguides are basically hollow pipes and used for transmission at ultra high frequency.

5. The waveguide may be of various shapes: rectangular, circular or elliptical.

6. The two important types of wave guide are coplanar waveguides and optical waveguides.

7. The electric and magnetic field configuration inside waveguide is called *mode*.

8. There are two types of modes-transverse magnetic (TM) mode and transverse electric (TE) mode.

❑❑❑

# Fibre Optic Communication

The Fiber optic communication is the science of transmitting data, voice and images by the passage of light through thin, transparent fibres. It is also spelt as *fiber optics*. In telecommunication, fibre optic technology has virtually replaced the copper wire in long distance telephone lines and it is used to limit computers within local area network (LAN).

The basic medium of fibre optics is a *hair-thin* fibre that is sometimes made of plastic, but more often of glass. A typical glass optical fibre has a diameter of 125 micrometers or 0.125 mm (0.005 inch). This is actually the diameter of the *cladding* or outer reflecting layer. The core or inner transmitting cylinder, may have a diameter as small as 0.10 mm. Through a process called *total internal reflection,* light rays beamed into the fibre can propagate within the core for great distances with remarkably little attenuation or reduction in intensity. The degree of attenuation over distance varies according to the wavelength of the light and composition of the fibre. When glass fibres of core/cladding design were introduced in early 1950s the presence of impurities restricted their employment to short lengths. In 1960 Electrical engineers working in England suggested using fibres for telecommunication, and within decades, silica glass fibres were being produced with sufficient purity that infrared light signals could travel through these for 100 km (60 miles) or more without having to be boosted by repeaters. Plastic fibres usually made of polymethyle/methacrylate, polystyrene or polycarbonate are cheaper to produce and are more flexible than glass fibres.

Optical telecommunication is usually conducted with infrared light in the wavelength range of 0.8–0.9 m or 1.3–1.6 m — wavelengths that are efficiently generated by light-emitting diodes (LED) or semiconductor lasers and that suffer least attenuation in glass fibres.

## 25.1 ADVANTAGES AND APPLICATIONS OF OPTIC FIBRE COMMUNICATION

### Advantages

In addition to the advantage of having extra information bandwidth using light as carrier signal, the optical fibre communication systems have several other advantages over the conventional systems:

- (*a*) Extra advantages of being light in weight and small in size, particularly in  underwater cables.
- (*b*) No possibility of internal noise and crosstalk generation along with the immunity to ambient electrical noise, echoes or electromagnetic interference.
- (*c*) No hazards of short circuits as in metal wires.
- (*d*) No problems when used in explosive environments.
- (*e*) Immunity to adverse temperature and moisture conditions.
- (*f*) Lower cost of cables per unit length compared to that of metal counterpart.
- (*g*) No need for additional equipment to protect against grounding and voltage problems.
- (*h*) Very nominal handling and installation costs, and
- (*i*) Lesser problems in space applications such as space radiation shielding and line to line data isolation.

### Applications

Because of these advantages, fibre optic communication is being currently utilised in telephones; computers, cable television, space vehicles, ships, submarine cable security & alarm systems, electronic instrumentation systems, medical systems, satellite ground stations and industrial automation & process controls.

## 25.2 BASIC PRINCIPLES INVOLVED IN OPTIC COMMUNICATION

The basic principle and law, governing optical communications are:

- (*i*) Total Internal reflection.
- (*ii*) Snell Law
- (*i*) **Total internal reflection (T.I.R):** If the angle of incidence is greater than the critical angle, the light is reflected back into the original dielectric medium; this is called *Total Internal Reflection* (T.I.R.)

Note that T.I.R. occurs at the interface between dielectrics of different (refractive) index, from higher to lower when ray of light is incident on the dielectric of lower index from the dielectric of higher index and the angle of incidence is greater than the critical angle. The ray in this case is also called as **meridional** ray; as that passes through the axis of the fibre core.

The Fig. 25.1 shows, total internal reflection in a multimode G.I.F. (Graded Index Fiber) cable with a parabolic profile core. The meridional rays are shown following curved path through the fibre core.

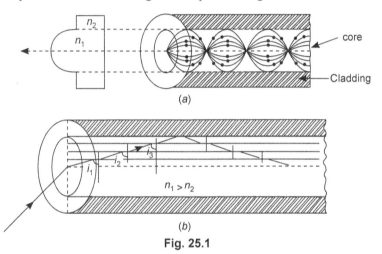

(a)

(b)

**Fig. 25.1**

The gradual decrease in refractive index from centre to boundary causes many refractions. The ray goes through increasing angle of incidence ($i_1$, $i_2$, $i_3$) through layers of the core and decrease in refractive index till total internal reflection is achieved and the ray returns towards the axis. [See Fig. 25.1 (b)].

The value of refractive index of the core ($n_1$) decreases as the radial distance from the centre of the core increases. The value of $n_1$ decreases till it approaches the value of refractive index of cladding ($n_2$). For proper mode propagation, the $n_1$ should be higher than $n_2$.

(ii) *Snell's law:* Figure 25.2 the Snell's law gives relationship at the interface of two mediums, having different refractive index. Snell's law states, how a ray of light reacts when it meets the boundary between two mediums. When the ray of light encounters the boundary, a part of the light is *refracted* to the other media, while other part is *reflected* back to the first (same) media.

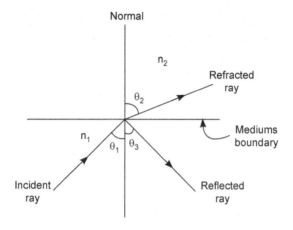

**Fig. 25.2**

## 25.3 PROPAGATION THROUGH OPTIC FIBRE CABLES (FIG. 25.3)

The propagation in optical cables is similar to that in the waveguides. Multimode propagation is also possible. Reflection in case of wave guides is obtained through the walls which are made of conducting materials whereas in optical cables, reflection is obtained through *total internal reflection*. Advance fibre manufacturing techniques ensure very low dispersion of light from the optical fibre.

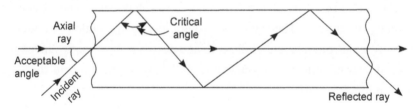

**Fig. 25.3**

## 25.4 BASIC FIBRE OPTIC COMMUNICATION SYSTEM

Essentially, fibre optic transmission system arose from a merging of two unrelated technologies: semiconductor technology and the optical waveguides technology. The first supplied the raw materials for the light sources and detectors needed in an optical transmission system, the second gave a transmission medium, the optical fibre. Rapid progress has been made in the past two decades in both: lowering attenuation in fibre and increasing the wavelength that it can handle. Early interest in the fibre was in the 800 – 900 nm (1 m = $10^9$ nm) wavelength region, where fibres exhibited a minimum local attenuation. Later fibre curves extended over a wider range of wavelengths, reflecting in the 1300 nm regions with their low *chromatic dispersion* and *low attenuation*.

Typical simple fibre optical communication systems have been shown in Fig. 25.4. In conventional optical systems Fig. 25.4 (*a*), a single light source transmits to a single detector over a single optical fibre. The disadvantage of such a conventional system is that the fibre or the cable cost may be greater than that of the other components and therefore, techniques are needed to make better use of a single fibre by increasing its information carrying capacity. This is being achieved by *wavelength division multiplexing* (WDM). In WDM technique, optical signals from different light sources are simultaneously transmitted through the same optical fibre while the message integrity of each signal is preserved from subsequent conversion to corresponding electrical signals [See Fig. 25.4 (*b*)]. As a matter of fact, the WDM techniques make use of LEDs or lasers as light sources but each source must emit at a different wavelength. The light from each source is combined for transmission down the same fibre using a device called *optical multiplexer.* The *photodetectors* at the far end of the fibre are broad band devices with respect to wavelengths being received, that is they respond to many different wavelengths but cannot distinguish one from another. Another device is, therefore, needed to separate the light into its component wavelengths. Such a device is called *optical demultiplexer.*

In another mode of WDM technique [Fig. 25.4 (*c*)], bi-directional transmission is being used. Here the multiplexing devices operate in two directions: the first device acts as a multiplexer for the signal from the light source 1 and as a demultiplexer for the signal from the light source 2. The second multiplexing device does the reverse. Thus, the signal from light source 1 travels in one direction down the fibre while the signal from the other light source travels the fibre in the opposite direction. The use of WDM is quite important in both *loop and trunk (telephonic) system* applications.

In the Fig. 25.4, λ stands for wavelength, MUX for multiplexer and 'DEMUX' for demultiplexer.

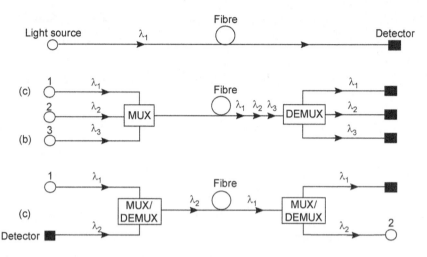

Fig. 25.4

## 25.5 THE OPTICAL FIBRE

The fibre is the core material used for making optical cables.

An optical fibre is a piece of very thin, highly pure glass with outside cladding that is similar but of slightly different composition and also with different refractive index. The fibres may be of two types:

(*i*) *Step Index Fibre* (SIF): A Step Index fibre has core diameter of 2 to 200 µm. The Fig. 25.5 shows two views of a Step Index Fibre.

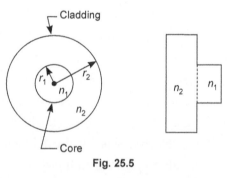

Fig. 25.5

(*ii*) *Graded Index Fibre* (GIF): The optical fibre, with refractive index reducing gradually from centre to outside of the fibre is called a **Graded Index Fibre**. The Fig. 25.6 shows its two views.

However, Step Index Fibres are more in use because of their higher *attenuation*. This can also be used for multimode or singlemode transmission, which yields best performance. The core diameter required for single mode transmission is 2 to 10 µm and for multimode transmission it is 50 to 200 µm. This can also be successfully used for digital transmission.

Remember that in metal wires, the signal is carried by **electrons** but in optic fibre cables, the signal is carried by **Photons**.

It may be mentioned again, that the core and the cladding both of an optical fibre are made of glass but with different composition and refractive index. The glass should be of highest purity and quality. The optical fibre covered with protective layers gives an *optical fibre cable*. The protective layers provide the cable extra mechanical strength and protection against environment.

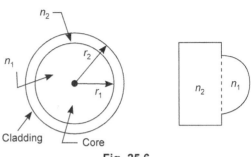

**Fig. 25.6**

## 25.6 TYPES AND STRUCTURE OF OPTICAL FIBRES

Depending upon the requirements and of particular usage, optical fibre may be of various types such as (*a*) active fibre, (*b*) conical fibre, (*c*) multiple fibre, (*d*) passive fibre, (*e*) glass coated glass fibre, (*f*) lasing (laser) fibre and (*g*) luminescent fibre. The **active fibre** is one that emits light as well as guides part of it. **Conical** fibre is used for light condensing and in aligned assemblies for magnification and demagnification purposes. Here the conical condensing property is of considerable value. The **multiple fibre** is a kind of fibre configuration that consists of a multitude of smaller diameter fibres. These have made possible the availability of very small fibres that are capable of high resolution and can be easily handled. The **passive fibre** is one that guides light incident on it from an external source. The glass coated glass fibre has a glass fibre core with coating of low refractive index glass. **Lasing fibres** are small diameter fibres where light amplification by stimulated emission of radiation (LASER) takes place. These fibres are capable of better mode selection, higher *pumping efficiency* and lower threshold. Finally luminescent fibres are those which are capable of emitting luminescent radiation when excited by ultraviolet, X-ray or high energy particles.

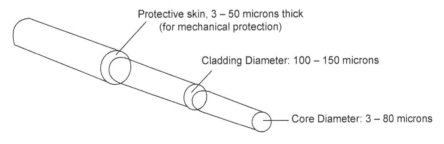

**Fig. 25.7**

Basically, optical fibre is a cylinder of transparent dielectric material which may be surrounded by a second dielectric.

The optical fibre cables (Fig. 25.7) make use of *passive fibres which* are as a matter of fact the glass coated ones. Hence the communication fibre has a core surrounded with a cladding where the refractive index of cladding has got to be smaller than that of core in order to have satisfactory guidance of light in the core. The thickness of cladding is normally one or two wave lengths of the light to be guided so that satisfactory optical isolation of fibre may be achieved. To have mechanical protection of the fibre, a protective skin is also used. It should be noted that the refractive index for core and cladding materials in fact decides the properties of communication fibres. The size of core and cladding also determines to some extent characteristics of these fibres.

## 25.7  PLASTIC FIBRES

The plastic fibres may consist of glass core with plastic sheath (cladding) or plastic core with plastic sheath or plastic core with glass sheath. Of these the plastic core with plastic sheath (cladding) fibres are popularly used because in this, both the core and cladding have similar *softening points* and that simplifies the production process. Plastic fibres have the advantage of more flexibility than glass fibres and can be used as single fibres with diameters of 2.0 nm, in systems that require flexibility. The attenuation characteristics of plastic fibres are worse as compared to that of glass and silica material systems. But even then they are frequently used for short distance computer applications with information capabilities of about 6 megabite per second over distance of 50–200 metres. In addition to flexibility, the mechanical handlings of plastic fibres such as gripping, crimping, etc. is easier. Plastic fibres are more easily damaged than glass and show a 'fatigue' effect. This is caused by the crazing of the plastic at the core sheath interface and it seriously reduces the transmitting efficiency of these fibres. Plastic fibres and cables have lesser operational temperature ranges, less than those of wire and co-axial cables and hence for high temperatures or long distance applications, plastic fibres and cables cannot be used. Typical continuous temperature range for operation of plastic fibre is 50°C and intermittent range is 90°C.

## 25.8  FIBRE MATERIALS

Three materials are normally used in the manufacture of fibres, namely, high content silica glass, multicomponent glass and plastic. A careful selection of raw materials is desired to obtain highest quality of optical fibres. Silica system consists of silicon dioxide with other metal oxides to establish a difference in the refractive index between core and the cladding. Dopants such as $TiO_2$, $Al_2O_3$, $GeO_2$ and $P_2O_5$, etc. are used to increase the refractive index of silica. It is essential that compound glasses used for core and cladding purposes in optical fibres must have similar coefficients of thermal expansion, similar viscosities at the drawing temperature, low material scattering (low compositional and density fluctuations), low melting temperatures and long-term chemical stability. Since cladding is normally having lower refractive index, it has got higher coefficient of thermal expansion than the core. It should be noted that attenuation and dispersion levels of multicomponent glass and plastic fibres are too high to be practical for most telecommunication purposes. Therefore, the highest quality fibres in mass production are made from high silica glass that has been synthesized by chemical vapour deposition technique. High silica glasses are drawn at a temperature of about 2100°C and contamination of the fibre surface must be minimized in order to preserve fibre strength. It should be noted that many dopants and materials used in fibre production facilities are *highly toxic* and so special precautions are needed to handle and apply them safely.

## 25.9 FIBRE BUNDLES

Generally speaking, fibre bundles may be grouped into three categories, namely, (*a*) aligned bundle, (*b*) fused bundle and (*c*) unaligned bundle. The *aligned bundle* is an *assembly of fibres* in which the coordinates of each fibre are the same at the two ends of the bundle. Therefore, these are also sometimes termed as 'coherent bundle'. *Fused bundle* is a flexible bundle of fibres in which fibres at the two ends are fused to achieve higher packing efficiency and surface quality. The *unaligned bundle* is an assembly of fibres in which fibres are randomly positioned, used primarily for conducting light around corners. Thus, basically fibre bundles may be coherent or noncoherent. In the noncoherent bundles little attention is paid to the optical isolation of a fibre from its neighbours (*i.e.*, here the light output from an optical fibre is a function solely of the light input to that fibre). In coherent bundles, the fibres are assembled in such a way that the relative positions of the ends of any fibre within the bundle are simply related. It should be noted that manufacturing of coherent bundles is much more difficult than that of noncoherent bundles so the range of components which can be made is much smaller. As a matter of fact, vast majority of applications involve the use of *image transferring coherent bundles* which can be used either by direct viewing by the eye, or the transferred image can be processed electrically or photographically. *Noncoherent bundles* are simpler to design and manufacture and hence, they are put to several uses such as (*a*) illumination for medical viewing instruments, (*b*) instruments illumination in general, (*c*) electronic switching, (*d*) calorimetry, (*e*) visual monitoring, (*f*) beam splitting or combining, (*g*) shape changing, (*h*) punched-card and tape reading, (*i*) photometric systems, (*j*) digital displays, (*k*) multi legend displays, and (*l*) high constant displays. *Conclusively* applications of these fibre bundles are limited to short distances and are better used with process control links, traffic control signals and certain computer applications.

## 25.10 OPTIC FIBRE CABLE (OFC)

In the case of telephone, trunk and trans-oceanic communication systems instead of fibre bundles, fibre optic cables are required which must have some competitive advantages over the properties of their metallic counterparts. Basically these *fibre optic cables* must be such that (*a*) their handling might be done in the same way as most ordinary communication wire and coaxial cables, (*b*) their mechanical and electrical properties are compatible for their specific use and surrounding conditions, including delivered powers to repeaters and resistance to applied static and dynamic stresses, (*c*) permanent splicing and connecting in the field may be done with ease and in reasonable time, (*d*) these are competitive economically with existing wire and coaxial communication cables, (*e*) their attenuation is quite low up to to the order of the individual fibre's and (*f*) fibres within the cables must be distinguishable.

Based on these particular requirements, a fibre optic cable can be designed. The constructional details of fibre optic cables used in telephone and trunk systems are shown in Fig. 25.8. As evident from this figure, there are seven major components in it.

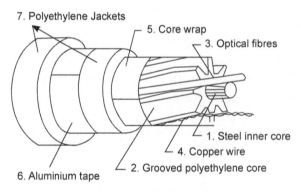

**Fig. 25.8**

1. *Steel Inner Core:* This core is made of aircraft steel and runs through the centre of the optical cable to absorb tension  applied during cable installation in ducts.

2. *Grooved Polyethylene Core:* Polyethylene pellets are melted and forced through a mould, forming grooves around the inner core; the core is twisted so the grooves spiral around the steel core first in one direction and then in the other. Thus the inner core wire gets surrounded by an extruded plastic core with six slots running in a reverse helix

pattern. The helix ensures that all fibres experience the same path increase around and bends. Each groove in the core contains six fibres or a copper pair; the latter may be used as an order wire or for other purposes.

3. *Optical Fibres:* Colour coded strands of fibre are laid in the spiral grooves. The structure is designed to *prevent pinching and squeezing* which could break the fibres or affect their performance. The colour aids in identification of each fibre during installation.

    The optical fibre is made by depositing many thin layers of glass from gaseous raw materials on the inside of a pure glass tube. The hollow form is then collapsed into a solid rod heated to high and precisely controlled temperature in a special furnace and 'drawn' into a fibre as thin *as a human hair* (approximately 0.125 millimetre in diameter) and several kilometers in length.

4. *Copper Wires:* A pair of plastic insulated copper wires are added for testing and to allow installers to "tap in" and talk to each other during installation.

5. *Core Wrap:* The bound cable is further wrapped in a protective film.

6. *Aluminium Tape:* A thin layer of aluminium is applied to further protect the cable from moisture and potential damage.

7. *Polyethylene Jackets:* The cable is coated with a final thick jacket of polythylene to add strength, keep water out and to prevent damage.

    In addition,

    1. *Powder Fill:* A specially mixed powder is sometimes packed into the grooves around the fibre. If water should enter, the powder expands into a jelly-like substance and prevents the water from moving along the cable.

    2. *Rodent Protection Sheath.* Cable manufactured for underground installation can have a thin layer of coated steel in addition to the aluminium sheath and polyethylene jackets. The steel prevents rodents, from biting through the jacket and damaging the cable.

**Note:** The cable is given a "buffer" coating which gives protection against mechanical and environment problems. Generally, hard plastic is used for the buffer coating. In addition to a buffer, the cable is also provided with a wrap of steel wire or it may be another layer of plastic or fibre glass. This provides sufficient mechanical strength to the cable, this layer is also called as "strengther". Moreover, the cable is also provided with outmost *sheath* for extra protection. An arrangement is also made not to allow penetration of water into the cable.

## 25.11  TYPES OF FIBRE CABLES

According to the construction, the cables may be

   (*i*)  Single fibre cable

   (*i*)  Double fibre cable

  (*iii*)  Multiple fibre cable.

   (*i*)  *Single Fibre Cable.* This is shown in the Fig. 25.9. It has a single buffer, while the fibre itself acts as "strengthener". This cable is more suitable for indoor applications.

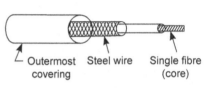

Fig. 25.9

  (*ii*)  *Double Fibre Cable.* The cross section of a double fibre cable is shown in Fig. 21.10. This is also provided with a steel wire fencing for better mechanical strength and environmental protection.

 (*iii*)  *Multiple Fibre Cable.* The cable has multiple buffer as shown in Fig. 25.11.

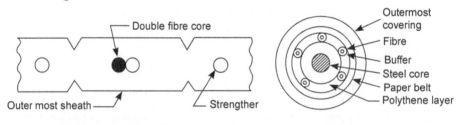

Fig. 25.10                                  Fig. 25.11

## 25.12  SPECIAL FIBRE OPTIC CABLES

Fibre optic cables developed so far may be grouped into four types, namely, (*i*) slotted core type, (*ii*) the loose tube type, (*iii*) loose fibre bundle, and (*iv*) ribbon.

   (*i*)  The slotted core cable has already been discussed.

  (*ii*)  In the loose tube cable, a centre strength member forms the cable core around which multiple tubes are stranded. There are many such filled loose tubes each having 12 fibres. The tube is filled with a soft thixotropic material allowing for free fibre movement in the tube. Nylon, polypropylene or dual extrusions of polymers are most common tube materials. The cable is completed with polyethylene (P.E.) sheath. See Fig. 25.12.

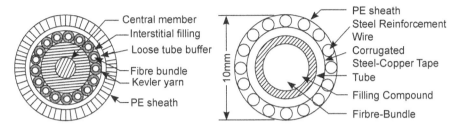

Fig. 25.12                              Fig. 25.13

(*iii*) In loose fibre bundle cable (Fig. 25.13), about 12 fibres are assembled in a unit or bundle which is, identified with a colour coded binder. Several such units are then assembled in the core of cable. A large diameter filled tube is extruded over the core. Thus, the fibres have access to free movement in the filling material. The fibre length is

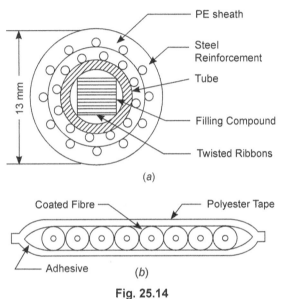

in little excess as compared to that of tube. This provides excellent mechanical performance. The core is completed with a reinforced sheath which in fact is composed of steel wires.

(*iv*) The ribbon cable [Fig. 25.14(*a*)] has a core consisting of stacked ribbons (about 12 in number). Each ribbon has 12 fibres which are packed between two adhesive backed polyester tapes in the manner as shown in Fig. 25.13 (*b*). The ribbon stack is usually twisted and a loose plastic tube is extruded over the core.

## 25.13 OPTICAL FIBRE CABLE VS CONVENTIONAL COMMUNICATION SYSTEMS

The optical fibre cables are replacing the other electronic communication systems due to following:

(*i*) Light radiation is same as the radio frequency radiation, but at a very high frequency say ($3 \times 10^8$ GHz), the information carrying capacity of a fibre transmission is much greater than microwave radio transmission.

(*ii*) Due to the above reason, a fibre cable can offer more no. of channels. If a conventional system duct has a diameter of 1.20 cm, the equivalent fibre system has only a diameter of 36 mm.

(*iii*) The material used for making fibre cables is silica glass, or silicon oxide, which is easily available, hence the cost of such lines is very less.

(*iv*) Since fibres are electrically insulators they can be used in places, where problems of electrical isolation and interference exist. The optical systems are also immune to crosstalk losses.

Note that, *Crosstalk* is the far and near ends coupling losses between the wires going close to each other.

(*v*) In optical fibre system, there is no fear of short circuits, sparking, etc.

(*vi*) The optical fibre cables are much lighter in weight and smaller in size.

## 25.14 LOSSES IN FIBRE CABLE

Following losses occur in fibre cables

(*i*) *Scattering Losses*. The glass used for optical fibres is an amorphous (non crystalline) solid, which is formed by melting the glass and then cooling it. While in plastic form, it is drawn into fibres. This is the reason of scattering losses occurring in the fibres. This loss can be reduced by adopting careful manufacturing techniques.

(*ii*) *Absorption Losses*. These may be ultraviolet or infrared absorptions. The absorptions occur in walls of the fibre cables.

(*iii*) *Bending Losses*. The losses may be due to "micro" bending or "constant radius" bending.

## 25.15 SUBMARINE OPTICAL CABLES

The fibre optic cables discussed so far are used in land based communication systems. For submarine or deep sea fibre optic communication, the design consideration of fibre optic cable is quite different as it depends first, on various required communication system design parameters and second, on mechanical requirements.

The mechanical design of a submarine cable must take into account the laying and recovery conditions, long term reliable operation of optical fibres under high water pressure and the protection of cables from sharks.

The Fig. 25.15 shows design of optical fibres submarine cables.

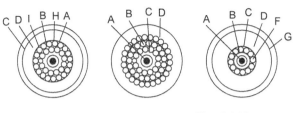

A Fibre unit
B Steel strand
C Copper tube
D Polyethylene
F Steel pipe
G Polyethylene sheath
H Inner pipe (Aluminium)
I Outer pipe (Aluminium)

**Fig. 25.15**

## 25.16 FIBRE OPTICAL SOURCES

Fibre optic communication systems require light sources which must have high efficiency, low cost, longer life, sufficient power output, ability to give desired modulation and the compatibility with the fibre ends.

The light sources used for fibre cables act as " light transmitters" . They should meet the following requirements:

- (*i*) Their light should be "monochromatic" (*i.e.*, of single frequency). The gas ionisation lamps, LEDs and laser diodes have lights of comparatively narrow band of frequencies though not pure monochromatic.

- (*ii*) They should have light of high intensity to overcome and compensate the various losses (scattering loss, bending loss, etc.) during transmission of the signal.

- (*iii*) The light source should be capable to be "modulated". Preferably *pulse code modulation* (PCM) is carried out due to lesser noisy reception.

- (*iv*) They should be small, compact and can be easily coupled to the fibre.

Two kinds of light sources are frequently used in fibre optic communication system namely, the semiconductor light emitting diodes (LEDs) and injection lasers. However, the latest ones are LEDs because of their simplicity and lower cost. Table 25.1 gives a brief comparison of LED and laser light sources. It should be noted that lasers can launch more power than LEDs. But the power output from laser is temperature sensitive as compared to LEDs. However, this temperature dependence can be controlled by the use of special feedback or electronic drive circuits.

**Table 25.1** Comparison of LED and LASER Diodes as Light Sources

| Particulars | Light Sources | |
|---|---|---|
| | LED | LASER diode (LD) |
| Wavelength | 1300 nm | 1300 nm |
| Coupled power to fibre (graded index 50 micron core) | – 13 dBm | 0.0 dBm |

Contd...

| Spectral Width | – 100 nm | < 2nm |
|---|---|---|
| Modulation band width, Mb/s | 30 – 100 | |
| Temperature sensitivity | – 15°C – 20°C | |

## 25.17  OPERATING PRINCIPLE OF LED AND LASER DIODE

The Fig. 25.16 (*a*, *b*) shows cross section and operation of LED and Fig. 25.17(*a*, *b*) show cross section and operation of laser diode. When current is passed through these structures, electrons are injected into the active layer.

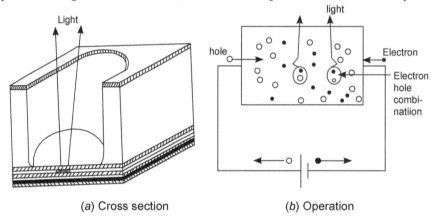

(*a*) Cross section                     (*b*) Operation

**Fig. 25.16**

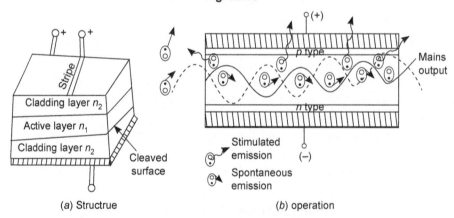

(*a*) Structrue                     (*b*) operation

**Fig. 25.17**

Light is generated when these injected electrons return from their excited (higher) energy levels to their normal (lower) energy levels. The energy difference is released as light. The wavelength of the light can be varied by changing the material of the active layer. The electrons release more energy to produce 'short wavelength' light than they do to produce 'long wavelength' light. In LED, the light is produced at right angles to the active layer. In the laser, the light originates from the plane of the confining layer.

Semiconductor *p-n* junction LEDs and lasers use essentially the *same physical recombination* mechanism for conversion of electrical energy into light. However, microscopic examination of such devices reveals large differences in the overall device construction and in the detail of the *p-n* junction region. These differences are evident with respect to typical surface emitting LEDs and stripe geometry lasers. Because the light emitted by lasers is *more directional* than that emitted by LEDs, larger fraction of their power can be coupled into optical fibres. For higher bit rate system lasers are used, whereas for lower bit rate systems LEDs are used.

The semiconductor materials chosen for the light generating (active) region and the adjacent (confining) layers must meet certain requirements, (*a*) the material that forms the active layer must be highly efficient in converting electrical energy into light energy, (*b*) to achieve good confinement of the injected electrons and thereby ensure the efficient creation of light, the material that forms the confining layers must have a higher energy bandgap and a lower refractive index than that in the active layer.

In view of these requirements, the choice of possible semiconductor materials is somewhat restricted. Nevertheless for the *short wavelength* region, devices based on the GaAs—AlAs alloy are most appropriate, whereas for the *long wavelength* region, alloys formed from the Ga In As P are the best.

## 25.18 DESCRIPTION OF VARIOUS LEDs

Important types of light emitting diodes are described below:

(*a*) *Surface emitting LED* (*SLED*) (Fig. 25.18): SLED operates at 850 nm wavelength. SLED is a five layered double heterojunction device consisting of GaAs and GaAlAs layers. The plane of the active light emitting region is oriented perpendicularly to the axis of the fiber. From the substrate of the device, a well is etched. Fibers are cemented in the well to accept the emitted light. The circular active area in practical surface emitters is normally 50 μm in diameter and upto 2.5 μm thick. SLED has a low thermal impedence in the active region which allows high current densities and give high radiance emission into the optical fiber.

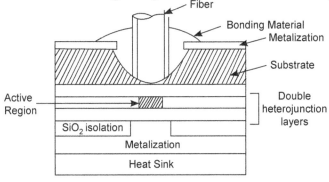

Fig. 25.18

The isotropic pattern from a surface emitter LED is lamberitian pattern in which source is equally bright when viewed from any direction, but the power diminishes as cos θ where θ is the angle between the viewing direction and the normal to the surface. Surface emitting LEDS are also known as "Burrus" because they were first described by The scientists, Burrus and miller.

(b)  *Edge emitting LED (ELED)* (Fig. 25.19): It is similar to surface emitter LED, and has "stripe geometry" having high radiance. The stripe geometry allows a very high carrier injection density for high currents.

It has transparent **guiding layers** with a thin active layer (50 – 150 μm). The light spreads into the guiding layers, reducing self absorption of the active region. The "wave guiding" also narrows the beam divergence to a half power width of around 30° in the plane perpendicular to the junction. However, lack of waveguiding in the plane of the junction gives a *lambrated* output with a half power width of around 120°.

The enhanced waveguiding of the "edge emitter" enables it to couple 7.5 times more power into low "numerical aperture (NA)" fiber than a "surface emitter diode".

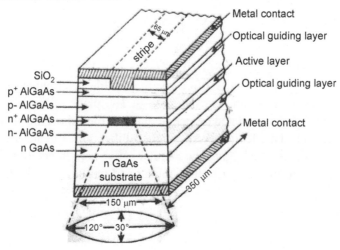

**Fig. 25.19**

In consists of an active region, (which is a source of **incoherent light**) and two guiding layers. The refractive index of both guiding layers is lower than the active region, but higher than of the surrounding material.

(c)  *Super luminescent LED (SLED)*: A superluminescent LED (SLED) is an edge emitting diode based on principle of superluminescence. It combines the high power and brightness of "laser diode" and low coherence of "conventional LED".

An SLED is similar to laser diode biased on an electrically driven *p-n* junction. When "forward biased", the SLED becomes optically active and generates amplified **spontaneous emission** over a wide range of wavelength.

The Fig. 25.20 (*a*) shows schematic diagram of SLED. An SLED is an elongated version of stripe geometry edge emitter. It has double heterogen (DH) structure with a long stripe (1000 μm). Its radiation pattern from the edge is well directed to within 10°. It has an anti-reflection (AR) coating or both facets.

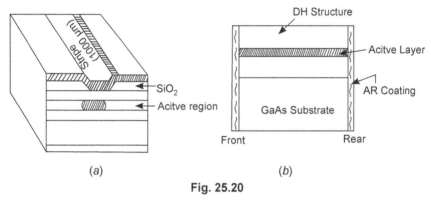

(*a*)                                              (*b*)

**Fig. 25.20**

When a forward voltage is applied to SLED, an injection current across the active region is generated. The current flows from *p*-section to *n*-section and across the active region, which is sandwiched between *p* and *n* sections. During this process, broadband light is generated through spontaneous and random combination of holes and electrons, which is amplified when travelling along the waveguide of the SLED, but because there is high loss at one end of the device, no optical feedback takes place.

Although there is amplification of the spontaneous emission, no laser oscillations are build up.

## 25.19 DESCRIPTION OF VARIOUS LASER DIODES (LDs)

Important types of laser diodes are described below:

(*a*) *Stripe geometry DH (double hetrogen) Laser:* In this laser, the *problems of the broad area DH* laser have been removed. Figure 25.21 shows a stripe geometry *DH* laser.

The active region is within the **stripe,** that acts as a **guiding mechanism,** which overcomes the problem of broad area device. The output beam divergence is 45° perpendicular and 9° parallel to the plane, thus the beam is quite narrow.

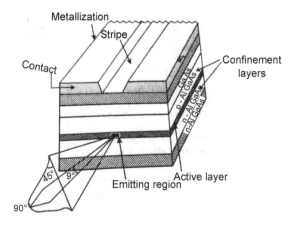

**Fig. 25.21**

The active regions are planar and continuous, therefore the stimulated emission characteristics of this laser is determined by the carrier distribution along the junction plane.

(b) *Distributed feedback (DFB) laser*: The Fig. 25.22 shows construction of distribution feedback (DFB) Laser. The optical grating is applied over the entire active region, which is **pumped.** The grating is etched into the substrate prior to the deposition of semiconducting layers.

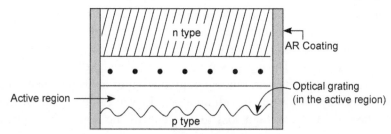

**Fig. 25.22**

*The both ends are anti-reflection (AR) coated.* The two types of structures are used:

(i)  *High-Low (Hi-Lo) structure*: In this, a high antireflection coating is applied on one end facet and low antireflection coating on the other end facet. This is done to improve the power output for single frequency operation.

(ii) *Double channel planar buried heterostructure*: This technique is used to improve the performance. It modifies the grating at a central point to introduce an additional optical phase shift, typically a *quarter wavelength.*

In DFB lasers, the main design objective is to generate a single line spectrum at the output under high data rates of modulation. The grating leads to an effective spatial modulation, which contributes to the feedback mechanism so that single mode is produced and the undesired modes are supressed.

(c) *Distributed Bragg Reflectors (DBR) Laser* (Fig. 25.23): The fundamental difference between DFB and DBR lasers lies in their grating mechanisms. In DFB, the grating is along the cavity length while in DBR, the grating is at both ends of the active region. When it

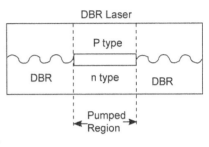

Fig. 25.23

is at both ends, it can act as a perfect optical mirror because of the difference between the constant refractive index of the active layer and the continuously changing refractive index of the grating layer. This provides a required feedback mechanism for optical power generation and spectral purity. DBR devices require higher threshold current than DFB structures.

(d) *Vertical cavity semiconductor (VCS) Lasers* (Fig. 25.24): VCS lasers emit light from their surface instead of the edge as in DFB lasers. The difference between an edge emitting laser and a VCS laser is that the laser cavities or the resonators are placed above or below the active layer, so that the light is emitted perpendicular to the active layer.

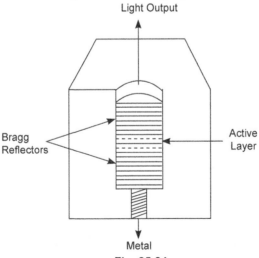

Fig. 25.24

(*e*) *Superfluorescent Fibre Laser Diode*: The output from a Superfluorescent laser grows from the spontaneous emission generated by the launched pump light and so the process can alternatively be termed as *amplified spontaneous emission* (ASE). Superfluorescent fibre lasers are quite useful as an alternate to superluminescent diodes in many sensor applications particularly in fibre gyroscopes and in signal processing fibre systems. The commercially available superluminescent diodes exhibit a short life time, a poor wavelength stability, a relatively low output power, high wavelength sensitivity to temperature, inefficient coupling to single mode fibres and a lack of immunity of optical feedback. The disadvantages of superluminescent diode (SLD) are overcome in Superfluorescent fibre lasers (SFL).

Figure 25.25 indicates the basic diagram of a typical SFL. It consists of a conventional $GeO_2 - SiO$, fibre of length 16.8 m doped with $P_2O_5$ and $Nd^{3+}$ (300 p.p.m.). The ends of fibre are mounted in capillary tube. One fibre end is polished at an angle of 10° to prevent any reflection and the other end is butted against a dielectric mirror.

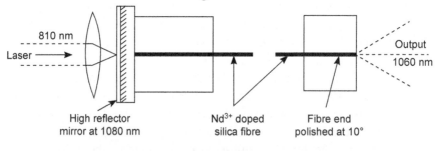

**Fig. 25.25**

The spectrum gets narrower as the output power increases and its linewidth approaches some asymptotic value at very high output levels. Thus the superfluorescent fibre laser requires a little high pump intensity of about 45 mW to enable the fibre gain to become high for strong broadband fluorescent emissions.

## 25.20  MODERN DIODE LASERS

In addition to the diode lasers described earlier, a variety of diode lasers have recently come up. These all are based on the materials and structure of semiconductor junctions. The application requirements such as power, speed, efficiency of coupling, temperature range, lifetime and cost were main factors responsible for the development of such devices. Table 25.2 gives a typical performance of some commercially available modern diode lasers.

**Table 25.2**  Performance of Some modern diode lasers as Light Sources

| Modern Diode Lasers (DL) | Material | Wavelength Range (µm) | Power (dbm) | Maximum bandwidth (GHz ) |
|---|---|---|---|---|
| DL Multimode | GaAlAs | 0.8–0.9 | 1 | 1 |
| | InGaAsP | 1.3 | 1 | 1 |
| | InGaAsP | 1.55 | – 2 | 2 |
| DL Single Mode | GaAlAs | 0.85 | 0.8 | 2 |
| | InGaAsP | 1.3 | –4 | 0.14 |
| | InGaAsP | 1.55 | –8 | |
| DL Distributed Feedback | InGaAsP | 1.5–1.65 | – | 0.24 |
| In the Table : Ga: Gallium, Al: Aluminium, As: Arsenic, P: Phosphorus, In: Indium, | | | | |

## 25.21  BLOCK DIAGRAM OF FIBRE OPTIC COMMUNICATION

Fibre optic communication systems would involve a series of components which are analogous to those in radio frequency communication, but are different in design. A basic block diagram of fibre optic communication system is shown in Fig. 25.26. It is apparent that the main blocks of the system are transmitter, repeater and receives. The Transmitter includes modulators, switches and drive circuits to lasers. It also has couplers to couple the transmitted light signals with the fibre. Fibres also may require connectors to increase the length of fibre medium. Repeaters detect the transmitted signals, regenerate, amplify and then retransmit them. Receivers amplify the signal and then convert them into the desired output.

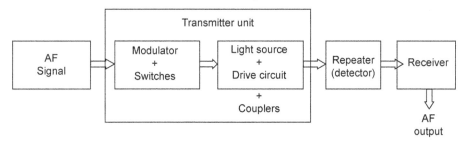

**Fig. 25.26**

Miniaturised version of above optical components like modulators, couplers, switches, detectors, etc., have also been developed. These involve guided wave optics into thin film waveguides. The principle is called *"integrated optics"* in analogy with the integrated semiconductor electronics. The main importance of such integrated optical components is that these are rugged, require less space and consume very little power. In addition, these are quite efficient and fast.

## Brief Description of Components/Blocks of Optical Communication

Below is given brief description of components:

(*i*) *Electronic Transmitter*: The information to be conveyed enters into Electronic Transmitter, where it is converted into electrical form, modulated and multiplexed. (Fig. 25.27)

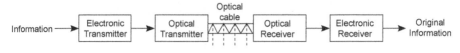

**Fig. 25.27**

(*ii*) *Optical Transmitter*: The electrical signal then goes into *optical transmitter*, where it is converted into optical form *i.e.*, light. The optical transmitter consists of drive circuit *optical source, modulator* and *a channel coupler*. [Fig. 25.28]

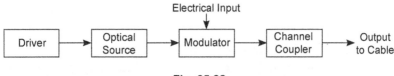

**Fig. 25.28**

Either LED or a Laser diode is used as optical source. The input signal **modulates** the intensity of light (called intensity modulation) coming from the optical source. The intensity modulation can be carried out electrically or optically. Analog or digital modulation is possible. In **coherent modulation** ASK, FSK or PSK is used, in that case, the receivers are more complex. Now-a-days **Electro-optic modulators** are used, which modulate the light by changing its refractive index through the electrical signal. The **channel coupler** couples the modulated optical signal to the optical cable. The coupler is a **microlens**, that focusses the optical signal on to the entrance plane of the cable with maximum efficiency.

(*iii*) *Optical Cable*: Through the optical cable, the signal travels **by total internal reflection.**

(*iv*) *Optical Receiver*: The main function of optical receiver is to convert the received optical signal to original electrical signal. The optical receiver consists of a channel coupler, optical photo detector, and a demodulator (Fig. 25.29).

**Fig. 25.29**

The coupler focusses the received optical signal on to the photodetector. The photo-detector converts the optical pulses into electrical pulses. The demodulator demodulates the received electrical signal.

(*v*) *Electronic Receiver*: The electrical signal is then processed in the electronic receiver to get the exact original information.

## 25.22 DIGITAL TRANSMISSION AND RECEPTION THROUGH OPTICAL CABLE

The Fig. 25.30 shows digital transmission through optical fibre cable.

(*i*) The digital data is sent with the help of "amplitude shift key" (ASK) technique.

(*ii*) The optical transmitter converts the electrical signal into optical signal. Any time-varying current can be used to **modulate** an optical *source* to produce optical (light) signal.

(*iii*) The optical signal is coupled to optical fibre. The optical signal gets attenuated and distorted, as it propagates through a fibre.

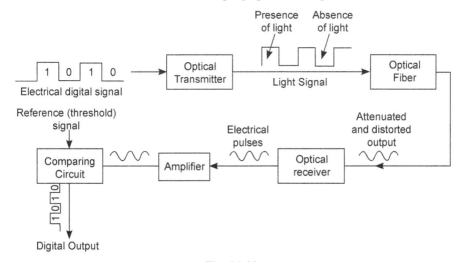

**Fig. 25.30**

(*iv*) The optical signal, now is coupled to optical receiver which uses a PIN or Avalanche photodiodes as a detector. It converts the optical signal back to original electrical signal (pulses).

(*v*)  A "comparer circuit" then compares this amplified signal in each "time slot" with a certain "reference" signal known as "threshold level."

If the received signal is more than the threshold level, an output binary '1' is obtained. If the received signal is less than the threshold level, a binary '0' is obtained.

## 25.23  VARIOUS DIGITAL MODULATION SCHEMES USED IN OPTICAL COMMUNICATION

In optical communication the various digital modulation schemes used are: Amplitude, Frequency and Phase shift key. (ASK, FSK, PSK)

The Fig. 25.31 (*a*) shows binary ASK, the Fig. 25.31(*b*) shows FSK, in which the binary 1 is transmitted at higher optical frequency than the 0 bit. The Fig. 25.31 (*c*) shows PSK, which shows 180° phase shift between Binary 1 and 0 bits.

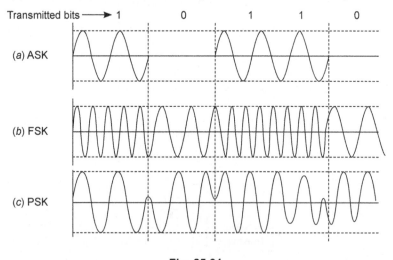

**Fig. 25.31**

It should be noted that in ASK, the amplitude of the optical carrier is effectively switched ON or OFF whereas, the amplitude of the optical carrier remains constant in other two modulation (FSK, PSK) schemes. This is the reason that ASK modulation is preferred. However, "continuous phase FSK" and "differential PSK" can also be used. Moreover, an alternative digital scheme has more recently come which is known as "polarization shift keying (Pol.SK)" .

## 25.24  OPTICAL (PHOTO) DETECTORS

An optical (photo) detector is a transducer, that converts an optical signal into an electrical signal. In other words, it generates an electric current proportional to the intensity of the incident optical radiation.

In communication, the photoconductors are the semiconductor devices used as optical/photodetector, generally there are three processes carried out by photoconductors:

(*i*) Generation of carrier by incident light.

(*ii*) Carrier transport and multiplication by current gain.

(*iii*) Interaction of current with external circuitary (load) to provide output signal.

**The photodetectors used in communication systems are:**

1. Photoconductor (light depending resistor)

2. P-N junction semiconductor photodetector

3. PIN diode

4. Avlanche photodiode (APD)

5. Photo-transistors

6. Photo emmissive detector

7. Metal semiconductor metal (MSM) detector.

The PIN photodiodes and APDs are popularly used as detectors are described below:

## 25.25 PIN PHOTODIODE

(*a*) The P-N junction is extended by an addition of a very lightly doped layer called intrinsic (I) layer between P and N type layers. The device so obtained is called PIN photodiode. Usually, it is made of silicon.

(*b*) The Intrinsic (I) layer has the following advantages:

(*i*) Because it makes the junction wider (*i.e.* wide depletion layer), it reduces the capacitance across the junction. Lower the capacitance of the junction, faster is the device's response.

(*ii*) It increases the chances of an entering photon being absorbed, because the volume of absorbant material is significantly increased.

(*iii*) Addition of I layer increases responsivity.

(*iv*) It decreases response time to picoseconds.

(*c*) The Fig. 25.32(*a*) shows PIN photodiode and Fig. 25.32(*b*) shows its operation.

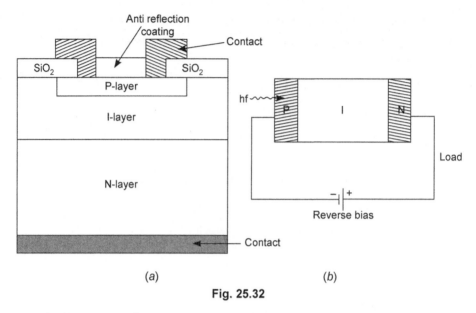

(a)                                             (b)

**Fig. 25.32**

(*d*) *Operation*: There are two ways of current carnage (conduction) across the junction:

   1.  Diffusion

   2.  Drift.

Increasing the width of the depletion layer favours current carriage by drift, which is faster than diffusion.

When the diode is biased, the Intrinsic region is flooded with charge carriers from P to N regions. The diode conducts current, when the number of electrons equals to the number of holes in the intrinsic region.

The PIN diode operates at any wavelength shorter than "cut off" wavelength. This suggests the use of a material, with low band gap energy ; But the low band gap energy increases the "dark current" and thus the noise.

A PIN diode has a *reverse recovery* time.

(*e*) *Applications*: A few applications of PIN photodiodes are given below:

   (*i*)  A PIN photodiode, if biased suitably, acts as a "variable resistor".

   (*ii*)  The diode is used as RF (radio frequency) and microwave variable attenuator.

   (*iii*)  The PIN diodes are used as input protection devices for high frequency test probes. If the signal is large, the diode starts to conduct and becomes a resistor and "shunts" most of the signal to ground.

(*iv*) The PIN diodes are used in fibre optics as a *photodetector*.

(*f*) The PIN diodes as compared to *pn* junction diode are widely preferred used due to the following:

(*i*) The problem of extreme thinness (thin depletion layer) of the *pn* junction diode is rectified by making the depletion layer thicker. The junction is extended by addition of a very lightly doped layer called **Intrinsic** layer between *p* and *n* layers. The wide intrinsic layer has only a small amount of dopant and acts as a very wide depletion layer.

(*ii*) Due to above, chances are increased for an entering photon being absorbed, because the volume of absorbed material is significantly increased.

(*iii*) Due to more width (*d*) of the depletion layer, it reduces the capacitance (C) across the device, $C \propto \dfrac{1}{d^2}$.

(*iv*) Lower the capacitance of the device, faster is its "response".

## 25.26 AVALANCHE PHOTODIODE (APD)

(*a*) An avalanche photo diode (APD) is a highly sensitive semiconductor electronic device that uses "*photo electric effect*" to convert light into electricity. The APDs can be considered as *photodetectors,* that provide a **built in** first stage of gain through **Avalanche multiplication**.

By applying a high reverse bias (upto 200V), the APDs show an internal current gain of 100 due to impact ionization (Avalanche effect).

Some silicon APDs employ alternate doping compared to traditional APDs that allow a greater voltage to be employed (upto 1500 V) before breakdown is reached; and hence a greater gain of more than 1000 can be obtained.

In general, higher the reverse voltage, higher is the gain. The APDs amplify the signal during detection process. Their working principle is same as that of a "Photo multiplier tube."

(*b*) *Construction of APD*: The Fig. 25.33 shows basic structure of APD. This is similar to PIN diode but with a very high reverse bias. The reverse bias may be from 50 to 200 Volts.

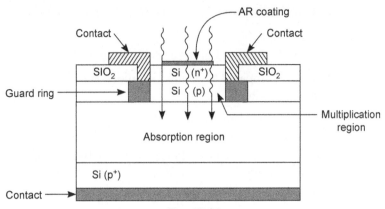

**Fig. 25.33**

(c) *Structural difference from PIN diode* : The main structural difference between APD and PIN diode is that the I layer, which is lightly *n* doped in PIN diode is lightly *p* doped in APD. It is also thicker. The APD is carefully designed to ensure a uniform electric field across the whole layer. The guard ring serves to prevent unwanted interaction around the edges of the multiplication region.

(d) *Operation*: The Fig. 25.34 (a), shows operation of an APD. It is reverse biased with 50 – 200V. The Fig. (b) shows the "gain region".

The arriving photons generally pass straight through $n^+p^-$ junction, as it is very thin and are absorbed in n layer. This absorption produces a "free electron" in the conduction band and a "hole" in the valence band. The potential across the device is sufficient to attract electrons towards +ve terminal and holes towards the –ve terminal.

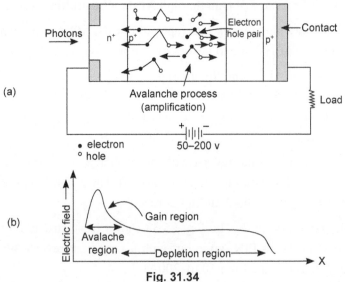

**Fig. 31.34**

Around the junction between $n^+$ and $p^+$ layers, the electric field is so strong that the electrons are strongly accelerated. When these electrons collide with other atoms, they produce new electron hole pairs. This process is called **Impact Ionization.** The new released electron/holes accelerate in opposite directions and collide again.

If electron/hole pairs have equal "**propensity**" to ionization, we can get an **uncontrolled avalanche,** which will never stop. The result of the above process is that single arriving photon can result in the production of 10 to 100 electron pairs.

If a very high gain is needed ($10^5$ to $10^6$), certain APDs (called single photon Avalanche diodes) can be operated with a reverse voltage above the breakdown voltage of the device. In this case, the APD needs to have its signal current limited and quickly diminished.

## 25.27 OPTICAL NETWORKS

An optical network may be defined as a "collection of transmission links and other equipment, which provides a means of information interchange within a group of end users". The term **network** also implies a geographical separation between the end users. Its concepts can be listed as below:

1. A network usually contains some shared resources *i.e.*, links and switching nodes are shared between many users.

2. Most networks have a centralized management system.

3. The information may be exchanged between any of the users.

4. A single user may communicate with multiple users simultaneously.

**A few examples of optical networks are:**

1. Couplers - splitters & combiners

2. Wavelength division multiplexers and demultiplexers

3. Optical amplifiers

4. Local area network (LAN)

5. Metropoletan area network (MAN)

6. Wide area network (WAN)

7. Personal area network (PAN)

8. Campus area network (CAN)

9. National area network (NAN)

We will here discuss first three networks.

## 25.28  OPTICAL FIBER COUPLERS

(*a*) The optical fiber couplers are the devices, that distribute light from a main fiber into one or more branches. These are also called **directional couplers** and are used in Local Area Networks (LANs), computer networks and distribution systems etc.

These may be **active or passive devices**. The transfer of power takes place through **core interaction** (through fiber cross section) or through **surface interaction** (through fiber surface and normal to its axis).

The most important is that the "coupler" should distribute light among all branches with no scattering loss or generation of noise.

The optical couplers are the key components used in **optical networks**.

(*b*) The optical (or opto) couplers are used for routing signals from one waveguide to another and/or for splitting optical signals into two independent signals at a predetermined power ratio to be transmitted over two different waveguides.

The opto coupler may be either an **active** or a **passive** device. The "active couplers" either **combine or split** the signal electrically and use optical detectors and sources for input and output. The "passive couplers" redistribute the optical signal without "optical to electrical" conversion.

The opto couplers usually include a LED and a light sensor (photodiode or photo transistor). The electrical isolation occurs because information is transmitted using light emitted by LED and received by the sensor. When the current driving the LED is changed, the emitted light also changes proportionately and consequently the electrical resistance of the sensor.

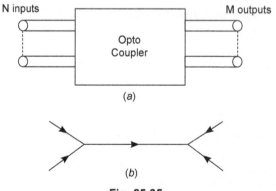

**Fig. 25.35**

See Fig. 25.35, the opto coupler has N input ports and M output ports. The N and M may range from 1 to 65. This number depends upon application of the couples.

(c) An opto coupler may include.

    1. Optical splitter       2. Optical combiner

    3. X coupler            4. Tree coupler

Here we describe optical splitters.

## 25.29 OPTICAL SPLITTER

This is a passive device, that splits the optical power carried by a single input fiber into two output fibers.

The Fig. 25.36. shows a "Y-Coupler", which splits the input power **evenly** between two output fibers.

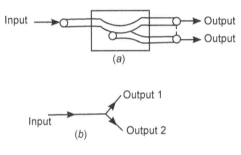

(a)

(b)

**Fig. 25.36**

However, an optical splitter may also distribute the input power **unevenly.** Sometime, a small amount of power is coupled into the output fiber, which is called a "T-splitter".

## 25.30 WAVELENGTH DIVISION MULTIPLEXERS AND DE MULTIPLEXERS

The couplers and splitters combine/split light signals at one wavelength, but the wavelength multiplexers and demultiplexers can combine/split the signals at different wave lengths.

1. *The wave division multiplexers*: These combine several wavelength channels into one channel (Fig. 25.37)

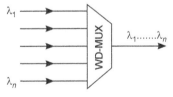

**Fig. 25.37**

2. *The wave division demultiplexers*: These split one wavelength channel into several channels. (Fig. 25.38)

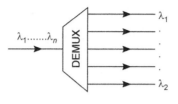

**Fig. 25.38**

## 25.31  WAVELENGTH DIVISION MULTIPLEXING (WDMUX)

(*a*)  *Wavelength division multiplexing (WDMUX)*:  In fiber optic communication, wavelength division multiplexing (WDMUX) is a technology, which multiplexes a number of optical carrier signals onto a single optical fiber by using different wavelengths of laser light. This enables "bidirectional" communication over one strand of fiber, as well as multiplication of capacity (Fig. 25.39).

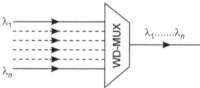

**Fig. 25.39**

(*b*)  The main features of WDM technology are:

    (*i*)  Capacity is increased.      (*ii*) Transparency also increases

    (*iii*)  Wavelength routing.      (*iv*) Wavelength switching.

(*c*)  A WDM system uses a multiplexer at the transmitter to join the signals together. Each of the wave is punched into the fiber and the signals are **demultiplexed** (splitted) at the receiving end. With the right type of fiber it is possible to have a device that does both jobs simultaneously.

    The most WDM systems operate on single mode fiber optic cables, which have core diameter of 9 μm. Certain forms of WDM can also be used in multimode cables, which have core diameter of 50 μm or so.

(*d*)  *WDM Vs TDM*: The WDM carries each input signal independently of others, this means, that each channel has its own bandwidth, all signals arrive at the same time rather than being broken up and carried out in "time slots" as is done in TDM (Time division multiplexing).

(*e*)  *WDM Vs FDM*: The term "WDM" is commonly applied to an optical carrier, which is typically described by its wavelength, whereas the term FDM (frequency division multiplexing) typically applies to a radio carrier, which is described by frequency. Since "wavelength" and "frequency" are related to each other (wavelength = 1 /frequency) the two terms describe the same concept.

## 25.32 TYPES OF WDMUX/DEMUX

Basically they are of two types (See Fig. 25.40):

    1. Unidirectional       2. Bidirectional.

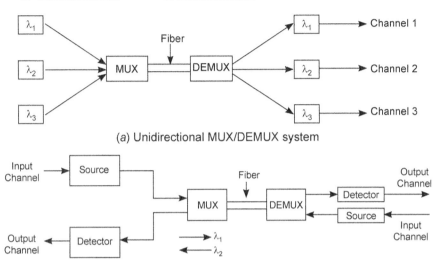

(a) Unidirectional MUX/DEMUX system

(b) Bidirectional MUX/DEMUX System

**Fig. 25.40**

## 25.33 OPTICAL AMPLIFIERS

(a) An optical amplifier is a device, which amplifies an optical signal directly without converting it into electrical signal.

As optical amplifier may be thought of a laser without an optical cavity; or in which the feedback from the cavity is suppressed. The stimulated emission in the amplifier *gain medium* causes amplification of the incoming light signal.

They are very important devices in the fiber optic communication.

(b) The optical amplifiers may be classified as:

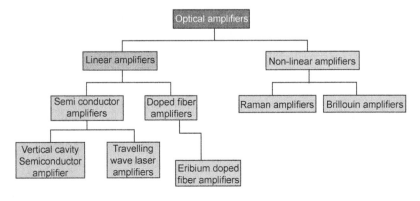

Here, we describe two linear optical amplifiers namely, semiconductor amplifiers and duped fiber amplifiers.

## 25.34  SEMI CONDUCTOR OPTICAL AMPLIFIERS (SOA)

(*a*) The semiconductor optical amplifiers (SOAs) use a semiconductor to provide the gain medium.

   The SOAs are made up of semiconductor compounds such as GaAs, AlGaAs, InP, InGaAs etc. These are operated at signal wavelength between 0.85 µm to 1.5 µm, producing gain upto 30 dB.

(*b*) The Fig. 25.41 shows a **vertical cavity semiconductor optical amplifier (VCSOA)**

   The active region has a width W, length L and as thickness. They have anti-reflection (AR) coating on facets ; which should ideally exhibit zero reflectivity.

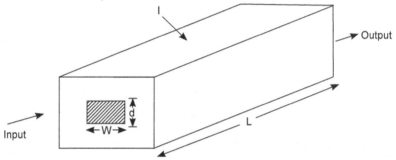

**Fig. 25.41**

## 25.35  DUPED FIBER AMPLIFIER (DFA)

(*a*) The doped fiber amplifier (DFA) uses a doped optical fiber as a "gain medium" to amplify an optical signal. They are related to "Fiber lasers". The Fig. 25.42 show a DFA.

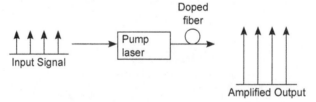

**Fig. 25.42**

   The signal to be amplified and the *pump laser* are multiplexed into the doped fiber. The signal is amplified through interaction with the doping ions.

(*b*) In **Eribium DFA (EDFA),** the core of a silica fiber is doped with trivalent **Eribium ions** and can be efficiently pumped with a laser at a wavelength of 980 and 1500 nm and exhibit gain in 1550 nm region.

**Table 25.3** Communication Systems at Different Frequency Ranges

| Range | Wavelength Range | Designation | Frequency Range Hz | Transmission Media | Application |
|---|---|---|---|---|---|
| 1. Optical | 50 nm to about 300 nm | Ultraviolet | Optical $10^{15}$ | Experimental fibres and laser beams | Telephone, data and video |
| | 400 nm to 700 nm | Visible | Between $10^{15}$ and $10^4$ | | |
| | 700 nm to 100 nm | Infrared to far infrared | $10^{14}$ to $10^{12}$ | | |
| 2. Extended Optical Ranges | 0.3 mm to about 1 mm | Sub mm waves | $10^{12}$ to $3 \times 10^{10}$ | Waveguides | Satellite |
| | 1 nm to 1 cm | mm waves | | | |
| | 1 cm to 10 cm | microwaves | $3 \times 10^{10}$ to $3 \times 10^9$ | Waveguides, microwave-radio | Experimental, navigation, satellite to satellite, microwave-relay earth satellite, radar |
| | 10 cm to 1 m | UHF | $3 \times 109$ to $3 \times 10^8$ | upper half bandwidth with wave guides and microwave radio, lower-half with coaxial cable and short wave radio | Radar and UHF TV |
| | 1 m to 10 m | very high frequency (VHF) | $3 \times 10^8$ to $3 \times 10^7$ | coaxial cable, short wave radio | Mobile, aeronautical, VHF TV and FM mobile radio |
| | 10 m to 100 m | High frequency (HF) | $3 \times 10^7$ to $3 \times 10^6$ | Coaxial cable short wave radio | Business amateur radio, international citizen's band |
| | 100 m to 1 km | Medium frequency (MF) | $3 \times 10^6$ to $3 \times 10^5$ | Coaxial Cable, long wave radio | AM Broadcasting |
| | 1 km to 10 km | Low Frequency (LF) | $3 \times 10^5$ to $3 \times 10^4$ | Pair of wires, long wave radio | Aeronautical Submarine cable navigation, trans-oceanic radio |

## SUMMARY

1. The optical fibre communication has higher bandwidth and uses light as carrier signal. The fibre cables have light weight, small size and have no risk of short circuit.

2. The optical fibre is the core material for making optical fibre cables. The fibre may be step index fibre or graded index fibre.

3. The materials used for making fibre are silica glass and plastic.

4. The optical fibre cable may be single, double or multiple fibre type.

5. The propagation in the optical fibre cable is obtained through *total internal reflection*.

6. The important fibre optical sources are: LED and LASER diodes.

# Propagation of Radio Waves

The radio waves are electromagnetic (e.m.) waves which propagate with a velocity of $3 \times 10^8$ m/s in free space. The e.m. waves consist of electric and magnetic fields. The direction of electric field, magnetic field and propagation are mutually perpendicular to each other. In this chapter we shall study the various modes of propagation of Radio waves.

## 26.1 ELECTROMAGNETIC WAVE

A simple e.m. wave is shown in Fig. 26.1, the electric field lines are perpendicular to earth and the magnetic field lines are horizontal to earth. The arrows indicate the instantaneous direction of fields, while the wave is travelling towards the reader.

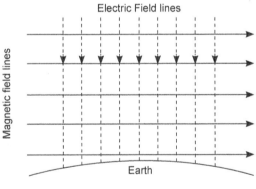

Fig. 26.1

## 26.2 PROPAGATION OF RADIO WAVES

The radio waves to some extent can be related to the waves produced in water due to some disturbance. The only difference is, that radio waves are transverse whereas water waves are longitudinal. By transverse, we mean that the oscillations produced are perpendicular to the propagation, also, as mentioned above, the direction of propagation, direction of magnetic field and direction of electric field all the three are perpendicular to each other, while travelling in a free space. Note that a *free space* is a space that does not interfere with the propagation of radio waves. The Fig. 26.2(*a*) shows propagation of a Transverse electromagnetic (TEM) radio wave in free space. Here, the $E$ is electrical field and $I$ is magnetic field. The Fig. 26.2(*b*) shows its field representation.

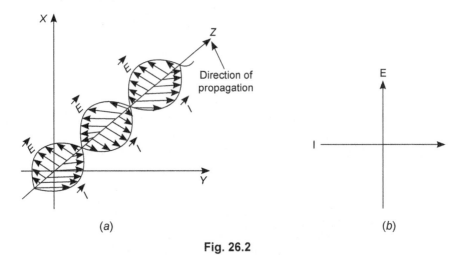

(a)                                                      (b)

**Fig. 26.2**

## 26.3 STRUCTURE OF ATMOSPHERE

The important regions (layers) of atmosphere are described below (Fig. 26.3):

1. *Troposphere*. It is that portion of the atmosphere which extends upto 15 km from earth surface. The entire belt is constant with increase in height. Water vapour components fall with rise in height.

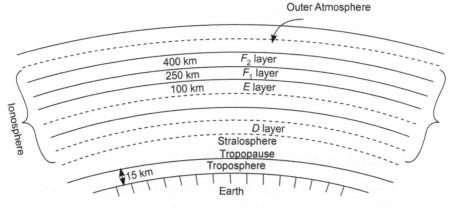

**Fig. 26.3**

2. *Tropopause* and *Stratosphere*. After troposhere, there is tropopause which ends before stratosphere, no intermixing due to air currents take place and composition of the atmosphere varies with height.

3. *Ionosphere*. The upper part of atmosphere where the ionization is appreciable is known as ionosphere. This part receives large amount of radiant energy in the form of cosmic rays and UV (ultraviolet) light. This not only heats the atmosphere but also ionizes *i.e.*, produces positive and negative ions in a gas under very low pressure. It is possible to knock off one or two electrons out of gas molecule. So the molecule which donates becomes positively charged and which accepts becomes negatively charged. These charged particles are affected by electric fields.

There are also recombinations which depend upon average distance between particles. As UV rays are absorbed in upper parts of atmosphere so there is relatively low ionization in lower part below 50 km. Since sky waves of different frequencies are found to return from different heights, it means ionosphere has several layers. Different layers exist as different gases in it get ionized at different pressure. The number of layers, their height and the amount of sky waves that can be bent by them will vary from day to day, month to month and year to year. There are three principal layers during day time called $E$, $F_1$ and $F_2$ layers. Below $E$ there is $D$ layer which lies between 50 to 90 km and is responsible for day time attenuation of high frequencies. During night, $F_1$ and $F_2$ layers remain and $D$ region vanishes.

Thus during night the layers existing are $E$ and $F$. The $E$ layer is generally found at 100 km and $F$ lies between 250–400 km. The $F$ layer is also called *appletion layer*.

| Layer | Ionic density per c.c. |
|-------|------------------------|
| $F_2$ | $10^6$ |
| $F_1$ | $5 \times 10^5$ |
| $E$ | $4 \times 10^5$ |
| $D$ | 400 |

## 26.4 TERMS RELATED TO PROPAGATION

Below few terms related to the propagation are given:

1. *Power Density*: The power density is defined as "radiated power per unit area" . It can be found by " Inverse Square Law" which states that "power density is inversely proportional to the square of distance from the source." Mathematically,

$$\rho = \frac{P_T}{4\pi r^2}$$

where                                   $\rho$ = Power density

$P_T$ = Total power radiated or transmitted

$r$ = distance from the Isotropic source. The Isotropic is that source (or antenna) which radiates power uniformly in all directions in the free space.

**Problem 26.1.** Calculate the power density

(*a*) At 500 m from a 500 W source.

(*b*) At ground from a satellite, stationed at 36000 km transmitting 4 kW power.

**Solution.** (*a*)        $\rho = \dfrac{P_T}{4\pi r^2} = \dfrac{500}{4\pi \times (500)^2} = 159.2 \; \mu W/m^2$              **Ans.**

(*b*)                      $\rho = \dfrac{4 \times 1000}{4\pi \times (36000 \times 10^3)^2}$

$= 2.5 \times 10^{-13} \; W/m^2$   **Ans.**

2. *Polarisation*: This defines orientation of radiated waves in space. The waves are said to be polarised "linearly" if they have the same alignment throughout the space. Mostly all antennas radiate linear polarised waves. Sun and other heavenly bodies radiate waves in haphazard alignment and said to be " randomly" polarised. If electrical lines are perpendicular to earth, the wave is said to be "vertically" polarised. If the lines are parallel to earth, they are "horizontally" polarised. Similarly they can be "elliptically" polarised.

3. *Attenuation*. The energy of e.m. waves diminish rapidly with distance. This is called "Attenuation" . The attenuation is inversely proportional to the square of the distance travelled by the waves. This is measured in *decibel*.

4. *Absorption*. In free space, absorption of radio waves do not occur, as there is nothing to absorb them.

However, when e.m. waves travel through atmosphere, some energy from these waves is transferred to the atmosphere and is lost as heat. The atmospheric absorption of the e.m. waves is quite insignificant below 10 GHz. This is a plus point, as most of our communication works under this frequency. The main components of atmosphere responsible for absorption are oxygen and rain water. It is to be noted that the rainfall used is the *peak intensity* and not the total rainfall around the year.

5. *Reflection*. The reflection of the electromagnetic waves is similar to that of the light waves. The two laws of light reflection are also applicable

to the electro magnetic waves. With each reflection, a part of energy of the waves is absorbed.

6. *Refraction*. The refraction in radio waves is similar to the light refraction. It occurs, when a radio wave passes from one propagating medium to the another propagating medium having different density. The velocity of the waves also changes, when it changes the medium.

7. *Interference*. The interference occurs when two waves from the same source meet at a point while travelling through separate paths. This mostly occurs in HF sky wave propagation or in microwave propagation. In the latter case, the antenna is located near the ground and the waves reach the receiving point not only directly but also indirectly *i.e.*, after reflecting from the ground. If in any case frequency and height of antenna is such that the difference between direct wave and the reflected wave is equal to half the wavelength, the two completely cancel each other at the receiving point, provided the ground is perfectly insulator. If the ground is not perfect insulator, partial cancellation of waves occur. If the above difference is equal to *one wavelength*, reinforcement of waves takes place. A succession of such points may be found lying one above the other giving rise to a "pattern" having alternate cancellation and reinforcement. In Fig. 26.4, the curve joins points of equal electric intensity. The pattern is due to the location of antenna at a height above the ground at about a wavelength. The three parts in the form of flower petals are called "Lobes" and represent the *reinforcement*. The empty spaces between lobes are called "Null" and represent *cancellation*.

In the UHF region and above, the interference dominates the propagation of waves. A radar will be completely blind to a target located in the Null zone.

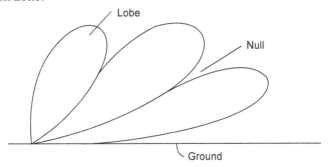

**Fig. 26.4**

8. *Diffraction*. The property describes the behaviour of the waves, when they strike a conducting plane with small slots, or encounter sharp edged obstacles. The behaviour is some-what like light, when a radio

wave strikes a plane with slots, instead a single wave passing through the slot, the wave spreads out on the other side of the slot radiating in all direction.

When a wave encounters an edge of an obstacle, diffraction takes place. It can be seen that radiation away from the main direction of propagation is also obtained. The " shadow zone" also gets some radiation. After the edge, the radiation gradually dies down.

The diffraction has some advantage also. It sometimes helps in obtaining radio signals behind tall buildings, mountains, etc.

9. *Virtual Height.* As the wave is refracted from the layer, it bents down gradually rather sharply. The actual path of the wave in the ionized layer is a curve and this is due to the refraction of the wave. Since it is more convenient to think of the wave being reflected rather refracted therefore, the path can be assumed to be straight lines TD and RD as shown in Fig. 26.5. Virtual height is always greater than the actual height.

Virtual height of an ionosphere layer may be defined as the height to which a short pulse of energy sent vertically upward and travelling with the speed of light would reach taking the same time as does the actual pulse reflected from the layer.

In the measurement of virtual heights, the transmitting point (T) and receiving point (R) are usually placed very close together so that the wave sent is nearly vertically upward.

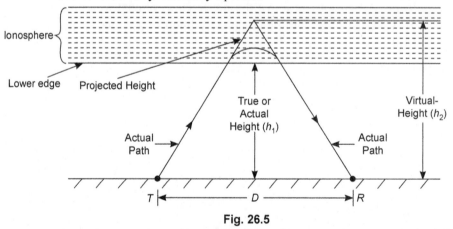

**Fig. 26.5**

**Problem 26.2.** A pulse of a given frequency transmitted upward is received back after a period of 6 milli seconds. Find the virtual height of the reflecting layer.

**Solution.**                    $t = 6$ milli seconds $= 6 \times 10^{-3}$ seconds.

Velocity of the pulse, $C = 3 \times 10^8$ metres/sec

$$h = \frac{Ct}{2}$$

$$h = \frac{3 \times 10^8 \times 6 \times 10^{-3}}{2} = \frac{10 \times 18^5}{2} \text{ metres}$$

$$h = 9 \times 10^2 \text{ km.} \quad \textbf{Ans.}$$

10. *Critical Frequency.* The critical frequency of an ionized layer of the ionosphere is defined as the highest frequency which can be reflected by a particular layer at vertical incidence. This highest frequency is called critical frequency for the particular layer and is different for different layers. It is usually denoted by $f_c$.

Mathematically    $f_c = 9\sqrt{N_{max}}$         $[N_{max} = $ Max. Electron density$]$

where $f_c$ is expressed in MHz and $N_{max}$ in per cubic metre. Thus if the maximum electron density $N_{max}$ is known, the critical frequency can be calculated.

**Problems 26.3.** The observed critical frequencies of E and F layers at Chennai at a particular times are 3 MHz and 9.9 MHz respectively. Calculate the maximum electron concentration of the layers.

**Solution.** (*i*) For E layers $f_c = 3$ MHz

We know                    $f_c = 9\sqrt{N_{max}}$

$$N_{max} = \frac{f_c^2}{81}$$

$$\rho = \frac{(3 \times 10^6)^2}{81} = \frac{9 \times 10^{12}}{81}$$

$$= 0.111 \times 10^{22} \text{ m}^{-3} \quad \textbf{Ans.}$$

(*ii*) For F layers          $f = 9.9$ MHz

$$N_{max} = \frac{(9.9 \times 10^6)^2}{81} = \frac{98.01 \times 10^{12}}{81}$$

$$= 1.21 \times 10^{12} \text{ m}^3 \quad \textbf{Ans.}$$

**Problem 26.4.** Calculate the critical frequencies for the $F_1$, $F_2$ and E layers for which the maximum ionic densities are $3 \times 10^6$, $4 \times 10^6$ and $1.7 \times 10^6$ electrons per c.c., respectively.

**Solution.**          $N_{max} = 3 \times 10^6$ cm$^{-3} = 3 \times 10^6 \times 10^{-6} = 3$ m$^3$

$$f_c = 9\sqrt{N_{max}} = 9\sqrt{3} = 9 \times 1.7 = 15.3 \text{ MHz (For } F_1)$$

$$f_c = 9\sqrt{4 \times 10^6 \times 10^{-6}} = 9\sqrt{4} = 9 \times 2 = 18 \text{ MHz (For } F_2)$$

$$f_c = 9\sqrt{1.7 \times 10^6 \times 10^{-6}} = 9\sqrt{1.7} = 9 \times 1.307$$

$$= 11.736 \text{ MHz.} \quad \textbf{Ans.} \text{ (For } E\text{)}$$

11. *Maximum Usable Frequency* (*MUF*). The Maximum Usable Frequency (MUF) is a limiting frequency, which can be reflected back to earth but for some specific angle of incidence. The maximum possible value of frequency for which reflection takes place for a given distance of propagation, is called as the maximum usable frequency (MUF) for that distance, and for the given ionosphere layer. If the frequency is higher than this, the wave penetrates the ionized layer and does not reflect back to the earth.

Normal values of MUF vary from 8 MHz to 35 MHz. However, it may be as high as 50 MHz. At the same time the highest working frequency between two particular points on the earth is obviously a bit less than MUF.

For a sky wave to return to earth, angle of refraction *i.e.*, $\angle r = 90°$,

The Fig. 26.6 shows curved path of a wave in the ionosphere, entering at oblique incidence ($\angle r = 90°$)

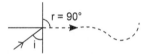

**Fig. 26.6**

Mathematically          $F_{muf} = fc\sqrt{r + \left[\dfrac{D}{2h}\right]^2} = f_c \cdot \text{secant } (i)$

where, $f_c$ = critical frequency, $D$ = distance between points, $h$ = Ionospheric height

This also shows that *muf* is greater than $f_c$ by a factor sec *i*, where *i* is the angle of incidence.

**Problem 26.5.** A high frequency radio link has to be established between two points at a distance of 2500 km on earth's surface. Considering the ionospheric height to be 200 km and its critical frequency as 8 MHz, calculate the MUF for the given path.

**Solution.**          $D$ = 2500 km, $f_c$ = 8 MHz and $h$ = 200 km.

$$F_{muf} = f_c\sqrt{1 + \left[\frac{D}{2h}\right]^2}$$

$$F_{muf} = 8\sqrt{1 + \left(\frac{2500}{2 \times 200}\right)^2}$$

$$= 8\sqrt{1 + \frac{625}{16}}$$

$$= 50.62 \text{ MHz.} \quad \textbf{Ans.}$$

## 26.5 RADIO WAVES

Electromagnetic Wave after being radiated by the transmitting antenna may be divided into various parts. One part travels along the surface of the earth and is called *Ground Wave*. The remainder parts, move upwards called *Skywaves* (Fig. 26.7)

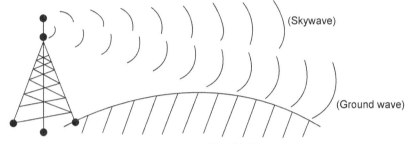

**Fig. 26.7**

Ground Wave further consists of:

(*i*) Surface Wave. (*ii*) Space Wave.

Skywave propagation is used for long distance communication. Ground Wave Propagation is effective for short distance communication at low frequency. Surface Wave is a part of ground wave, which travels in contact with the surface of earth. Space wave travels in the space just above the earth. The space wave is composed of (Fig. 26.8).

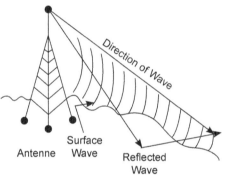

**Fig. 26.8**

(*i*) Direct Wave.

(*ii*) Reflected Wave. Here we give classification of radio waves.

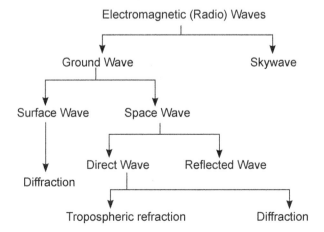

## 26.6 MODES OF PROPAGATION OF RADIO WAVES

From above discussion, There may be following modes of propagation. (Fig. 26.9)

1. Ground wave or surface wave propagation.
2. Skywave or ionospheric propagation.
3. Space wave propagation.
4. Scatter propagation.

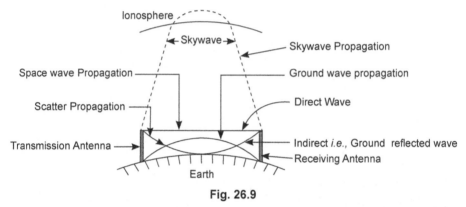

**Fig. 26.9**

Propagation of radio waves is used in radio communication for transmission over short and long distance. Now we will study above four types of propagation:

(a) **Ground Or Surface Wave Propagation (Upto 2 MHz):** The ground wave is guided along the surface of earth. Surface wave permits the propagation around the curvature of earth. The mode is used when the transmitting and receiving antennas are close to the surface of earth. Any horizontal component of electric field in contact with earth is short circuited by the latter. The ground wave propagation along the surface of earth induces charges in the earth, which travel with the wave and hence constitute current. Thus the earth behaves as *leaky capacitor*. The Fig. 26.10 shows equivalent circuit of earth. Wave while passing over the surface of earth, looses some of its energy by absorption. Energy so lost is replenished to some extent by the energy diffracted downward from upper portion of the wave front present immediate above earth surface. The ground wave therefore suffers variable attenuation while propagation along the curvature of earth. It is a more for higher frequencies. The mode is suitable upto 2 MHz. It is medium wave propagation and used for local broadcasting. All the broadcast signals received during the day time are due to ground wave propagation.

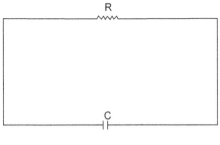

**Fig. 26.10**

## Groundwave propagation at high frequencies

Since surface wave is greatly attenuated at high frequencies, the only alternative left is to use direct and reflected waves. To avoid their cancelling effect, their path difference may be considerably increased by increasing the height of transmitting and receiving antennas.

(b) **Skywave Or Ionospheric Propagation (2 MHz To 30 MHz):** This is used for long distance communication and for medium and short waves. After propagation the waves reach the receiving end after reflection from the ions in the ionosphere of upper atmosphere situated between 50 km to 400 km above the surface of earth. It is used for electromagnetic waves between 2 to 30 MHz. (Fig. 20.11)

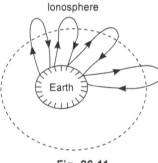

**Fig. 26.11**

The e.m. waves of frequency greater than 30 MHz are not reflected back but penetrates through it. This propagation is also called *short wave propagation*.

As long distance point to point communication is possible so it is also called *point to point wave propagation*. Round the globe communication is possible with multiple reflections of sky waves. In a single reflection, it covers a distance of 4000 km.

Following information is important regarding sky wave propagation:

(a) *Skip distance and Skip zone (Fig. 26.12)*: The skywave after reflection is received at a point much farther from the transmitting antenna. There is no reception in the area between the transmitting antenna and the point where the skywave first reaches the earth. The only source for reception in between the area is the ground wave. The distance between the transmitting antenna and the point where the skywave first reaches the earth is called " skip-distance."

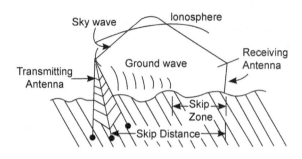

**Fig. 26.12**

The ground wave becomes lesser and lesser significant as we move away from the transmitting antenna. A point comes after which there is no reception due to ground wave. If the point lies somewhere in the skip distance, then in the region between this point and the point where sky wave is received first, there is no reception at all. This region is termed as 'skip-zone.'

(b) *Single and Multihop Transmission (Fig.* 26.13): When a sky wave is received on earth after being reflected from the ionosphere only once, the transmission is termed as single hop transmission. If there is enough power in the signal when it reaches the ground, it may get reflected from the earth and starts moving towards ionosphere to get reflected for the second time.

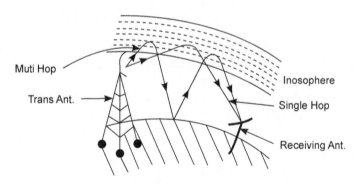

**Fig. 26.13**

This process may be repeated a number of times. Such a transmission is called *multihop transmission*.

Since the circumference of the earth is about 40,000 km, multihop propagation paths are generally occurring (Fig. 26.14). Further there is no problem in south-north multihop propagation paths. However, care is taken far East-West propagation paths.

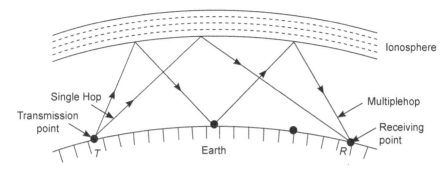

**Fig. 26.14**

With multihop transmission, extremely long distance communication is possible. However, its demerits are that the wave is rendered very weak due to successive reflections and refractions, since each time it undergoes a reflection or refraction, some energy is lost. Another disadvantage is that single hop and multihop waves both of them are arriving at the same point.

(c) **Space Wave (or Tropospheric) Propagation (30 MHz – 300 MHz):** (Fig. 26.15 ) Space wave propagation is of practical importance at VHF (30 MHz to 300 MHz) and for communication like television, radar, etc. In this mode the electromagnetic waves reach receiving antennas in two ways, directly from antenna to antenna and by reflection through ground thus inducing a phase shift of 180° in ground reflected wave. Both the waves leave the transmitting antenna at the sametime but reach the receiving antenna either in phase or out of phase as the path travelled is different. At receiving point the signal strength is sum (or addition) of direct and indirect waves. In space wave propagation the reflection occurs from ground in troposphere region so it is also called Tropospheric propagation.

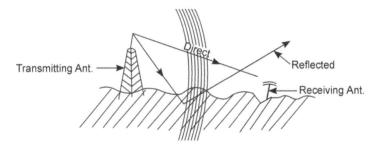

**Fig. 26.15**

The space wave propagation is also known as *Line of sight* propagation. Since the direct and reflected waves travel is straight lines, their use for communication is limited to line of sight conditions otherwise the waves are ineffective. Line of sight means that the transmitting and the receiving antenna must be facing each other without any obstruction in between.

The line of sight propagation has now extended to what is known as "*satellite communication*". This has made possible trans-oceanic propagation of micro waves with their potentiality for large bandwidth. The space communication is applied to radio traffic between a ground station to satellite or between a ground station via natural space bodies (*i.e.*, sun, moon, venus) to man made space bodies (*i.e.*, communication satellite). The propagation in this case is by direct wave, which travels through an isotropic and homogeneous medium notably in outer space in straight line.

(*d*) **Scatter (Or Duct) Propagation (Above 300 MHz):** The UHF and microwaves (e.m. waves above 300 MHz) were found to be propagated much beyond the line of sight through forward scattering in the troposphere. This propagation is possible in VHF (30–300 MHz) as well a UHF (above 300 MHz) region.

In addition, depending upon the condition of troposphere, there may be formation of "Inversion layers" and propagation to a larger distance with lower attenuation in the frequency range of 300 MHz to 30 GHz is possible. This is also called "*Duct propagation*".

Note that, troposphere is an ionized region at a distance of about 15 km from earth's surface. High frequency waves (in UHF range) if beamed at the troposphere get scattered in different directions. Some come back towards the earth (called back scatter) while some are lost in the undesired direction (called lost scatter). A part of it also reaches the receiving antenna placed symmetrically opposite on the other side of the horizon. (See Fig. 26.16). It may be mentioned that the ratio of received power to the radiated power is extremely small and that is why large transmitting powers are required.

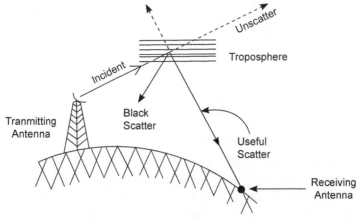

**Fig. 26.16**

The Fig. 26.17 shows two UHF antennas-one transmitting and other receiving and their beams intersect each other at point $O$ above the horizon. The irregularities in the troposphere scatter radio waves due to turbulence and this happens when they are situated in the common volume facing the antennas. At low frequency, the scattering occurs in the back direction and for high frequency forward scattering dominates into a cone of angle θ. Both ionospheric and topospheric scatters sometimes produce an undesirable noise, which may be reduced by diverting reception. The duct *propagation* takes place under the communication path with less attenuation in frequency range 300 MHz to 30,000 MHz.

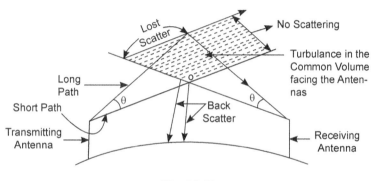

**Fig. 26.17**

## 26.7 FADING

The loss of signal is called fading:

(*a*) In troposphere, Fading is due to

    (*i*) Variation in electrical composition of air.

    (*ii*) Local wind eddies

    (*iii*) Variations in refractive index.

    (*iv*) Variation in earth radius factor.

(*b*) In ionosphere, Fading is due to:

    (*i*) Phase interference

    (*ii*) Change of polarization of incident wave relative to the receiver antenna.

    (*iii*) Absorption of energy by the ions.

## SUMMARY

1. The radio waves are electromagnetic waves which propagate at a velocity of light.

2. A radio wave has electrical as well as magnetic fields.

3. The direction of propagation of radio wave, direction of electrical field and direction of magnetic field all the three are mutually perpendicular.

4. The regions of atmosphere are: Troposhere, Stratosphere and Ionosphere.

5. The important properties of radio waves are: polarisation, attenuation, reflection, refraction, interference, diffraction, etc.

6. When a radio wave is reflected: virtual height is always greater than actual height.

7. The critical frequency of an ionised layer of ionosphere is the highest frequency, which can be reflected back by a particular layer at vertical incidence.

8. The maximum usable frequency (MUF) is the limiting frequency, which can be reflected back to earth for a specific angle of incidence rather than vertical.

9. There are following modes of propagation:

   (*i*)   Ground or surface wave propagation (upto 2 MHz)

   (*ii*)  Skywave or ionospheric propagation (2 MHz to 30 MHz)

   (*iii*) Space wave propagation (above 30 MHz)

   (*iv*)  Scatter propagation (above 300 MHz)

10. When a skywave is received on earth after being reflected from ionosphere only once, this is called single hop transmission. When skywave is reflected repeatedly, this is called the multihop transmission.

□□□

# Radio Navigation

In olden days, long journeys over sea were accomplished with the knowledge of movement of Sun, Moon and other Stars. These days a large variety of navigational systems have been developed for use in different navigation applications.

In this chapter, few navigational aids have been discussed.

## 27.1 RADIO NAVIGATION

Radio navigation is a process or technique used for directing the movement of a vehicle (aeroplane, ship, missile, etc.) from one position to another along the desired trajectory (direction).

## 27.2 DEAD RECKONING NAVIGATION AND INERTIAL NAVIGATION

(*a*) DEAD RECKONING NAVIGATION is one of the earlier techniques used for navigation of ships. In this, the position of the ship is determined from speed and elapsed time since the last position.

(*b*) INERTIAL NAVIGATION is a self contained navigation system and is associated with the use of an *Inertial Measurement Unit* (IMU) to help to determine state of propagation. The IMUs contain accelerometers mounted on a gyro stabilised platform. An accelerometer is required for each direction of possible motion. The accelerometer output can be integrated to determine position. Gyros in the IMUs' can be used to provide information about orientation of the vehicle. The gyros are either conventional electromechanical type or of the laser type.

## 27.3  GENERALISED NAVIGATIONAL SYSTEM

Figure 27.1 shows the basic block diagram of a navigation system. A processing unit known as the *Guidance, Navigation and Computer (GN* and *C)* is the heart of the system. The processing unit receives inputs from the on-board vehicle motion sensors and other navigational aids. It also receives pilot's commands. The processing unit provides information to the pilot in the form of displays and also generates required commands for the control system to keep the vehicle along the desired trajectory.

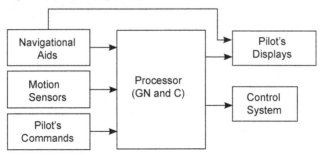

**Fig. 27.1**

## 27.4  IMPORTANT NAVIGATION AIDS

The important navigation aids are:

1.  Long range Navigation (LORAN) system.

2.  Visual Aural range (VAR) system.

3.  Visual Omni range (VOR) system.

4.  Radio direction finder (RDF) system.

5.  Instrument landing system (ILS).

6.  Distance measuring equipment (DME) system.

Here we will discuss above systems in brief.

## 27.5  LONG RANGE NAVIGATION (LORAN)

*Loran* is short-form of 'Long range navigation aid'. It is based on measurement of the difference in the time of arrival of electromagnetic waves from two transmitters to the receiver in the craft (plane, ship etc.). It is also called *hyperbolic system of navigation* (Fig. 27.2).

In this, signals are transmitted from a pair of transmitters based on the shore. One of the transmitters, called *master station*, sends out a pulse in the direction of the plane or ship and also in the direction of *slave station*. The pulse reaches the slave station and triggers it, which sends out its own pulse towards ship.

The receiver finds out the time difference between the time taken by master pulse to reach the receiver and time taken by slave pulse to do the same.

This time difference establishes a line called *line of position* and the ship or plane may be at any of the infinite points on this line. This line of position is a hyperbola because the locus of the point is such that the difference between its distance from two fixed points remains same. Time taken by the pulses is proportional to nothing but the distance only.

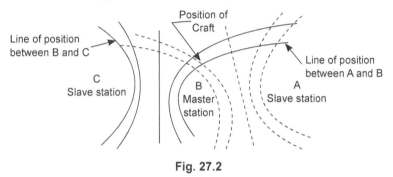

**Fig. 27.2**

The exact location may be determined by drawing another line of position due to a second pair of transmitters. The point of intersection of two lines of position gives the exact location of the ship.

Standard loran operates in the frequency range of 1.8 to 2.0 MHz with a transmitter pack power of about 100 kW. The slave stations are normally at about 325 to 650 km away from the master station and are synchronized by means of ground wave propagation. Ground wave reception is mainly operative in day time and is much better over sea. However, at night, both ground and skywave receptions are possible. The average range for ground waves over sea is about 600 km. The range over land is reduced to about one half. If sky wave receptions are used, the range is appreciably higher.

Accuracy achieved by a loran system is dependent on two factors. The first factor is the accuracy of measurement of time interval, which in turn is dependent on the signal strength. Another factor affecting the accuracy is the synchronization of transmitter stations.

## 27.6 VISUAL AURAL RANGE (VAR) SYSTEM

The VAR system, as the name, it gives Visual as well as Aural (audio) indication.

The operation of VAR system is based on transmitting four different signals on separate carriers by directional antenna each having same frequency. Each antenna transmits the signal in a 180° arc [Fig. 27.3 (*a*)]. Two of the four signals represent letters *N* and *A* respectively, *i.e.*, the signal for *N* is repetition of the signals corresponding to Morse-code for *N*' and signal for *A* is repetition of signal corresponding to Morse-code for A. [Fig. 27.3 (*b*)]. The other two signals are modulated carriers by 90 Hz and 150 Hz tones.

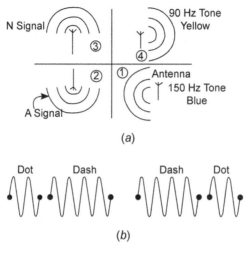

(*a*)

(*b*)

**Fig. 27.3**

If the pilot hears the morse-code symbol for letter *A* and see *yellow* indication, he is in fourth quadrant. Similarly, if he sees a blue indication and hears Morse-code symbol for letter *N*, then he is in first quadrant.

The system cannot give us the exact location of the plane. However, it tells us about the quadrant, the plane is in. The plane can be anywhere in that quadrant. Then the pilot has to take help of other navigational aids to find the exact location.

## 27.7 VISUAL OMNI RANGE (VOR) SYSTEM

In this system, two signals are transmitted on separate carriers and by separate directional antennas. One of the signals has got a variable phase depending upon the direction of transmission, whereas the other has a fixed phase. (Fig. 27.4)

In the aircraft, both the signals are received and their phase difference, which depends upon direction of aircraft from transmitter, is used to indicate the exact bearing position of the plane.

The airborne equipment used to avail of the VOR facility consists of a broad band omnidirectional antenna, a multichannel AM receiver which can be tuned over the desired band and also instrumentation for processing the receiver output for display unit. The frequency band over which the receiver works extends from 108 to 135 MHz covering 560 channels with a separation of 50 kHz each. The receiver directly recovers the 30 Hz amplitude modulation due to pattern rotation. The 9960 Hz subcarrier is selected by a suitable filter and applied to an FM detector to recover the 30 Hz reference signal.

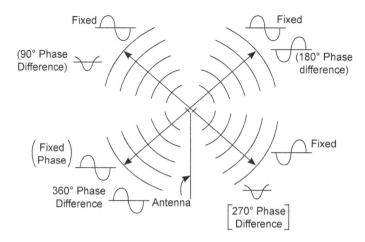

**Fig. 27.4**

A VOR transmitter radiates between 50 to 200 watts. As the frequency of operation is in the VHF range, the range of VOR facility is limited to line of sight only. This range is thus determined by the height of the VOR antenna and the aircraft using it. For an aircraft flying at about 6000 m (20,000 ft), the range of this facility is about 340 km.

VOR system is not entirely free from errors. A few errors can be listed as:

1. Error due to site irregularities.

2. Error due to ground station and airborne equipment.

3. Error due to terrain features.

4. Error due to polarization.

## VOR Vs. VAR

The VOR gives the exact location of the plane or ship, whereas VAR does not. In VOR, indication is visual whereas in VAR, the pilot has to be attentive towards aural (audio) indications also.

Another major defect in VAR is that when the plane lies on the vertical line or the horizontal line dividing the quadrants, there will not be a visual indication or aural indication respectively. In this case signals cancel each other because both are equally dominant. On the vertical line both yellow and blue are having same strength, whereas on horizontal line, morse-coded signals for $N$ and $A$ combine to give a continuous tone.

## 27.8 RADIO DIRECTION FINDER (RDF)

The radio direction finding is the process of using radio signals to find out the relative position of the receiver (on the plane) with respect to the transmitter.

## Principle of Operation of RDF

In RDF system, there is a directional antenna whose directional characteristics are such that it picks up different amplitudes of signals from different directions and when the plane of the antenna is perpendicular to that of the direction of the signal, it picks up zero signal. Therefore, the antenna is rotated after tuning in a particular station so that the signal received drops to the lowest or zero level. In that case, plane of the antenna is normal to the station transmitting the signal (Fig. 27.5).

If the direction of the plane or ship from two known stations is known, the exact bearing can be determined.

Each RDF measurement gives a straight line and the plane can lie anywhere on that straight line. The point where the two lines meet, is the exact place the plane is situated at the time of measurement. (Fig. 27.6).

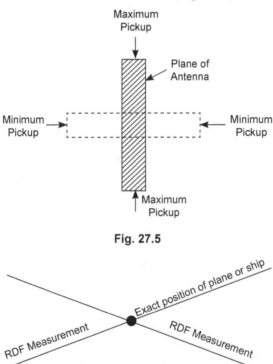

**Fig. 27.5**

**Fig. 27.6**

## VOR Vs. RDF

In VOR (Visual Omni Range), the signals transmitted from the transmitting station are meant for providing navigational aids and nothing else. On the other hand, RDF makes use of any signal, may be from a commercial broadcast station or from some radio beacon or another transmitting station.

For example, if a pilot who is lost, tunes his RDF receiver to signal, which he knows is coming from a particular broadcasting station, he can find out his location.

In early days, the direction of arrival of electromagnetic waves at the receiving point was used for direction finding purposes. Direction finding can be carried out at any frequency, but specific frequencies have been allotted for this service. A number of different types of direction finding systems are available, although the basic principle involved is the same.

## 27.9 INSTRUMENT LANDING SYSTEM (ILS)

This system comprises of three units namely the *localizer, the glide slope* and marker (beacons). A sketch showing the location of these facilities with respect to the runway is given in Fig. 27.7 (*a*).

**The Localizer** gives a vertical equisignal plane which passes through the centre line of the runway while the **glide slope** gives an equisignal plane inclined to the horizontal at the desired angle which generally lies between 2° to 5°. (See Fig. *b*)

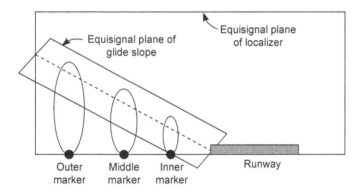

**Fig. 27.7**(*a*)

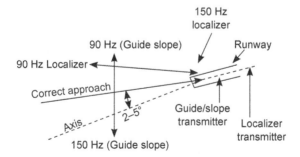

**Fig. 27.7**(*b*)

The intersection of these two equisignal planes gives the *approach path*. Three **markers (beacons)** are installed at certain specified distance from the touchdown point on the runway, helping the pilot to check his position in the approach path.

Thus the pilot is able to find out the exact horizontal and vertical position of plane and makes the adjustments accordingly.

A localizer transmitter is located at about 300 m from the end of the runway. It operates in the VHF band (108–112 MHz). There are two antenna systems. One transmits a carrier, modulated by two sinusoidal signals of 90 Hz and 150 Hz. The other antenna system consists of two arrays. To these arrays, only sidebands of 90 Hz and 150 Hz are applied.

The guide slope operates at about 330 MHz. Glide slope facility provides an equisignal path type guidance in the vertical plane. The transmitter of glide slope system is located on one side of the runway near the touch down point at a distance which varies from 130 to 140 m.

Marker beacons are used to indicate position of the aircraft along the localizer path from the touch-down point. Three beacons are used for each runway. All the three operate at 75 MHz.

## 27.10 DISTANCE MEASURING EQUIPMENT (DME)

Distance measuring equipment is a secondary radar system. A block diagram of DME system is shown in Fig. 27.8.

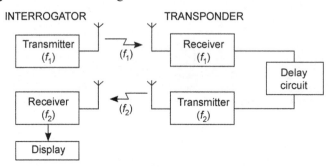

**Fig. 27.8**

An Interrogator is installed on the aircraft while transponder is installed on the ground. The interrogator transmits RF pulses periodically at a frequency $f_1$. These pulses are received by the transponder and are amplified, demodulated and again modulated to be transmitted by its transmitter at another frequency $f_2$. This retransmission is done after introducing some delay in the received signal. In the aircraft the receiver is tuned to frequency $f_1$ and $f_2$; this receiver then calculates the delay between the transmitted and the received pulses to measure the distance between the aircraft and ground station transponder.

The interrogator operates in the frequency band 1025 to 1150 MHz. The transponder operates in the band 962 to 1213 MHz. The frequencies $f_1$ and $f_2$ are fixed and they differ by 63 MHz. The output power of a DME interrogator ranges from 50 W to 1 kW.

The transponder, also called beacon, transmits identification signals. The identification is given in the form of 3 letter Morse code.

The distance calculated is displayed in the aircraft on a meter.

## SUMMARY

1. The radio navigation is the process for directing movement of planes, ships, missiles, etc. The radio waves are used for this technique.

2. The two techniques of navigation used are : Dead reckoning system and Inertial system.

3. The Loran is used to find location of a ship.

4. The VAR does not give exact location but only the quadrant, in which the vehicle lies. The VAR system gives aural as well as visual indication.

5. The VOR gives exact location of the vehicle but the system gives only the visual indication.

6. The RDF can find direction *i.e.*, relative position of a vehicle w.r.t. the transmitter.

7. The ILS helps safe landing of plane on the runway.

8. The distance measuring equipment (DME) is a secondary radar system.

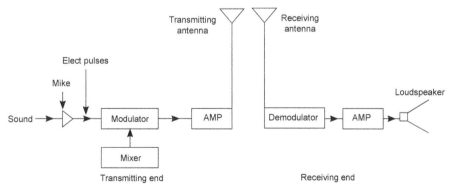

# 28

# Consumer Communication

In this chapter, modern communication will be discussed such as

(*i*) Wireless telephone     (*ii*) Cordless telephone

(*iii*) Pager     (*iv*) Mobile phone

(*iv*) TV remote     (*v*) Video phone

(*vi*) E-mail     (*v*) Internet

These are discussed below one by one.

## 28.1 PRINCIPLE OF WIRELESS/RADIO TELEPHONY

The ordinary telephone system is an example of wire telephony, whereas a pager, mobile, etc. are the examples of wireless or radio telephony, in which signal does not flow through wires but flows through space in the form of radiowaves.

**Fig. 28.1**

When a person speaks before the microphone of his telephone (see Fig. 28.1), his sound is converted into electrical pulses and modulated, amplified and relayed in space. At the receiving end, the same signal is demodulated and amplified. The loudspeaker converts the electrical pulses again into original sound.

## 28.2 CORDLESS TELEPHONE

A cordless telephone basically is a wireless (cord means wire) telephone. We can make/receive a call, while moving in a radius of 30 to 100 m from the point of telephone connections. The wire of the conventional telephone has been replaced by a low power radio transmitter contained in the "portable unit" of the telephone.

A cordless telephone has two parts: (Fig. 28.2)

   (*i*)  *Base*. The Base is connected with the telephone exchange of the area. The line supplying power to the *base* also acts as its transmitting antenna. The *base* has also a receiving antenna.

   (*ii*)  *Portable unit*. The portable unit has a keyboard for dialing. It is supplied power by a chargeable cell which gets charged, while the unit is resting on the BASE. The unit contains a transmitter (1.8 MHz), microphone and a loudspeaker. Like base, the unit also has transmitting and receiving antennas.

*Working*. When we dial a number from keyboard of the portable unit, the signal so produced modulates a carrier and the modulated signal is transmitted through its transmitting antenna to the *base*. The base receives this transmitted signal through its receiving antenna, demodulates and sends to the telephone exchange of the area. Thus a bell rings on the telephone set of the person called.

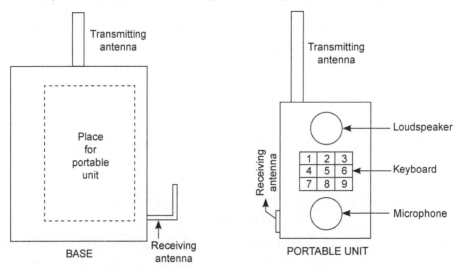

**Fig. 28.2**

When the bell rings, the person picks up his portable unit from the *base* and presses a switch which turns ON the transmitter of his unit. When he speaks before his mike, the signal so produced is modulated and is transmitted through the transmitting antenna. The receiving antenna of his *base* receives the same, demodules and sends to the telephone exchange of the area.

In this way, both the subscribers are connected through the telephone exchange.

The radio frequency of cordless telephone is about 1.8 MHz.

The cordless telephone helps in keeping privacy, as we can move away from *the base* of the telephone. In conventional telephones, such privacy/secrecy is not possible.

## 28.3 PAGER/RADIO BELL OR SUBSCRIBER RADIO

The literal meaning of "paging" is to call a person by sending a "page" (*i.e.*, massage). The pager is a very convenient method of "one way communication" *i.e.*, sending messages. By one way, we mean that only message can be given that some body wants to talk or wants to meet him at some place etc. No immediate reply is possible.

It is a form of one way mobile communication facility and is useful in communicating with a person with an unknown location. The person wanting to communicate with another person with unknown location dials the number alloted to the paging receiver being carried by the person to be located. The call is delivered to the paging transmitter through service provider. The paging transmitter sends out a coded radio signal which is received and decoded by the paging receiver. The moving person then contacts a prearranged telephone number and gets the detailed message.

For this purpose, we can dial the *pager number* of the person (to whom we have to send message) on any ordinary telephone and thus get connected with the "Base station" . The base station then searches that person and sends him the message to contact the caller by telephone or by meeting personally. In emergency, through pager we can call a doctor to attend a patient. Similarly, if a person is within the range of the base station, he can be contacted through pager.

With very economical use of radio spectrum, paging system can keep thousands of people in touch with a small, light weight body worn equipment *i.e.*, pager. When the call is sent on a paging system, the pager carried by the person sounds a "beep" tone.

### The Paging System

The Fig. 28.3 shows a simple block diagram of the "paging system". It basically consists of "Paging control terminal" (PCT). It is heart of the system and is installed at the exchange. The PCT contains CPU (Central processing unit), Input/output processor, coder etc.; moreover a number of radio transmitters are installed at suitable places to take advantage of antenna height and to cover a wide area. The frequency bands available for paging cover VHF as well UHF range (*i.e.* from 30 to 900 MHz).

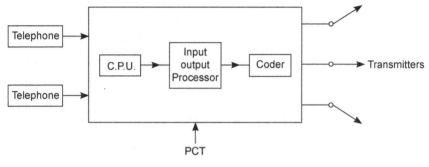

**Fig. 28.3**

## Operation of Paging System (Fig. 28.4)

When a subscriber dials a pager number from any telephone, the PCT (paging control terminal) receives call from PSTN (Public switching telephone network) through the exchange. After decoding, the system asks the subscriber to tell the message.

Note that PSTN makes use of telephone network for switching and data transmission. The digital data is first converted into an analog signal in a *modem* before it is transmitted over the circuit estabished by dialling the number associated with the host or terminal line. The analog signal is converted back to digital data in the modem at the receiving end.

The message received from telephone is modulated (usually frequency modulation is used) and is relayed by the transmitter of the pager system through its transmitting antenna into space. The pager (whose number is dialled) gives a "beep" and the message is displayed on the small screen of his pager. The message should be brief, in one or two lines.

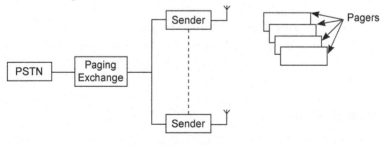

**Fig. 28.4**

## Pager Receiver

The pager receiver is always to be worn on the body. It has a marked effect on its antenna performance (the pager is provided with a loop antenna). It may be kept in the pocket also, however, in all cases it should maintain a contact with the body of the subscriber. A pager is basically a receiving system *i.e.,* it can only receive a call but it cannot make a call.

The Fig. 28.5 shows a simple block diagram of the pager receiver. It has double heterodyne receiver, demodulator, decoder, tone generator, microcomputer, display, etc.

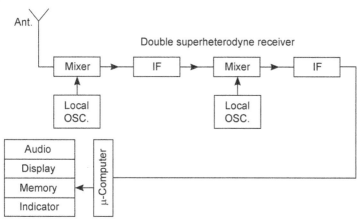

**Fig. 28.5**

The pager is a robust pocket device (Fig. 28.6) and uses generally a single pencil cell for its working. The power consumption is very low, rechargeable cells may also be used. The sensitivity of the pager is about 10 μ.V/m.

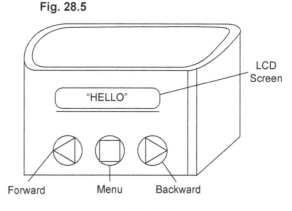

**Fig. 28.6**

The modern pagers include many useful features like:

1. Battery economiser

2. Multitone alert.

3. Out of range indication.

4. Low battery alarm.

5. Message store.

6. Auto or Manual reset.

## Advantages of Paging

Following are the advantages of paging:

1. Persons with pagers are free to move within the paging zones and clients can be in contact randomly.

2. It provides proper professionalism and thus helps in competitive business.

3. It aids in time saving and quick communication with rapid response.

4. It reduces the risk of miscommunication and thus provides better efficiency.

5. It is an economical way of communication with only disadvantage that it is a one way communication.

   Now the Pager is obsolete. Due to one way communication now. It is not in use.

## 28.4  MOBILE/CELLULAR PHONE/CELL PHONE

The mobile phone has replaced cordless telephone, pager etc.

This is an advanced telephone system in which we can call/can be called by a person while moving. Today we have world wide roaming mobiles. The mobile may be a *handset* or fitted in a vehicle, It is also called cellular phone/ cell phone, as the total area is divided into number of small *cells*.

The working principle of this system is same as of a cord less telephone with the difference that it uses high power devices, covers a bigger range and provides service to many subscribers at a time.

Cellular mobile services offer telephone and data communication services even when the calling party or the called party or both are mobile (in motion).

## Terms Related to Cellular Phone Service

Below are given terms which are used regarding the cellular service:

AMPS  —  Advanced mobile phone service

CTS  —  Cellular telephone service

MMC  —  Master mobile centre

MTSO  —  Mobile telecommunication switching office

MSC  —  Mobile switching centre

NCS — Network control system

PSTN — Public switching telephone network

RSG — Remote switch group

SMC — Satellite mobile centre

SIM — Subscriber identity module (smart card)

SAT — Supervisory audio tone.

## Distribution of Area into Cells (See Fig. 28.7)

The total area is splitted in to small parts known as "cells". Each cell covers an area of 15 km and has its own transmitter, receiver and other control equipment and can send/receive calls from the system, moreover each cell has a particular channel to work, however the same channel can be used for more than one conversations. Further, each cell can handle frequencies for two way operation. However, the adjacent cells use different frequencies to avoid interference.

Within cells, all communication is performed using a given bandwidth and centre frequency. We arrange such that in Cell $A$, centre frequency $f_A$ is used in Cell $B$, centre frequency $f_B$ is used, etc. If two cells are widely separated so that a receiving antenna in one cell cannot detect the signal transmitted from the other cell, both cells can be given the same centre frequency allocation.

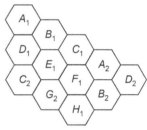

**Fig. 28.7**

In Fig. 28.7 each pair of cells $A_1$ and $A_2$, $B_1$ and $B_2$, $C_1$ and $C_2$ and $D_1$ and $D_2$ use the same centre frequency. This technique which is designed to conserve the spectrum is called *frequency reuse*.

The bandwidth allotted to each cell is divided into $N$ channels thus if the bandwidth of a cell is $B_C$. The bandwidth alloted to each user is

$$B_u = \frac{B_C}{N}$$

In the AMPS (Advanced mobile phone service), the communication is by FM and each user channel is alloted a bandwidth $B_u = 30$ kHz. The total Spectrum allocation per cell is $B_c = 40$ MHz so that $N$ can be calculated.

The mobile unit first communicates with a near by cell site. The cell site is linked to the telecommunication switching office by conventional voice and data circuits (Cell sites and telecommunication switching offices are fixed installations). The telecommunication switching office switches the traffic between the cellular system and rest of the telephone network.

## Base Station

The base station has a high power transmitter, receiver, control unit and other equipment. It has also an elevated antenna. The base system may be owned by government itself or by private companies.

The **transmitter** of the base station contains modulators, amplifiers and the *receiver* contains demodulators, amplifiers, etc., for processing the sound signal. The control unit has logic circuits to maintain a total control on all operations.

## The Handset

Technically it is called a "cell phone" which can be used from any place. But when a cell phone is used in a moving vehicle or it is fitted in a vehicle itself it is called a mobile phone/unit. Note that "mobile" means " in motion" .

A mobile unit has its own transmitter, receiver and controlling equipment. It works on the same way as an ordinary/conventional telephone except that it is through the base station.

A cell phone (handset) consists of—the mobile terminal (phone itself) and the subscriber identity module (SIM), which fits into the handset. Sometime if we do not have our own mobile, with this card, we can use any other

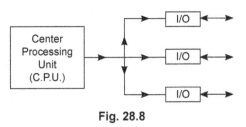

Fig. 28.8

person's mobile, without affecting his billing. Note that the mobile handset should be rugged enough as it may even drop on the ground. The availability of channel is indicated by a visual indication. The mobile unit has 10 digit Keyboard like other telephones.

The Fig. 28.8 shows the block diagram of the hand set. Note that it is essentially a microcomputer. The blocks marked as I/O are input/output devices.

## Working of a Cell Phone System

For making a call, when a cell phone is switched ON, a request automatically reaches to the *cellsite*, which interprets the coming call and sends digital signal to MTSO (Mobile Telecom Switching Office) which connects the subscriber to the destination. If the user is travelling while calling, the cellular system automatically switches the cell to another cell as and when the caller moves from one area to the another.

The receipt of the call is just opposite. The central office acknowledges the caller and connects the concerned MTSO. A channel is made available and a link is established between "the caller" and "the called".

## Cell SiteTransmitter

The Fig. 28.9 shows block diagram of the cell site transmitter.

According to the instructions obtained from CPU of the cell phone, the frequency generator (FG) generates the desired channel frequency. This signal then reaches to the exciter, where the process of modulation occurs. The output is then amplified in stages—finally it goes to the power amplifier (P.A.) and to the antenna as shown.

**Fig. 28.9**

## Receiver Block Diagram

The Fig. 28.10 shows the block diagram for the receiver.

The signal received from antenna goes to the high frequency unit (HFU) where it is preamplified and shifted to the frequency generator (F.G.). It is then lowered by intermediate frequency unit (I.F.U.) to I.F. range and lowered further by Audio Frequency Unit (A.F.U.) to the A.F. range for its further transmission to the MTSO (Mobile Telecom Switching Office).

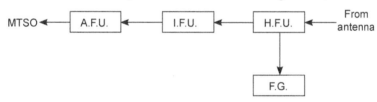

**Fig. 28.10**

The modem mobiles provide the following (and many other) facilities:

**Video:** Video clips of sports, movies, news, cartoons and a lot more.

**Audio:** Songs on your handset in Hindi, English and regional languages.

**Surf:** Access a wide range of interesting sites like Yahoo and Rediff and experience the difference.

**News:** The latest business, sports and weather news in 6 major Indian languages.

**Messaging:** Use SMS, e-mail and a range of greetings. Download from hundreds of pictures, add your text and send to loved ones.

**Panchaang:** Astrology, Hora, Rahukalam and Numerology.

**Convenience:** Access, whenever you need to.

**Games:** Access vast library of games, download and play.

**Finance:** Keep a tab on you stocks through a frequently updated ticker.

All modern modern mobile phones have internet facility.

Figure 28.11 shows, few design of mobile phones.

**Fig. 28.11**

**Wireless loop (WLL):** This is a mobile with limited mobility. Generally, this does not have "roaming" and S.M.S. (Short massaging service) facilities. The W.L.L. works under C.D.M.A. (Code division multiple access), while a mobile works under G.S.M. (Global system for mobiles) technology.

## 28.5 TV REMOTE CONTROL

The TV remote was invented by Mr. Eugene polley. The different functions to be controlled in TV receivers are:

1. ON/OFF control

2. Channel selection

3. Volume control, etc.

**Note that,** the **contrast** and **brightness** are not included in remote control, because the AGC circuit automatically changes the gain of the TV receiver to maintain desired control of contrast and brightness.

The Fig. 28.12 shows working of remote control. It has two parts:

(a) *Transmitter* (*Remote Control Handset*): The Fig. 28.12 (a) shows the basic principle of remote control. The remote control handset which acts as a transmitter is used from a distance from the TV receiver and it sends out an electrical signal which is intercepted by a transducer (microphone) provided in the TV receiver. The nature of signal received is intercepted by a "sensor" for the particular control required. The received signal is processed to control electronic units. In electronic system, d.c. voltages are used to vary the controlling function. For example, for volume or channel control, operating bias of the amplifiers is varied, which forms part of the circuit.

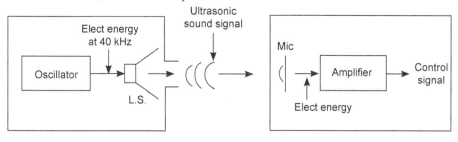

(a) Transmitter (Remote control handsel)                    (b) TV receiver

**Fig. 28.12**

The remote control handset generates **ultrasonic** waves of frequencies around 40 kHz. In the air, these waves are directed towards the TV receiver. The use of ultrasonic frequencies has the advantages that they are out of reach of the human ears and do not interfere TV receivers in neighbouring rooms.

Each controlling function needs a different frequency. An oscillator can be used to generate the required ultrasonic frequency. The typical (approximate) value of the frequencies for few controls are :

| Channel up ↑ | 42 kHz |
|---|---|
| Channel down ↓ | 41 kHz |
| Volume up ↑ | 37 kHz |
| Volume down ↓ | 43 kHz |

The transmitter may use a **Hartley oscillator**. When any push button at the remote is pressed, different capacitors get connected across the secondary of its output transformer, which changes the resonant frequency of oscillator's "tank circuit" and the required frequency output is obtained from the oscillator for the particular function to be controlled. The output of the oscillator is fed to the ("Piezo electric") loudspeaker, which converts the output into ultrasonic sound energy and radiates it towards the TV receiver.

(b) **TV Receiver** [Fig. 28.12 (b)]. In the TV receiver, the ultrasonic energy is picked up by a ("Piezo electric crystal) microphone", which converts the sound energy into electrical energy. The microphone uses "Barium Titanate" crystal which when "strained" by the striking sound along one "axis", a proportional voltage of the same frequency is generated along the other "axis" . The output of the microphone is given to the amplifier. The output of the amplifier is used to provide various controls (Volume, Channel selection, etc.)

Now-a-days for remote control, digital I.C.s are used. The recent development in high speed *I.C. counters* have made the remote controls very sophisticated.

## 28.6 VIDEO PHONE

The Video (or picture) phone is nothing but a telephone plus T.V.

The Fig. 28.13 shows a typical video phone, which is an I.S.D.N. (Integrated Services Digital Network) phone and is a combined unit of a $h_i f_i$ (high fidelity) telephone and video conferencing: By this unit, we can telephone on I.S.D.N. (Integrated Services Digital Network) as well as P.S.T.N. (Public Switching Telephone Network) facilities.

**Fig. 28.13.**

It has a 5 inch L.C.D. screen, on which we can see the picture (photograph) of the person called as well our own picture while calling *i.e.*, it has "picture in picture" facility. It is a mobile unit and can be shifted to any place. It has a simple key board. We can also install a large size monitor on the unit, if required.

Now a days this facility is provided in mobile phones itself.

## 28.7 E-MAIL

It is an electronic mail often referred to as E-mail. It is the transmission of textual data from one place to another using electronic means for capture, transmission and delivery of information. In simple words, an electronic mail system is a method of sending messages, mail or documents. It is also known as electronic delivery system or electronic documents distribution/communication system.

Electronic Mail refers to transmission of messages at high speed over communication channels. The simplest form of computer based mail system allows one user of the service to send a message to another. The second user at his or her convenience retrieves the message on his or her display monitor or by taking a hard copy of it. The mail can be duplicated, revised, incorporated into other documents, passed on to new recipients or filed like any other document in the system.

E-mail is rather a service with various facilities like:

(a) *Storing and Forwarding messages.* Messages are given to the recipient on request and person to person contact is not required.

(b) *Advice delivery.* The sender can be confirmed of the reception of message to the recipient and an immediate reply can also be demanded.

(c) *Off time working.* During one's absence the incoming messages can be stored.

(d) *Gateway.* E-mail services include access to other facilities like Telex system, on-line information services, etc.

(e) *Closed user group.* The use of E-mail service can be restricted to smaller or larger areas.

Apart from the above services, E-Mail also offers **Radiopaging** *i.e.,* you will hear a beep (audible tone) while receiving an urgent message, **Telemessages** *i.e.,* replacement of old Telegram system. *Message translation i.e.,* message can be translated to the recipient's own tongue and at the recipient's end.

We can conclude from the above that E-mail system is actually a message handling system. The Fig. 28.14 shows block diagram of E-mail.

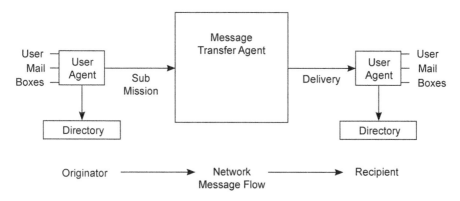

Fig. 28.14

## Advantages of E-mail

1. Message can be sent at whatever time or day suits to the user. These can also be prepared in advance for subsequent despatch at convenient times.

2. Messages will be in the recipient's mailbox within minutes.

3. No need to speak to the recipient in person.

4. Delivery of messages can be confirmed.

5. Incoming messages can be saved.

6. E-mail reduces the volume of paper that is to be processed.

7. Telex services can be provided as part of the E-mail facility.

8. On-line information services may also be available.

## Drawbacks of E-mail

1. Recipients must also be E-mail users.

2. Only text can be sent.

3. Text formatting is restricted to the basic punctuation and alphanumeric characters.

4. Until a mailbox is checked, there is no way of knowing that a message has arrived.

As told the modern mobile phone has E mail facility.

## 28.8 INTERNET

The Internet was conceived in 1969.

By 1987, it was called as internet because by this time it was made available to all who want to access it.

Thus, the Internet has grown into an immense network of computers and wires interconnecting them. There are millions of computers connected to this "network of networks", sepanning thousands of computer designs, operating various kinds of connections including coaxial cable, optical fibre, etc.

The Internet is a computer network made up to thousands of networks worldwide. No one knows how many computers are connected to the internet, or how many people use it, though it is certain that both are expanding at a rapid rate. No one is **incharge** of the Internet. There are organizations which develop technical aspect of this network, but there is no governing body in control.

An Internet is a group of computers connected mutually for exchanging information and sharing equipment. In case something goes wrong with any part of the internet, information finds an alternative route around the crippled computers in order to reach its goal. That is why the internet is also called the **Information Superhighway** or **Cyberspace**.

## Equipment Needed for Internet (Fig. 28.15)

We need the following equipment for internet.

1. *Computer*. A compatible PC (Personal Computer) can be used for an Internet connection.

2. *Programs*. We require special programs to use the Internet. These programs are given free of charge by most service-providers.

3. *Modem*. A modem serves as a medium to exchange information between a computer and the Internet.

   Modem is a device by which computers exchange information over telephone lines. The word "Modern" is short for modulator/demodulator.

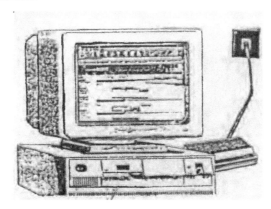

**Fig. 28.15**

4. *Telephone line*. All Internet-information travels over telephone lines.

5. *ISDN line*. Integrated Services Digital Network (ISDN) is a line by which data is transmitted over digital telephone lines. ISDN is two to four times quicker than the most speedy modems. Use of ISDN lines on the Internet's World-Web helps to transmit or receive text, graphics, sound and video. Many telephone companies offer ISDN lines.

## Utilities of Internet

1. *E-Mail (Elcctronic-Mail)*. The internet enables us to exchange messages throughout the world with people, friends, colleagues, relatives and even strangers, we happen to meet on the Internet.

2. *Information*. Any information on any subject can be obtained easily on the *Internet*. We can have a good information from *newspapers, magazines, academic papers, government documents, famous speeches, recipies, works o f literary figures* and what not.

3. *Entertainment*. The Internet offers hundreds of simple games to various people free of cost. Children as well as grown-ups can enjoy *chess, football, snooker, carrom-board, cards, lawn tennis* and such other games. We can have a good look at current movies or listen to over 1,000 television theme-songs. We can have several conversations even with famous personalities of the world.

4. *Programs*. The Internet offers thousands of free programs. Some of these programs are World Processors, Spreadsheets and Games.

5. *On-line shopping*. We can order desired goods and services on the Internet just sitting comfortably in our home. Items *like flowers, books, cars, computer programs, music, CD's, pizza* and many others can be brought.

6. *Finding people*. If you have lost track of some body, you can find him or her anywhere in the country.

7. *Finding businesses products and services*. The yellow Page directory services enable you to search the type of company you're looking for. You can indicate a code to specify the location. People are shopping for that hard-to find special gift item.

8. *Education*. School teachers coordinate projects with classrooms all over the globe. College students and their families exchange e-mail to facilitate letter writing and keep the cost of phone calls down. Students do research from their home computer. The latest encyclopedias, are online.

9. *Healing*. Patients and doctors keep up-to-date with the latest medical findings, share treatment, experience, and give one another support around medical problems.

10. *Investing*. People buy stock and invest money. Some companies are online and trade their own shares.

The significance of the internet lies not so much in where it is today, but in where it will be in the next five or ten years of time, it is still a technology in its high future prospects of extensive usage in its remarkable early success.

**Difference between the Internet (NET) and the World Wide Web (WWW).**

The Internet is the maze of phone and cable lines, satellites and network cables that interconnect computers around the world. The www is all that which can be accessed on the Internet using a Uniform Resource Locator, (URL). For example, www.goodreading.com. You can use the Internet without using the Web, for example, when you send e-mail.

## SUMMARY

1. Cordless telephone is a wireless telephony which works on the principle of radio telephony. The radio frequency of cordless phone is 1.8 MHz.

2. Pager is a convenient device of one way communication, *i.e.*, only a message can be given that some body wants to talk, no immediate reply is possible. Now the pager obsolete.

3. The mobile is an advanced telephone by which we may call a person while moving around. The roaming mobile may have theoretically an infinite range.

4. The E-mail is a transmission of data/message from one place to another using electronic means.

5. The Internet is a computer network made up to thousands of networks worldwide. Infinite no. of computers are connected to the internets. It is the largest computer system in the world.

6. All modern mobile phones are equipped with Internet.

□□□

# Appendices

## A. ABBREVIATIONS

| | |
|---|---|
| AF | Audio Frequency |
| AM | Amplitude modulation. |
| Amp | Ampere or Amplifier |
| A3 | Double-sideband, full carrier AM. |
| A3A | Single-sideband, reduced-carrier AM. |
| A3B | Independent-sideband AM. |
| A3H | Single-sideband, full carrier AM. |
| A3J | Single-sideband, suppressed-carrier AM. |
| A5C | Vestigial-sideband AM. |
| AIR | All India Radio. |
| ADF | Automatic Direction Finder. |
| AFC | Automatic Frequency Correction (or control) |
| AGC | Automatic Gain Control. |
| AOS | Atlantic Ocean Satellite. |
| ARC | Automatic Request for Correction. |
| ARQ | Automatic Request for Repetition. |
| AVC | Automatic Volume Control. |
| ADM | Adaptive Delta Modulation. |
| ASK | Amplitude Shift Keying |
| ASCII | American Standard Code for Information Interchange. |

| | |
|---|---|
| BBC | British Broadcasting Corporation |
| BD | Breakdown |
| BE | Base emitter |
| BW | Band Width |
| BFO | Beat-Frequency Oscillator. |
| BMEW | Ballistic Missile Early Warning. |
| BWO | Backward-Wave Oscillator. |
| Baiun | Balance-to-unbalance transformer. |
| Bit | Binary digit. |
| B & W | Black & White (monochrome) TV. |
| CB | Common Base or Collector Base |
| CE | Common Emitter or Collector Emitter |
| CC | Common Collector |
| CCi | Current Gain |
| CT | Centre Tap or Current Transformer |
| Ckt. | Circuit |
| Ch. | Characteristic |
| CCTTT | Consultive Committee for Integrational Telephone & Telegraph (Code). |
| CFA | Crossed-field Amplifier. |
| CRT | Cathode-Ray Tube. |
| CRO | Cathode-Ray Oscilloscope. |
| CCIR | Comite Consultative International Radio. |
| COMSAT | Communications Satellite (Corporation). |
| CTE | Channel Translating Equipment. |
| CW | Continuous Wave. |
| Codan | Carrier-operated device antinoise. |
| Coho | Coherent Oscillator |
| Compander | Compressor-expander. |
| CW | Continuous Wave |
| CATV | Cable TV. |
| CCTV | Close Circuit TV. |

| | |
|---|---|
| DC | Direct Current or Direct Coupled |
| DPCM | Diff. pulse Code Modulation. |
| DF | Direction Finder. |
| DECI | Decimal Digit |
| Domsat | Domestic Satellite (System). |
| DM | Delta Modulation. |
| EB | Emitter Base |
| Eq. | Equivalent or Equation. |
| ESBAR | Epitaxial Schottky Barrier (diode) |
| ECCM | Electronic Counter-Counter Measures. |
| EHT | Extra High Tension |
| E. mail | Electronic mail. |
| FB | Feedback or Forward Bias |
| FOM | Figure of Merit |
| FDM | Frequency-Division Multiplexing. |
| FDMA | Frequency-Division Multiple Access. |
| FHP | Fractional Horse Power. |
| FM | Frequency Modulation. |
| FMVFT | Frequency-Modulated Voice-Frequency Telegraph. |
| FSK | Frequency-Shift Keying. |
| FCC | Federal Communications Commission. |
| Fax | Fascimile |
| GCA | Ground Controlled Approach. |
| Ge | Germanium |
| GTE | Group Translating Equipment. |
| HW | Half Wave |
| HF | High frequency |
| IC | Integrated Circuit |
| IF | Intermediate Frequency |
| I/P | Input |
| IR | Infra Red |
| IMPATT | Impact Avalanche and Transit-time (diode). |

| | |
|---|---|
| INSAT | Indian National Satellite. |
| INTELSAT | International Telecommunication Satellite (Consortium). |
| ITU | International Telecommunication Union. |
| ISRO | Indian Space Research Organisation. |
| IFF | Identification Friend or Foe. |
| ILS | Instrument Landing System. |
| IOS | Indian Ocean Satellite. |
| ISB | Independent Sideband (modulation) |
| ISDN | Integrated Services Digital Network (Computer Communication). |
| IMU | International Measurement Unit. |
| LORAN | Long Range Navigation. |
| LE | Line Equipment. |
| LSA | Limited Space-charge Accumulation. |
| LASER | Light Amplification by Stimulated Emission of Radiation. |
| LAN | Local Area Network. |
| MEWS | Missile Early Warning System. |
| MFC | Multi-frequency Coding. |
| MTBF | Mean Time Between Failures. |
| MTl | Moving-Target Indicator (radar). |
| MUF | Maximum Usable Frequency. |
| MV | Multivibrator |
| MESFET | Mesa Field-Effect Transistor. |
| MARISAT | Maritime Satellite (system) |
| Maser | Microwave Amplification by Stimulated Emission of Radiation. |
| MODEM | Modulator-demodulator. |
| NASA | National Aeronautics and Space Administration. |
| NTSC | National Television Standards Committee. |
| OFC | Optical Fibre Cable. |
| OOK | On-Off Keying. |
| OC | Open Circuit |
| OSC | Oscillator |

| op. | Operation or Operating |
|---|---|
| O/P | Output |
| OCS | Overseas Communication Services (India). |
| OTC (A) | Overseas Telecommunications Commission (Australia). |
| PLCC | Power Line Carrier Communication. |
| PAM | Pulse-Amplitude Modulation. |
| PCM | Pulse-Code Modulation. |
| PG | Power Gain |
| PA | Power Amplifier or Public Address System |
| PT | Potential Transformer / Picture Tube |
| PM | Phase Modulation. |
| PPM | Pulse-Position Modulation. |
| PSK | Phase-Shift Keying. |
| PTM | Pulse-Time Modulation. |
| PWM | Pulse-width Modulation. |
| PFN | Pulse-Forming Network. |
| PLL | Phase-Locked Loop. |
| POS | Pacific Ocean Satellite. |
| PPI | Plan Position Indicator. |
| PPM | Periodic-Permanent Magnet/Parts Per million. |
| PRF | Pulse-Repetition Frequency. |
| PIN | P-Intrinsic-N (diode). |
| PRR | Pulse Repetition Rate. |
| PSTN | Public Switching Telephone Network. |
| RB | Reverse Bias |
| RF | Radio Frequency or Rupple Factor |
| RDF | Radio Direction Finding. |
| R.F. | Radio Frequency. |
| Radar | Radio Detection and Ranging. |
| SC | Short Circuit |
| Si | Silicon |
| SLV | Satellite Launch Vehicle. |
| SAGE | Semiautomatic Ground Environment. |

| Stalo | Stable Oscillator (in radar). |
|-------|-------------------------------|
| SWR | Standing Wave Ratio. |
| SSB | Single Sideband |
| SIM | Subscriber Identity Module (mobile card). |
| TE | Transverse Electric. |
| TEM | Transverse Electromagnetic. |
| TM | Transverse Magnetic. |
| Tr | Transistor |
| T/F | Transformer |
| TR | Transmit-Receiver. |
| TRF | Tuned Radio Frequency. |
| TWS | Track While Scan. |
| TDM | Time-Division Multiplexing. |
| TDMA | Time-Division Multiple Access. |
| TED | Transferred Electron Device. |
| TEO | Transferred Electron Oscillator. |
| TRAPATT | Trapped Plasma Avalanche Triggered Transit (diode). |
| TWT | Travelling Wave Tube. |
| USB | Upper Side Band. |
| UHF | Ultra High Frequency. |
| UV | Ultra Violet |
| VG | Voltage Gain |
| VAR | Visual Aural Range. |
| VOR | Visual Omni Range. |
| VTM | Voltage-Tunable Magnetron. |
| VCO | Voltage-Controlled Oscillator. |
| VFO | Variable-Frequency Oscillator. |
| VHP | Very High Frequency. |
| VSB | Vestigial Side Band (transmission). |
| WB | Wein Bridge |
| WLL | Wireless Local Loop (Mobile phone) |

# B. GLOSSARY

### ACTIVE HOMING

In active homing, the missile has its own radar. The missile homes on to the reflected energy from the target.

### ACCURACY

The accuracy of a D/A converter is the difference between the actual analog output and ideal expected output when a given digital input is applied.

### ADAPTIVE RADAR

In an adaptive radar, the operational parameters of the radar such as PRR, Pulse width etc. are varied within limits as a function of radar's performance on variety of targets.

### AGC

AGC stands for Automatic Gain Control. In this technique, a control input to an amplifier is used to control its gain to keep the amplifier output constant irrespective of the changes in the input signal amplitude.

### AMPLITUDE MODULATION (AM)

In amplitude modulation, amplitude of carrier is varied in accordance with the amplitude variation of the intelligence to be transmitted. The carrier frequency remains constant.

### AMCW RADAR

The transmitted signal is amplitude modulated CW signal. It is a CW (Continuous Wave) radar that makes use of two slightly differing CW transmissions. The range is determined from the phase difference between the two received doppler signals. The phase difference is linearly proportional to the distance and is unambiguous upto a phase difference of 2.71 radians.

### AM DEMODULATOR

It is an electronic circuit that demodulates an amplitude modulated signal fed to it. It recovers the modulating signal from the modulated signal.

### AND-GATE

The output of an AND-GATE is logic '1' only when all of its input are in logic '1' state. For all other possible input combinations, the output is a logic 'O'.

### ANTENNA GAIN

The antenna gain is defined as the ratio of voltages produced at a given point by the actual and the hypothetical antennas. The hypothetical antenna is nothing but an omni-directional antenna.

### ANTENNA RECIPROCITY

The phenomenon of using the same antenna for transmission as well as for reception is known as antenna reciprocity.

### ASCII

Stands for American Code for Information Interchange. This is a 7-bit code and is the most commonly used code all over the world. It can be used to represent up to 128 characters including all alphabet (both lower case and upper case), numeric digits and a number of special characters.

### ASPECT RATIO

The ratio of width to height of the picture frame is called aspect ratio.

### ASIC

ASIC (an abbreviation for Application Specific Integrated Circuit) is an IC belonging to LSI or VLSI category that has been designed for specific applications.

### ASYNCHRONOUS TRANSMISSION

It is a transmission mode in which each information character is individually synchronised by the use of start and stop elements.

### ATM

ATM stands for Automatic Teller Machine. An ATM is a computer terminal that allows individual customers to carry out transactions without any human intervention.

### AUDIO TRANSFORMER

It is a transformer used to transform electrical signals spread out in the audio frequency range from one circuit to another.

### AVC

AVC stands for Automatic Volume Control. It is used to keep the strength of the received signal at the output of the detector as constant irrespective of the strength of signal received.

## BAUD

It is a unit of measurement of signalling speed and is equal to the number of signal events per second, or bits per second.

## BCD

Stands for Binary Coded Decimal representation. This is a system of representing decimal numbers in the form of binary numbers.

## BISTATIC RADAR

In the bistatic radar system, separate antennas are used for transmit and receive functions. In a typical bistatic radar set up, these antennas may even be miles apart.

## BIT

BIT stands for Binary Digit and is the smallest unit of information. It is either 0 or 1.

## BLIND RANGES (Radar)

When a target lies at such a distance from radar that the radar receives the echo while it is still transmitting, it fails to detect the target. In such a case, the target is said to be at a blind range.

## BODY STABILIZED (Satellite)

A satellite whose solar panels are fixed always to face the sun while antenna and body stabilization is provided through a spinning wheel.

## BOOLEAN ALGEBRA

It is an algebra which has been named after its inventor, George Boole. This algebra is quite similar to ordinary algebra but handles only two valued variables, known as Boolean variables, which can take on one of the two values TRUE or FALSE.

## BPI

Stands for Bits Per Inch. On a disc, data are recorded serially on tracks. On a tape, data are recorded in parallel on several tracks. BPI tells us about the storage density of any mass storage device.

## BYTE

A string of 8 bits is called a byte. Byte is the basic unit of data operated on as a single unit in computers.

## CAD/CAM SYSTEMS

CAD/CAM stands for Computer-Aided-design and Computer-Aided-Manufacturing systems. These systems usually have a CRT display, a keyboard, plotter and some graphics I/O devices.

## CAPTURE RANGE

Capture range is frequency range over which a VCO output can acquire lock with the input frequency.

## CARD READER

Card reader is an input device which is used to read the punched cards and the data read is then transferred to the computer for processing.

## CAT (or CT)

CAT stands for Computerised Axial Tomography. It is a type of non-invasive testing that combines X-Ray techniques and computerised processing to carry out medical diagnosis.

## CCD

The CCD (Charge Coupled Device) is a storage element structured to offer a serial access. It can be regarded as a stretched enhancement MOSFET with a string of gates between source and drain.

## CHIRP RADAR

See Pulse Compression Radar.

## CHROMATICITY

The chromaticity diagram is a graphical representation in space co-ordinates X, Y, Z, of all spectra! colours. The colours which are produced as a result of additive mixing of these colours based on values of the component colours are also represented in this diagram.

## CHROMINANCE SIGNAL

Chrominance signal is the electrical signal that represents colour information of the scene to be televised.

## CITIZEN BAND

The range of frequencies from 26.96 MHz to 27.41 MHz allocated to private citizens for short-range radio communication.

## CO-AXIAL CABLE

A co-axial cable is a two conductor transmission line with a center conductor surrounded by a braided shield. The inner conductor is supported by some form of dielectric insulation.

## O-AXIAL RELAY

A co-axial relay is terminated into some kind of RF connector. These are extensively director RF switching operations of equipment interconnected by cables.

## CLOCK

It is an electronic device that generates periodic signals and is used to control the timing of all the CPU operations.

## COMMAND GUIDANCE

It is a technique for missile guidance. In command guidance of missiles, there are two separate tracking radars, one tracking the missile and the other tracking the target. The computer sends guidance commands to the missile continuously till it intercepts the target.

## COMPANDING D/A CONVERTER

Companding type D/A converters are so constructed that the more significant bits of the digital input have a larger binary relationship than to the less significant bits.

## COMPANDING (PCM)

Companding mean compressing the signal at the transmitter and expanding it at the receiver. The signal is modified at the transmitter by artificially making the smaller signals immune to quantizing noise by increasing the amplitudes of smaller signals. The original amplitudes are restored at the receiver.

## CONVERSION TIME (A/D Converters)

It is the time that elapses from the time instant of start conversion signal until the conversion of complete signal occurs. It ranges from a few nanoseconds to microseconds.

## COPROCESSOR

It is a microprocessor that can be plugged into a microcomputer to replace or work with microcomputer's original (or main) microprocessor.

## CROWBARRING

Crowbarring is a type of over voltage protection for the power supplies. Whenever the output voltage reaches a certain preset limit, the power supply output is short circuited by an SCR which takes the short circuit current through it.

## CRT

Stands for Cathode Ray Tube. It is an electronic device, upon which the information can be displayed.

## dBM

dBM is nothing but decibels referred to 1 milliwatt. It is used as a measure of signal power in communication circuits. Zero 0BM equal 1 milliwatt into a specified impedance, often 600 ohms.

## DECIBEL (DB)

A logarithmic measure of ratio between two powers.

## DECODER

A decoder is a circuit that converts information from n-inputs to a minimum of 2 output lines.

## DIFFERENTIAL PCM

In differentia! PCM, as compared to the conventional PCM, only the relative amplitudes 01 various sample and not the absolute amplitudes are indicated.

## DELTA PCM

It is similar to differential PCM. In this technique, only one bit is transmitted per sample just to indicate whether the sample in question is larger or smaller in amplitude than the immediately preceding sample.

## DEMULTIPLEXER

It is exactly the reverse of a multiplexer. It receives data from one high speed line and distributes it to one of the low speed output line

## DIFFERENTIAL NON-LINEARITY(DNL-A/D Converters)

It indicates the worst case difference between the actual analog voltage change and the ideal LSB (Lower Side Band) voltage change. DNL is also expressed in percentage of full scale.

## DIGITIZER

'A digitizer (or scanner) is a direct entry input device that can be moved over text, graphics, maps, drawings, pictures etc, to convert them into digital data. This is used to find coded price of books etc.

## DOPPLER EFFECT

Doppler effect implies that the frequency of a wave when transmitted by a source is not the same as the frequency of the same wave when picked up by a receiver if there is a relative motion between the transmitter and the receiver.

## ECHO SUPPRESSOR

Echo suppressor is a device or a circuit that allows transmission only in one direction at a time. Echo suppressors are used in telephone circuits to attenuate echo on long distance circuits.

## EBCDIC

Stands for Extended Binary Coded Decimal Interchange Code. It is an eight-bit code.

## EDTV

EDTV stands for Enhanced Definition Television. It employs newer encoding techniques and a compatible transmission to achieve a resolution better than the conventional system.

## ENCODER

Encoder is any device which modifies information into the desired pattern or form for a specific method of transmission.

## EQUALIZATION

Equalization is the process of reducing the effects of amplitude, frequency and/or phase distortion of a circuit.

## FACSIMILE (Or Fax)

Facsimile is a method of transmitting pictures, printed pages or film to a remote location, where the transmitted information may be reproduced in the hard copy form.

## FADING

Fading phenomenon is the result of electromagnetic waves reaching the receiving antenna via two different paths.

## FIDELITY

The fidelity of a receiver represents the variation of output with modulation frequency when the output load impedance is a pure resistance.

## FLAT PANEL DISPLAY

It is an alphanumeric or graphic terminal that uses an LCD type or an electroluminescent type display.

## FLOPPY DISKETE

It is a thin, flexible circular plastic plate, coated with magnetic material and enclosed in a card board jacket. It is used as a secondary storage device.

## FM ALTIMETER

In the modulated signal, the instantaneous frequency depends upon the time that elapses alter the start of the pulse. The frequency of the received echo thus gives an indication of the height of the transmitter above ground.

## FMCW RADAR

In an FMCW radar, the transmitted signal is a frequency modulated CW (Continuous Wave) signal. It can be used for measurement of both the target velocity and range.

## FREQUENCY MODULATION

It is, the modulation technique in which frequency of the carrier varies in accordance with the amplitude of the modulating signal. The rate of change of frequency is proportional to the frequency of the modulating signal.

## FREQUENCY SHIFT KEYING

It is a form of frequency modulation commonly used in low speed MODEMS in which the two states of the signal are transmitted as two separate frequencies.

## FULL-DUPLEX

Full duplex refers to a communication system or equipment that is capable of simultaneous two-way communications.

## FM DISCRIMINATOR

It is a circuit that converts signal frequency variation to corresponding amplitude variation fed at its input. It demodulates a frequency modulated signal.

## FREQUENCY DIVISION MULTIPLEXING

Frequency Division Multiplexing is the process of utilising the frequency scale for simultaneous transmission of more than one signals on the same carrier.

## GEOSTATIONARY

It is a satellite orbit where the speed of rotation of satellite is equal to the speed of earth's rotation with the result that any spot on earth within the coverage of the satellite remains in the same relative position with respect to the satellite.

## GPS

The GPS (Global Positioning System) is a navigational aid implemented with a series of satellites. System can be used to fix user's position on earth with an accuracy of few meters.

## HALF DUPLEX

Half duplex refers to a communication system or equipment that is capable of providing communication in both directions but in only one direction at a time.

## HDTV

HDTV stands for High Definition Television. It is the new generation TV technology, displaying images with much higher resolution on screens having an aspect ratio of 16 : 9 instead of 4 : 3 at present.

## HOLLERITH CODE

A type of code used to represent alphanumeric data on punched cards.

## DITORMATION

Ditormation is data that is organised and is meaningful to the user. Data processor transforms data into deformation.

## INERTIAL NAVIGATION

Inertial navigation system is a self contained navigation system and is associated with the use of an Inertial Measurement Unit (IMU) to help determine state propagation.

## INSTRUMENT LANDING SYSTEM

It is the system that is used by aeroplanes to assist the pilots to make proper landing. The system comprises of two transmitters, the localizer and the Glide path.

## INTERACTIVE VIDEO

Interactive video refers to multimedia concept that merges computer text, audio and graphics by using a video disk, video disk player, micro computer and computer software.

## INTERLACED SCANNING

In the interlaced scanning, all the odd lines are scanned first followed by scanning of all the even lines dividing one complete frame into two parts.

## IRRIDIUM PROJECT

This is a satellite based global communication system.

## ISDN

ISDN stands for "Integrated Services Digital Network." it integrates different data and voice services so as to transmit them over a single communication channel.

## ISO-9000

The term ISO-9000 broadly refers to a series of international standards laid down by the International Organisation for Standardisation.

## LORAN

LORAN stands for Long Range Navigation. It is used to provide navigational aid to ships and planes.

## MAXIMUM FREQUENCY DEVIATION

It is the range of frequencies between the lowest earner frequency and the unmodulated carrier frequency.

## MODULATION INDEX (For AM)

Modulation index in amplitude modulation is defined as the ratio of peak amplitude of the modulating signal to the peak amplitude of unmodulated carrier signal.

## MODULATION INDEX (For FM)

Modulation index in FM is defined as the ratio of maximum frequency deviation and (he highest modulating signal frequency.

## MODEM

MODEM stands for Modulator-Demodulator. It is a type of data communication equipment that converts digital data to an analog signal (and vice-versa) for transmission on telephone circuits.

## MONOTONOCITY

The D/A converter is said to exhibit monotonocity, if its analog output either increases or remains same but does not decrease.

## MTI RADAR

MTI radar is used to determine both range as well as velocity of a moving target. Velocity determination is based on the doppler principle while range is determined as it is done incase of pulse radar

## MULTIPLEXER

It is an electronic device that receives input from many low speed lines and then multiplexes or concentrates and transmits a compressed data stream on a high speed and a much more efficient data transmission channel.

## MULTIPLYING D/A CONVERTER

In a multiplying D/A converter, the converter multiplies an analog reference by a digital word.

## NAVIGATION

It is a process or a technique used for directing the movement of a vehicle (aeroplane, ship, missile etc.) from one position to another.

## NEGATIVE LOGIC

In this, the more positive of the two voltage or current levels represents a logic 0 and the less positive of the two voltage or current levels represents a logic 1.

## NOISE FIGURE

Noise Figure (NF) of a receiver is a measure of the extent to which the noise appearing in the receiver output in the absence of signal is greater than the noise that could be present if the receiver were a perfect one.

## NON-LINEARITY

Non-linearity specification of an A/D converter describes its departure from a linear transfer curve. Non-linearity error does not include Gain, Offset and Quantization errors.

## OMNIDIRECTIONAL ANTENNA

An antenna that radiates in or receives equally from all directions is called on omnidirectional antenna.

## OVER MODULATION

Modulation percentage greater than 100 per cent is called over modulation.

## PASSIVE HOMING

Passive homing depends for its functioning on the natural energy emitted by or reflected from the target. The sensor in the missile makes use of this energy to track the target. Anti-radar missiles for instance may home onto the transmitted power produced by the radar.

## PERCENTAGE OF MODULATION (For FM)

The percentage of modulation in an FM signal is defined as the ratio of maximum frequency deviation to the standard deviation (75 kHz) fixed for the service.

## PHASED ARRAY RADAR

A phased array radar produces a beam that can be electronically scanned over a wide angle without any physical movement of the antenna.

## POINT CONTACT DIODE

Point contact diode is primarily intended for RF applications due to its extremely small internal capacitance.

## POLARISATION

Polarisation is that characteristic of the electromagnetic wave which gives direction of electrical component of the wave with respect to ground.

## PRIMARY COLOURS

Red, Green and Blue have been identified as the primary colours.

## PULSE AMPLITUDE MODULATION (PAM)

In a pulse amplitude modulated signal, the amplitude of the unmodulated pulse train varies in accordance with the instantaneous amplitude of the modulating signal

## PULSE CODE MODULATION

In Pulse Code Modulation (PCM), the signal to be transmitted is sampled at various instants. These samples are then transmitted.

## PULSE COMPRESSION RADAR

A pulse compression radar (also known as chirp radar) has the increased range detection capability of long-pulse radar system while retaining better range resolution

## PULSE POSITION MODULATION (PPM)

In a pulse position modulated signal, the time of occurrence of pulses in the unmodulated pulse train changes in accordance with the instantaneous amplitude of the modulation signal.

## PULSE WIDTH MODULATION (PWM)

In a pulse width modulated signal, the width of the pulses is changed in accordance with the instantaneous amplitude of the modulating signal.

## PULSE TRANSFORMER

A pulse transformer is used to transform pulses from one circuit to another.

## PUNCH CARDS

These are cards made up of thick paper (or card-board) and have 80 columns and 12 rows on which holes can be punched. Each column refers to a character and characters are stored by punching holes into the card.

## QUANTIZATION ERROR (A/D Converter)

The error resulting from the resolution limit is known as the Quantization Error.

## RADAR ALTIMETER

Radar altimeter is similar to a pulse radar and is used to determine height of the aircraft above ground

## RADAR BEACON

Radar beacons are stations to provide signals for use as navigational aids.

## RADAR CLUTTER

All those objects which reflect the radar signal and which are other than the target are termed as clutter.

## RADIX

Radix of a number system is defined as the number of different symbols used in the number system.

The radix of binary number system, for instance is 2.

## RASTAR

Rastar is illumination on screen in the absence of any video signal.

## RDF

RDF stands for "Radio Direction Finding." It is the process of using radio signals to find out the relative position of the aeroplane

## RECIPROCITY THEOREM

In a linear, active and bilateral network with a single source, the ratio of excitation to response remains unchanged when the excitation and response interchange their positions.

## RESOLUTION

The resolution of a A/D converter is the number of states that the full scale range is divided or resolved into.

## RESOLUTION

It is the ability of the radar to distinguish between two closely spaced targets.

## RF AND IF TRANSFORMERS

These are narrow band transformers that are used to couple signals in a narrow band around a center frequency from one stage to another.

## SCANNING (Horizontal and Vertical)

Picture elements in that line (called horizontal scanning) and then various lines may be scanned sequentially in order to cover the complete picture frame (called vertical scanning).

## SECONDARY COLOURS

Yellow (= Red + Green), Cyan (= Blue + Green) and Magneta (= Red + Blue) are known as secondary colours.

## SENSITIVITY

The sensitivity of a communication receiver is the ability of the receiver to get a standard output.

## SMART TERMINAL

A smart terminal has a microprocessor and primary storage to enable the user to do some processing, before data is actually sent to the computer.

## SVGA-(Super Video Graphics Array)

It is a type of video system offering upto 1024×768 resolutions producing a total of 7,86432 pixels.

## SYNTHETIC APERTURE RADAR (SAR)

SAR coherently combines signals obtained sequentially from different locations of a moving antenna for mapping or target identification purposes.

## TELECONFERENCE

Teleconference refers to a meeting that occurs via telephone/electronic facilities thereby eliminating need for travel.

## TELECOMMUTING

Telecommuting refers to working at home and communicate with the office or send data to the office via electronic machines.

## TELEX

Telex is 3 communication service used for transmission of plain messages. It makes use if teleprinters as terminals and telephone lines as communication medium.

## TELETEXT

Teletext service is the use of television sets for display of data, usually alphanumeric, in addition to receiving television broadcast.

## TRUTH TABLE

A truth table lists all possible combinations of input and the corresponding outputs of a logic system.

## TURN-AROUND TIME

It is the actual time required to reverse the direction of transmission from sender to receiver or vice versa when using a half duplex circuit.

## TRACKING

Tracking in a communication receiver is the ability of different tuned circuits to change their resonant frequencies in the same proportion when a common shaft is rotated.

## TRACKING RADAR

A tracking radar is used to continuously track a moving target.

## UART

UART stands for Universal Asynchronous Receiver/Transmitter.

## UPS-(Uninterrupted Power Supply)

UPS is a power supply that is not interrupted by mains failure. When the mains power failure occurs, the UPS acts as an auxiliary power source for the equipment. The UPS system draws its power from a storage source such as batteries.

## VAR

VAR stands for Visual Aural Range. VAR system can be used by aeroplanes and ships to determine their location.

## VIDEO TRANSFORMER

It is a wideband transformer with a frequency response spread over frequency range from a few Hertz to a few Megahertz.

## VIDEO DISPLAY ADAPTER

It is located in one of the expansion slots in the computer and acts as an interface between computer and the Video Display Unit.

## VIDEOTEXT

It is a general name given to two way interactive information system which allows for collection and retrieval of information through terminals which are specially adapted TV sets.

## VOR

VOR stands for Visual Omni Range, VOR can be used to give exact bearing of the plane.

## WAVEGUIDE

It is a specially designed and constructed metallic pipe for transmission of microwave electromagnetic radiations.

ㅁㅁㅁ